The Evolution of Insect Mating Systems

The Evolution of Insect Mating Systems

Randy Thornhill and John Alcock

Harvard University Press
Cambridge, Massachusetts, and London, England   1983

Library of Congress Cataloging in Publication Data

Thornhill, Randy.
　　The evolution of insect mating systems.

　　Bibliography: p.
　　Includes index.
　　1.　Insects—Evolution.　　2.　Insects—Reproduction.
I.　Alcock, John, 1942–　　.　　II.　Title.
QL468.7.T46　　1983　　　595.7′056　　　82-21351
ISBN 0-674-27180-7

*To Nancy, Albert Patrick R., and Aubri Thornhill
and to Sue, Joe, and Nick Alcock*

Preface

THIS PROJECT grew out of our conviction that there was need for a book that would review insect reproductive behavior and related topics from an evolutionary perspective, particularly with reference to sexual selection theory. The book reflects our fascination with the diversity of mating systems of insects and our appreciation for the ability of modern evolutionary theory to make sense of the diversity. We hope to persuade our readers to share this attitude.

We also wish to draw the attention of students of animal behavior to the delight of watching insects. Ethologists typically choose vertebrates to study; the larger the animal and the closer its phylogenetic affinity to the human species, the better. This is true even though several of the founders of the discipline, including Nobel Prize winners Niko Tinbergen and Karl von Frisch, worked extensively with insects.

There are numerous advantages to studying insect behavior, not the least of which is a breathtaking array of species and biological characteristics that puts birds and mammals to shame. Moreover, because insects often occur in dense populations, it is relatively easy to collect quantities of information in a short time. The task of the observer is made easier still by the tolerance of the average insect to human beings. One of us recently had three weeks in which to conduct a study on a population of territorial damselflies—hardly time enough to unpack the equipment needed for a comparable piece of work on a territorial bird or other vertebrate, yet more than sufficient time to accumulate data on the lives of over a hundred damselflies. Many individuals were captured *by hand,* and upon release returned at once to what they had been doing, as if the observer hardly existed. It is almost embarrassing at times for students of insect behavior to hear of the difficulties and discomforts endured by their counterparts, whose research on the mating behavior of a songbird or the social organization of a mammal requires hours of laborious fieldwork for each noteworthy observation.

But the primary reason for watching insects behave is not that they are easy to study, but that their lives are captivating and instructive. For every problem in evolutionary biology there are at least a thousand insect species (to hazard a guess) particularly well suited to provide an answer, yet in most cases only a tiny fraction of the potential studies have been exploited. The opportunities for biologists to test hypotheses about the adaptive value of insect behavior represent both a challenge and a reproach to entomologists and students of behavior.

The book has been prepared with two audiences in mind—professional entomologists and beginning students of animal behavior. The discipline of entomology has developed somewhat separately from the fields of zoology and biology. As a result, advances in evolutionary theory that were made by zoologists and biologists in the 1960s were not immediately incorporated into the mainstream of entomological thought. Entomologists today are increasingly aware of evolutionary theory, and growing numbers accept its importance for understanding insect behavior.

The impetus for starting the manuscript came from a contact between Harvard University Press and R.T., who is therefore shown as senior author; however, the final text represents a truly mutual endeavor, with both authors involved in the writing of each chapter.

Colleagues and friends helped in various ways as we wrote the book. Many individuals graciously gave us permission to use photographs or drawings from their own work. Several people read and constructively criticized the complete manuscript: T. H. Clutton-Brock, Gary Dodson, Darryl T. Gwynne, James E. Lloyd, Larry M. Marshall, David E. McCauley, and Bruce Woodward. Others commented on various sections: James Bandoli, William G. Eberhard, Teri Markow, and Mary Jane West-Eberhard. Linda DeVries, Becky Payne, Cynthia Schooler, and Joanne Tapia typed the manuscript (repeatedly), and we are grateful for their contribution. Jill Smith and Marilyn Hoff Stewart did the graphs and original art; the high quality of their work speaks for itself. We are indebted to William Patrick, Vivian Wheeler, and other members of the staff at Harvard University Press for their help in producing the book.

James Findley, former chairman of the Department of Biology, University of New Mexico, provided R.T. with an environment conducive to scholarship during the three years in which the book was being written. J.A. received similar assistance from Ron Alvarado and Kathy Church, chairpersons of the Department of Zoology at Arizona State University. The support of both institutions is gratefully acknowledged.

The National Science Foundation awarded R.T. grants DEB 79-10193 and BNS 79-12208, which funded much of his research that is discussed here.

R.T.

J.A.

Contents

1

Evolutionary Hypotheses and Tests

BECAUSE THIS BOOK applies an evolutionary approach to insect behavior, it is essential to recognize at the outset what an evolutionary hypothesis is and is not. Evolutionary theory cannot be used to explain all aspects of an animal's biology, only to account for why an animal may have evolved one set of characteristics rather than other alternatives. It is entirely possible to study the biology of a species without reference to the evolutionary basis for its traits. For example, if we observe male dung flies (*Scatophaga stercoraria*) flocking to a recently deposited pat of cow manure, we might seek to explain their behavior in terms of their ability to detect odoriferous substances drifting on air currents, the effect of light intensity or air temperature on their response, the role of experience in altering an individual male's behavior, and so on. Research on these factors could test our hypotheses about how developmental and physiological systems within individual flies provide them with the capacity to detect and react to cowpies. We would then better understand the immediate or *proximate* foundation of a fly's behavior, but we would be no closer to understanding why this species of fly had evolved one set of mechanisms rather than another.

Male dung flies come to fresh dung piles far more frequently than to older ones (Parker, 1970a,b). At the proximate level, the flies may do this because they can smell fresh manure more readily than more mature dung. Still, other insects are drawn to older pats by the cues they provide. Why have male dung flies evolved the odor preferences that lead them to recently deposited rather than to hours-old cowpats? To deal with this question we must consider how the force of natural selection has led to the evolution of a species whose current representatives have certain capabilities, but not others. This requires an analysis of the ultimate or evolutionary function of the behavior—which is, we shall argue, to propagate the genes of the individual.

The hypothesis that an animal has the ultimate goal of passing on its genes does not imply that the animal is consciously seeking to achieve that goal. Evolutionary entomologists sometimes write as though bees know who their relatives are, flies have strategies, and a bug can demand, "like some errant macho Californian, proof of his fatherhood before paying out paternity benefits" (J. E. Lloyd, 1980a:3). As Lloyd points out, anthropomorphisms of this sort do not imply that insects have the cognitive powers of human beings. A male fly drawn to fresh dung surely is not aware that its chances of encountering receptive females are better there than at older cow droppings; nevertheless, this is the result of its preference. It is sufficient that the fly's nervous system operates in such a way that certain odors in certain concentrations are perceived and elicit certain responses. In the past such proximate responses evidently were associated with reproductive success; the fly is descended from ancestors all of whom reproduced.

Evolutionary hypotheses about the ultimate function of behavior are not teleological. The "goal" of behavior is the logical outcome of a process based on natural selection, on past differences in the reproductive success of individuals. Even so, many entomologists (and other biologists) feel more comfortable with research on the proximate mechanisms of behavior rather than on its ultimate causes. Studies of developmental or physiological systems within an individual often lead to an understanding that has a satisfying air of stability and precision. In contrast, research on the adaptive value or evolutionary foundation of a behavioral trait is reputedly more speculative. We hope to show that hypotheses on the ultimate function of behavior can be developed in a rigorous way and tested in a manner no less scientific than the methodology of geneticists and physiologists. Without an attempt to solve questions about evolution and adaptive value, our understanding of a trait would be forever incomplete. The ability to behave has an evolutionary history, and exploration of the effects of this history is important, entertaining, and fully scientific.

The Mechanism of Evolution

Part of the confusion surrounding the development of an evolutionary approach to animal behavior comes from past failures to consider precisely what causes evolution to occur. Not long ago it was widely accepted that the traits of individuals had evolved because they contributed to survival of the species as a whole. The logic here was that traits that contribute to the survival of species will survive, whereas those that do not will be lost when a species becomes extinct. So deeply ingrained was the acceptance of this kind of species-level selection that its philosophy still pervades countless articles on insects (and other animals).

For example, the observation that males of many species tend to be dispersed or evenly spaced through the environment has frequently been interpreted as a device that ensures a more even distribution of males and thus a more efficient means of uniting the sexes (see, for example, Cazier and Linsley, 1963; King, Askew, and Sanger, 1969; Catts and Olkowski, 1972; Coulson, 1979). The assumption is that males behave in ways that speed the insemination of all females so that the reproductive potential of the entire species is maximized. This maintains high population levels and encourages the survival of species.

Perhaps ironically, the same kind of "group reproductive efficiency" hypothesis has been applied to swarming by males of various insects. Two kinds of species-preserving benefits have been proposed for swarming. The first is that by gathering in large and conspicuous groups, males make it easier for females to locate mates. To our knowledge, advocates of group benefit hypotheses have never specified the conditions under which swarming rather than even spacing of males provides the greater measure of "reproductive efficiency" for the species as a whole. In any case, there is yet another supposed advantage for species that swarm; by gathering together, individuals may assess local population density and use this information to regulate their reproductive output to maintain an optimal population size (that is, one that makes extinction least likely). This hypothesis of reproductive regulation is at the heart of Wynne-Edwards' (1962) proposal that group selection is responsible for the evolution of much of the social behavior of animals.

Implicit in many group selectionist arguments is the expectation that an individual's behavior should contribute to the welfare and survival of the group, even if this means that the individual sacrifices some (or all) of its own reproductive opportunities. For example, imagine that in a species with typically well-dispersed males there were some individuals that failed to move away from an area with a cluster of receptive females. These males might father relatively more offspring than those that abandoned such centers and distributed themselves more evenly in their habitat. But the reproductively "selfish" males might also reduce the overall mating efficiency of the population, in that their failure to disperse could mean that some isolated females would not mate. An advocate of group selection would predict that evolution would result in elimination of the selfish males that stayed near groups of females because they would tend to reduce the size of the next generation, making the species more vulnerable to extinction.

The same logic can be applied in the case of swarming males. According to the group selection argument, a dense swarm might communicate information to individuals that the local population was dangerously large, threatening the long-term survival chances of the group. In this case individuals should restrain reproduction, even if the species was one in which breeding occurred in a single season only (a semelparous species, like most insects).

The Problem with Group Selection

These examples illustrate the fundamental difficulty with Wynne-Edwardsian group selection, a problem identified by Williams (1966) in his book *Adaptation and Natural Selection*. Williams makes a powerful case for accepting natural selection at the individual level, not the group level, as the primary mechanism of evolutionary change. His argument begins with recognition that an animal is a product of the genes it receives from another individual or individuals, which in turn received their genetic information from others in the still more distant past. If individuals differ in the genes they possess, their relative reproductive success will determine the frequency of the various forms in the next generation. There is in effect unconscious competition among alternative forms of genes (alleles) for representation in the finite population of any generation. Therefore it is difficult to imagine how reproductively self-sacrificing behavior can persist over evolutionary time. Consider the reproductive contest between disperser males and selective nondispersers in a species whose males typically are evenly spaced. Let us accept the argument that those males that disperse improve the chances that the species will avoid eventual extinction at some point in the distant future. But if those that fail to move away from female clusters reproduce more often than those that space themselves evenly in their environment, the reproductively self-ish genes will tend to replace group benefit alleles. The genes of the self-sacrificing males may be beneficial to the species, but their bearers damage the chances of survival of these genes.

The same argument applies to interpretations of swarming behavior. Imagine a semelparous insect species with two genetically different types of reproducers, one that sacrificed a portion of its reproductive output when population density was high and swarms were large, and another that maximized its own genetic propagation even when this was contrary to group benefit. If, as seems likely, the reproductive altruists produced fewer progeny than the unrestrained reproducers, the genes for reproductive sacrifice would be reduced in frequency in the next generation and ultimately eliminated over time. Logic dictates that traits that promote selfish individual reproduction should spread through populations.

We do not deny that extinction will have an effect on the species composition of communities, and thus on the persistence of certain behavioral traits. If a particular characteristic does contribute to the disappearance of a species, obviously it also will disappear. The point is, however, that traits can become established and be maintained in populations by intraspecific reproductive competition, despite the fact that these same traits have deleterious effects at the level of a population or species. Even if a species is on the road to extinction (as most are), the attributes of its members are being shaped by individual selection. The absence of a group selection mechanism powerful enough to overcome this force (Lewontin, 1970) greatly reduces the attractiveness of hypotheses about the evolutionary function of an animal's behavior that are based on Wynne-Edwardsian ideas.

The controversy about group selection that followed publication of Wynne-Edwards' book led to efforts to develop more sophisticated models of group selection (nicely summarized in E. O. Wilson, 1975; Wade, 1978; Wittenberger, 1981a). These models have shown that what matters is the relative rates of extinction of individuals versus groups. Group selection is theoretically possible, but the requirement that the population be composed of genetically differentiated, well-isolated groups—some of which have a very high probability of going extinct because of their genetic constitution—is not likely to be met in most species. Thus there is little support for the proposition that group selection has been a major factor in shaping the behavior of animal species, although research on this question continues (D. S. Wilson, 1975, 1977, 1980; Wade, 1976, 1977, 1980.)

The debate on group selection has benefited the biological sciences.

There is now widespread recognition of a real difference between adaptationist hypotheses founded on individual selection and those derived from a group selectionist perspective. Furthermore, students of behavior have been forced to acknowledge the logical implications of natural selection, and this has had a revolutionary impact on the field. To a surprising extent, behavioral researchers prior to 1966 had ignored the utility of Darwinian thought. The original "evolutionary behaviorists," the European ethologists, often operated under the assumption that animals behaved to prevent species extinction. Thus Lorenz, one of the founders of the field, wrote *On Aggression* (1963), in which he concluded that animal aggression was designed to prevent overpopulation, to select the best members of the species for breeding purposes, and to accomplish all this without a great deal of bloodshed. All three "functions" supposedly evolved through group selection to preserve the species. This attitude has now been virtually replaced by an individual selectionist philosophy and by two new behavioral disciplines—behavioral ecology and sociobiology. These disciplines are founded on the working hypothesis that an animal's behavior should propagate its genes better than possible alternative traits. Adoption of this hypothesis has shifted the focus to reproductive competition among individuals and has stimulated a massive reevaluation of past data and old interpretations, at the same time revitalizing the entire study of behavior.

The Individual or the Gene?

The argument that the individual is the basic unit of selection is not, however, universally accepted. Dawkins (1976, 1978) proposed that the unit of selection is the gene, not the individual and not the group, and that competition within genotypes is central to the evolutionary process. The key question here is whether competition within the genome could result in a reduction of individual fitness. Dawkins argued that a gene which could induce its individual carrier to reproduce on its behalf could spread through a population, even if its effect were to damage the transmission chances of the other units in the genotype and to reduce the number of offspring produced by the individual relative to other members of its species.

Genes that fail to cooperate with other members of their genome have been labeled "outlaw genes" (Alexander and Borgia, 1978), and they apparently do exist (Hartung, 1981). The best-known examples are those that subvert the replicating process of meiosis for their own selfish ends, to the disadvantage of other components of the genome. Thus in *Drosophila* there are a number of alleles that distort meiosis so that they become overrepresented in the gametes produced by the individual (Crow, 1979). Such alleles can spread through a population even if they happen to be lethal in the homozygous condition.

A similar case is the driving Y chromosome. Males (XY) with this mutant Y chromosome make only Y-bearing sperm. All the progeny of these males will also be male. Theoretically, the trait can spread through a population until only male progeny are produced in a final generation and the species goes extinct (Fig. 1.1). The driving Y chromosome clearly reduces the reproductive success of a parent that would have more surviving offspring if it were able to produce equal numbers of male and female progeny (Chapter 3; W. D. Hamilton, 1967).

Other phenomena, such as the apparently functionless DNA found in many animals (Doolittle and Sapienza, 1980; Orgel and Crick, 1980) and the formation of eggs with two sets of certain chromosomes rather than one, have been attributed to outlaw genes or outlaw gene complexes (Axelrod and Hamilton, 1981). But the list is not long,

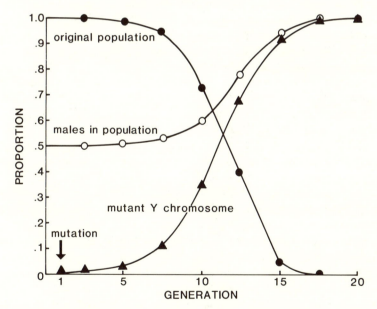

1.1 The effect of a driving Y chromosome on the population in which it occurs. After the first mutant appears the proportion of individuals carrying the driving Y goes steadily up and the sex ratio becomes more and more male biased, even as the population is sliding to extinction. (From W. D. Hamilton, 1967. Copyright 1967 by the American Association for the Advancement of Science.)

perhaps because it is difficult for outlaw genes to persist in populations.

First, if the presence of one gene in a genotype reduces the fitness of many other genes, selection will favor mutant alleles that arise at any of the disadvantaged loci which happen to suppress the meiosis-distorting or otherwise damaging effects of the outlaw gene (Alexander and Borgia, 1978; Hartung, 1981). Since there are a great many loci in most genomes, the probability that a suppressor mutation will eventually occur at one of these loci must be high. Evidence in support of this argument comes from studies of meiotic drive in *Drosophila*, which have shown that segregation-distorting alleles rarely succeed in completely subverting meiosis; they often appear in 55–65 percent of all gametes produced rather than 100 percent. Because the distorting effects are usually slight, modifier mutations must have occurred at other loci and have been selected for because they prevented the distorter alleles from reducing the fitness of other genes and of the individual that carries them (Crow, 1979).

The same conclusion, that conflict among the genes in a genome is resolved in favor of the majority, comes from an examination of the occurrence of haploid effects in sperm (Sivinski, 1980a). Haploid effects are phenotypic characteristics of sperm that are produced by expression of the haploid genotype of a gamete. Such effects might result in a sperm acting in its own genetic interests even if these ran contrary to the reproductive success of the individual (genotype) that produced it. But haploid effects are rare in sperm, and generally the attributes of gametes are controlled by the parent's diploid genotype (Cohen, 1971; Beatty, 1975; Sivinski, 1980a, forthcoming).

Thus selection should make it difficult for a few genes to take advantage of the majority. Moreover, genes do not exist independently of their carriers. Because the probability that they will be transmitted to the next generation depends upon proper functioning of the individual, all the genes in a genome have a common interest in producing a competent, successful individual. Conflict that damages the reproductive ability of the carrier individual will usually lead to a decrease in the frequency of the genes that are unable to cooperate (Hartung, 1981). To the best of our knowledge no one has advanced a detailed argument in support of the proposition that *any* behavioral character is due to the presence of an outlaw gene in an individual's genome.

Genes are not themselves "visible" to selection. They achieve their effect by influencing the development of a phenotype, which is a characteristic of the individual, not of the gene. Even Dawkins (1976:48) agreed that the "immediate manifestation of natural selection is nearly always at the individual level." In this book we shall therefore treat the individual as the *unit of selection.*

The long-term consequences in differences in the reproductive ability of individuals can be measured in terms of the rise and fall in frequency of competing alleles in the gene pool. Genes are therefore the *units of evolution;* only they persist over evolutionary time, provided that the individuals in which they temporarily reside can outreproduce other individuals in the population. For this reason it is helpful to think of the phenotypic differences among individuals as the product of differences in a single gene (gross oversimplification though this may be). The advantage of this approach is that it enables one to perform a simple "thought experiment" to examine hypotheses that purport to explain the evolutionary basis of a trait. When earlier in this chapter we criticized group selectionist hypotheses, we employed the one-gene, two-allele model. We asked what would happen to the (hypothetical) allele for group benefit if there arose a (hypothetical) alternative form of the gene whose carriers consistently outreproduced individuals with the other allele. Gene thinking of this sort has proved to be an extremely productive way in which to identify logically flawed hypotheses and to develop new interpretations of many behavioral traits.

Testing Evolutionary Hypotheses

It is one thing to propose that an individual's behavioral ability contributes to its chances of propagating its genes and quite another to demonstrate that this is true. For the confirmed individual selectionist, an animal is generally viewed as a constellation of component parts and attributes, each of which in some way promotes genetic success. This attitude has been vigorously attacked (Lewontin, 1978; Gould, 1978; Gould and Lewontin, 1979). Skepticism about adaptationist thinking is not new. For example, the German ornithologist Stresemann (1914:399) denounced a hypothesis about the adaptive nature of the resemblance between two bird species with the comment that "this [interpretation] belongs to the realm of unlimited imagination which has always characterized the extreme selectionists."

The opposition to "extreme" selectionism is based on recognition that not all aspects of the phenotype qualify as adaptations. For example, some traits are *currently* maladaptive because of the lag time in evolutionary adjustment. The attributes of living organisms are products of past selection pressures created by environmental conditions that may no longer exist. Thus creatures may exhibit behavior that was adaptive in the past but under current circumstances no longer promotes genetic success.

Conceivably, some traits have not been adaptive in the past or in the present. Genes often have multiple effects, some of them incidental to the primary functional product of gene action. These might have no significant positive influence on gene survival, or they might even have a weakly negative influence (presumably more than compensated for by the adaptive value of the major character related to the gene). Thus an enzyme that contributed to the construction of a useful component of an insect's nervous system might also have side effects—on pigment deposition or bristle number, say—that are incidental rather than adaptive. The pleiotropic effects could nevertheless persist over evolutionary time because of the generally advantageous outcome of the gene's influence on neural development. Selection will, however, continually favor mutant alleles that have the same positive results without negative pleiotropic effects (see Wittenberger, 1981a).

Lewontin and Gould also argue that some traits could "evolve" as a result of random changes in gene pools (in other words, genetic drift). Especially in small populations it is possible for a mutant allele to spread through the entire group in a short period of time because of chance events. Once fixed at 100 percent it may persist indefinitely if no reproductively superior allele appears in the population.

Williams (1966) preceded Lewontin and Gould in arguing that biologists should be aware of these possibilities and should exercise restraint in attributing adaptive value to a trait. Nevertheless, we suspect that evolutionary biologists are not often tempted to propose that the incidental or random characteristics of individuals are adaptive. Usually it is the complex and elaborate traits that arouse curiosity about their evolved functions. Development of these characters generally requires the integrated action of many genes. This is particularly true of behavioral traits, which typically have a complex polygenic foundation (Ehrman and Parsons, 1976). It is unlikely that the pleiotropic effects of several genes could accidentally generate an apparently well-coordinated behavioral ability. Thus the risk of proposing that a behavioral trait is adaptive, when in reality it is a trivial or random aspect of gene expression, is probably not great.

Just-So Stories

Lewontin and Gould argue that extreme selectionists typically assume that an interesting character is adaptive, and use whatever information is available to show how the trait might advance individual genetic success. Gould calls creating an account of the supposed adaptive value of a character the art of telling "just-so stories," *à la* Rudyard Kipling. He argues that only the ingenuity of the storyteller limits his ability to create a tale that is consistent with the assumptions of selectionist theory.

Consider the following just-so story about the behavior of males of the hippoboscid fly, *Lynchia hirsuta,* a blood-sucking parasite of quail (Fig. 1.2). In this species males are often found riding on the backs of gravid females (Tarshis, 1958). The mounted females typically have a well-developed prepuparium swelling the abdomen. (Females nurture a single fertilized egg within their body for several days until it has passed through the larval stage, depositing the offspring only when it is about to pupate.) Although a male may ride on the female for many hours, copulation does not take place until the female has placed the prepuparium in its resting place. She flies away from the quail to do this, with the male still mounted on her back, after which she mates with her rider and then returns to another host bird.

An extreme adaptationist would interpret this male behavior in the following way. The mounted male selects females that are reasonably close to becoming receptive because in this way he can minimize riding time and maximize the number of females he copulates with in his life. He rides on the back of an about-to-become-receptive female to guard a valuable potential mate from other males (he can prevent competitors from assuming the precopulatory position if he has a firm grip on the female). Both the preference for gravid females and the investment in guarding behavior increase his contribution of genes to

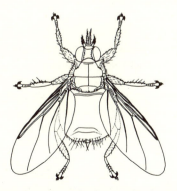

1.2 *Lynchia americana,* hippoboscid relative of *L. hirsuta,* showing the hooklike tarsal claws that make it difficult for a host to remove the parasitic fly. (From Cole, 1970.)

the next generation. (Note again that adaptationist hypotheses do *not* require that an animal be consciously aware that its behavior will increase its genetic success.)

Gould could point out that a host of similar stories might be developed to account for the male's behavior. For example, perhaps the males are not really guarding their mate but simply ride on the female in order not to lose contact with her (because females are hard to find) before she becomes receptive. Alternatively, it could be argued that females are able to lead their males from a depleted host to one that offers generally superior food (perhaps the female is a better flier). Critics of the adaptationist approach take the ease with which a variety of just-so stories can be devised as evidence that none of the stories is likely to be valid. Moreover, there is no guarantee that males do not ride on the backs of females as a simple holdover from a time when such behavior was adaptive, or because this trait is a pleiotropic effect of selection for some other attribute.

There is no question that a selectionist approach will not always produce a correct hypothesis. Proadaptationists have not always recognized that there is a significant difference between a "working hypothesis" and a "tested hypothesis." Gould himself seems not to appreciate the distinction, for he implies that it is improper even to advance ideas about the possible ultimate causes of behavior. "Storytelling," a pejorative synonym for "hypothesis making," implies that evolutionary working hypotheses are frivolous. Yet the educated guess, the speculation, the story is an indispensable first step in the solution of a scientific puzzle. Some of Gould's own ideas have an element of speculation, such as the arguments he has presented on the inconstant rate of speciation events over evolutionary time (Eldredge and Gould, 1972; Gould and Eldredge, 1977). (See also the contrast between Gould, 1980, and Stebbins and Ayala, 1981.) The idea that evolution proceeds more erratically than gradually does not deserve to be dismissed out of hand merely because it is an incompletely tested hypothesis. In the last analysis the utility of a hypothesis comes from the testable predictions it generates. The adaptationist approach has been a rich source of such predictions, whereas the approach that proposes that a trait is adaptive *unless* it has been caused by "allometry, pleiotropy, random gene fixations, linkage and indirect selection would be utterly impervious to test" (Lewontin, 1978:230).

The value of having testable hypotheses is shown by the history of how beetle horns (see Figs. 7.16 and 7.17 later) have been explained. Arrow (1951) made a major survey of the beetle groups in which males (and sometimes females) possess large and conspicuous

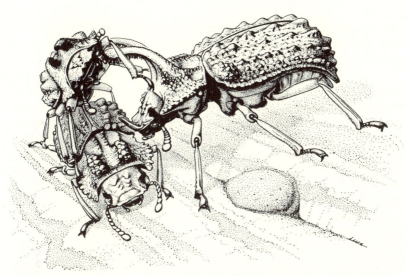

1.3 The use of horns by a male tenebrionid beetle, *Bolitotherus cornutus,* to displace a rival from a female. (Drawing courtesy of D. Luce.)

horned structures. He rejected the adaptationist approach by arguing a priori that in many cases the shape of the horns would preclude their effective use in aggressive disputes. He concluded that the horns were in all probability incidental by-products of the developmental process. This conclusion closed the door on the search for function and did not suggest useful predictions to Arrow or to others. But if one assumes that structures as complex and varied as beetle horns should be adaptive, then one can predict that they should be used in aggressive situations and that the shape of the horns should be correlated with the manner in which they are used (Fig. 1.3). Eberhard (1979, 1980) and L. Brown (1980) have shown this to be the case by demonstrating that males do use their horns in contests for valuable resources and for females. Some thick, powerful horns are employed as levers, whereas thin, delicate ones may be used as pincers to grasp an opponent, remove him from a perch, and drop him to the ground. The search for adaptive function has been infinitely more productive as a research approach than the pleiotropic effect hypothesis.

Working Hypotheses and Testable Predictions

A person armed with a knowledge of individual selection can produce hypotheses with two important virtues: first, they are shaped by the constraints of the logic of natural (individual) selection, a logic that has withstood close examination for more than a hundred years; and second, the hypotheses can generate testable predictions. Our working hypothesis about the adaptive significance of the behavior of *L. hirsuta*, for instance, can be made to yield some predictions.

If males are the product of individual selection, they should attempt to transmit as many of their genes as possible in reproductive competition with their rivals. In order for riding behavior to contribute to this unconscious goal, copulations with females that have already mated at least once (witness their recently gravid condition) must lead in at least some cases to fertilizations. Moreover, if males really are guarding their mates against rival males, there should be at least occasional competition for the opportunity to ride on the back of a female. Thus, simple predictions derived from our original hypothesis are that (1) males will sometimes succeed in fertilizing an egg of a gravid female on whose back they are riding, and (2) there will be contests for control of gravid females, in which mounted males will sometimes prevent an opponent from securing the female.

These predictions are testable. We could conduct a series of breeding experiments, pairing a female first with a male with a distinctive genetic marker and later with other males without the gene. If all the young produced by that female exhibited the attribute of her first mating partner, we would conclude that although females permit more than one male to copulate with them, they do not use the sperm they receive from any but the first male. This would compel us to reject the first prediction as incorrect. Likewise, if through observation of the flies we found (1) no evidence that male-male struggles were occurring, or (2) that the mounted male was consistently supplanted from the female's back by any male that challenged him, we would also reject the second prediction. Our confidence in the original hypothesis would be severely reduced and we would be required to consider other possibilities.

On the other hand, our predictions might be supported by our experiments and observations, in which case we might construct still more refined predictions. The likelihood and intensity of aggressive acts between a mounted male and rival should increase, for example, as the female nears the the time when copulation can occur. The logic of this prediction is that males could gain more fertilization chances overall if they were to accompany fully gravid females instead of females distant from the moment of receptivity. Generally in insects the more females a male mates with, the greater the number of offspring he will produce and the larger the number of genes he will transfer to the next generation. Females about to become receptive are more valuable to males because the time spent waiting on a female's back for the copulation is shorter and there is more time to find additional mates. As the genetic value of a particular female increases for males, they can make a greater and greater investment in aggressive defense of the female or aggressive thievery attempts and still, on average, enjoy a net gain as measured ultimately by the number of progeny produced relative to other males.

This prediction is also amenable to test by observing the behavior of a group of males enclosed with females at different stages of their reproductive cycle. Measures of the frequency and duration of aggressive interactions would enable an observer to confirm or reject the prediction, further strengthening or weakening the reliability of the original hypothesis about the adaptive value of riding behavior.

Basic to this approach is the recognition that behavioral traits have costs and benefits in terms of their effects on reproductive success. Male riding behavior may thwart rivals and increase the probability that the mounted male will copulate with and fertilize the female under his control (benefits of riding). But the trait also has costs, including the time spent riding on one female which cannot be used to find additional mates. The argument is that if a trait is adaptive it should have certain functional properties whose benefits exceed its costs. Otherwise it will be replaced over evolutionary time by an alternative trait that offers greater reproductive success for its bearers. Our predictions stem from considering what conditions have to be met if the hypothesized benefits of male riding behavior are to exceed its costs. We can then determine whether the necessary conditions actually apply (for instance, whether egg fertilization occurs from matings other than the initial copulation or whether riders win some contests for control of mounted females).

Ideally, in our initial speculations we would draw up a sizable list of reasonable alternative functions and consider how we might devise predictions whose tests would enable us to discriminate among the alternatives. Although it often happens, it is not desirable for an observer to become such a champion of a single hypothesis that he is blinded to other possibilities. In fact, many traits make more than one contribution to reproductive success. Whatever the length of our list, the logical underpinning of this approach is, first, to assume that individuals are attempting to maximize their genetic contribution to subsequent generations; second, to suggest how a trait might assist in achieving this goal; and third, to predict what conditions must apply if the hypothetical benefits of the trait are to exceed its costs.

Optimal Behavior and Evolutionary Prediction

The ultimate extension of this method of testing adaptationist hypotheses comes from the application of optimality theory to prediction making (Parker, 1978a,b,c; Maynard Smith, 1978a). In one sense the demonstration that the proposed benefits of a trait are likely to be greater than its costs is a fairly weak test of adaptation. Selection should theoretically result in the evolution of *optimal traits,* characteristics whose benefit-to-cost ratio (B/C) is greater than that associated with alternative phenotypes that have arisen over the history of the species. In theory at least, the most demanding test of a suggested adaptive function of a trait would be to require that the trait be optimal and then to determine if the evidence supported the prediction.

We shall once again use the hippoboscid *L. hirsuta* to illustrate this method. Male riding, we hypothesized, is adaptive because it enables males to guard potential mates. But if females vary greatly in the riding time required before copulation can occur, males do not necessarily reap the maximum benefit by riding on the back of any female they encounter. Our prediction from optimality theory is that males will mount and guard only those females whose defense yields the greatest possible number of egg fertilizations per unit of riding time. In order to test this prediction, one would need to know at a minimum (1) the number of eggs fertilized per copulation, (2) the probability of contacting females in various stages of their reproductive cycle, (3) the time before a female in a particular stage will deposit her prepuparium and accept a partner, and (4) the probability that a mounted male will be displaced from the back of a female in various stages of "pregnancy." Disregarding for the time being the difficulties in acquiring this information, we can, with the hypothetical data in Table 1.1, attempt to answer the following question: What is the optimal decision a male should make with respect to choosing a female to guard?

If, for example, a male finds a female in day 3 of her reproductive cycle, should he remain with her or continue searching in order to contact a female in day 4 or 5, closer to the time when she will become receptive? The fertilization rate or egg gain per day for a male that mounts a female in day 3 of the 5-day cycle is 0.27 (1 egg per 3 days of riding multipled by 0.80—the probability that the male will retain control of the female for the necessary time, 3 days). If the male abandons this female, he has a 15-percent chance of finding a female in day 4 and a 10-percent chance of finding one on the last day of the cycle. His fertilization rate per day from continuing the search would be 0.155 (0.15 × 0.43 plus 0.10 × 0.90). The gain from the "continue-

Table 1.1   Hypothetical data on the biology of the hippoboscid fly *Lynchia hirsuta* that are related to optimal decisions by males about mate choice.

	Female's position in reproductive cycle (day):				
	1	2	3	4	5
Probability of contacting an unmounted female in 1 day of searching	0.40	0.30	0.25	0.15	0.10
Riding time in days before copulation	5	4	3	2	1
Probability of displacement from female by rival male over entire riding time	0.20	0.20	0.20	0.15	0.10
Average egg gain per day of riding time[a]	0.16	0.20	0.27	0.43	0.90

[a] On the assumption that males gain 1 egg fertilized when copulation occurs.

searching" option is substantially less than the "remain-with-day-3-female" option, and thus we would predict that males encountering females 3 days or more along the cycle will not abandon them. If, however, a male encounters a female at day 2 of the cycle, his egg gain rate will be 0.20 per day by adopting the riding option. If he continues his search for a "better" female, his gain will be 0.223 (0.25 × 0.27 plus 0.155). Thus we can predict that males will reject females at this reproductive stage in order to search for one closer to depositing her prepuparium.

Problems with Optimality Theory

Our predictions could be tested against the observed responses of males in natural populations to females of different types. But there are a great many practical and theoretical difficulties with the application of optimality theory (Maynard Smith, 1978a; Oster and Wilson, 1978). In the first place, the mathematics required is usually far more complex and intimidating than in the simple illustration above. Moreover, the approach has been criticized because its users generally contrast a supposedly optimal trait with a very limited number of alternatives, rather than with a broad range of competing options. In our example we compared males that choose to ride on females at day 3, 4, or 5 versus those that select females at an earlier stage in the female's cycle. We did not consider the possibility that males search for deposition sites in order to wait there for mating opportunities with unguarded females. Because imaginations are finite, an observer will never identify every possible option—nor would it be practical to test every alternative against a proposed optimum solution. The prediction tester, therefore, may be stacking the deck in favor of one hypothesis by not considering certain other possibilities.

Still more troubling is the realization that in order to use this method of prediction testing we must make many simplifying assumptions and exclude many parameters from our analysis. For example, we have had to assume that our data on the probability of contacting a female hippoboscid in a given stage of her reproductive cycle would be duplicated at other times and places. If the probabilities vary greatly over time and space, any one sample is unlikely to reflect the past selection pressures that in actuality shaped male behavior.[1] We have, moreover, excluded a host of constraints that might affect the profitability of male riding behavior. For example, if riding is costly because it prevents males from feeding or exposes them more frequently to their predators, then males may very well not do what our calculations predicted they will. Perhaps most important of all, we have had to assume that the number of eggs fertilized by a male is an accurate measure of his fitness. Ideally, we should like to know how many of the eggs a male fertilizes actually reach the age of reproduction themselves, but these data are extremely difficult to collect. As a result, users of optimality theory are forced to settle for some measure that approximates fitness, although the degree of error is almost always unknown.

In our example, although it is not unreasonable that copulation frequency is related to egg fertilization frequency, which in turn is an indicator of the rate of production of surviving offspring, there is no guarantee that the relation between these variables is simple. Some

---

1. On the other hand, if there is a great deal of variation among different populations (or at different times in the same population) in the parameters affecting the profitability of mate guarding, an optimality approach can be used to predict how decisions on mate guarding should vary from population to population or from time to time. This has been done, for example, with a species of *Asellus,* an isopod. J. T. Manning (1980) has shown that altering the ecological conditions acting on males results in a modification of their guarding behavior. In an experimental setting in which there were more males than females, males guarded females that required a longer period before they would copulate than in experiments in which the sex ratio was even or skewed toward females. The increase in competition for mates made it advantageous for males to initiate guarding sooner, presumably because a male that rejected a potential partner was unlikely to find a superior one through continued searching. Essentially identical results have been secured for a spider mite (Everson and Addicott, 1983).

males may fertilize more than one egg as the result of a single copulation; some fertilized eggs may be more likely to develop successfully than others. Certain males may achieve a high copulation rate simply because their dominant rivals concede reproductively inferior females to them in order to have more time to defend and inseminate the rarer, high-fecundity females.

There is also confusion about what to do if a prediction derived from optimality theory is not supported. Does this require rejection of the view that natural selection shaped the behavior whose hypothetical function was analyzed by the test? No. The point of testing a prediction derived from optimality theory is not to prove or disprove the theory of natural selection (Maynard Smith, 1978a). The theory is assumed by the adaptationist to be true, for the reasons given above by Lewontin (1978). A negative result for a particular test *does* mean that we reject our initial prediction and it weakens or destroys our original hypothesis.

It is often possible to modify a rejected hypothesis to take into account new information or other possibilities not previously considered. We might for example devise a way to incorporate the constraint of predation pressure on the selection of females by males. It would not, however, be legitimate to accept the modified hypothesis as proven without subjecting it to a new test. Lewontin (1978) notes that the practice of explaining away negative results, although common, is not defensible. It *is* scientific to reject an original hypothesis, modify it, and then examine a prediction based on the altered hypothesis (or at least suggest how the test might be done).

This approach has been adopted by a number of behavioral ecologists who have been able to develop predictions that work. Despite the apparent odds against success, optimality models have been applied with good effect to a broad variety of behavioral traits, especially foraging behavior and reproductive tactics (see Krebs and Davies, 1978, 1981). Because many predictions based on optimality assumptions have been shown to be consistent with the evidence, the various assumptions of the optimality approach may not be so unrealistic, nor the practical limitations of the method so severe, that it cannot be used. At least it is not implausible that reproductive behaviors are intimately related to fitness and have been under continuous intense selection. If optimality theory is to be applicable at all, these traits should be reasonable candidates for examination.

For an actual example of application of optimality theory to insect reproductive behavior, let us briefly review G. A. Parker's analysis of searching behavior by male dung flies. Parker (1970b, 1978a,c) tested whether males behave in ways that give them the best possible chance to encounter receptive females. He measured fitness by assuming that the number of females found was correlated with fertilization opportunities, which in turn was linked directly to the number of surviving offspring produced by a male.

In the first 20 minutes after a cow has graced a pasture with a fresh mound of dung, a male dung fly's reproductive success is essentially a function of his competence in finding females drawn to the site to lay their eggs. (After 20 minutes have passed, there are copulating pairs on the dung and therefore opportunities for takeovers. This alters the optimal strategy of males, as they now have two potentially productive options.) During the initial period males hurry to the dropping; they station themselves on and about it in an effort to intercept the incoming females, which either fly directly onto the dung or walk through the grass to it after landing nearby. In many situations searching males do not fight with one another and so males can distribute themselves freely. If they disperse in the optimal manner, they will take into account both the density of competitors in an area and the likelihood of females' arriving on the dung and in zones about the cowpat. As the number of males builds up in the prime areas for contacting females, there comes a point at which a male gains more by leaving to search for females at secondary sites. In his study Parker divided the area around a dropping into five zones; most males searched on the dung itself and in the first zone about the pat, with ever-declining numbers farther from the oviposition resource. If males distribute themselves in the predicted way, the probability of encountering and capturing a female (which equals the number of receptive females in a zone divided by the number of males searching there) should be the same in all zones for any individual male.

Notice that Parker's prediction did not take into account the possibility that the females found in different zones are not equally fecund or that they differ in their willingness to accept another male whose sperm would supplant that of the original discoverer. Parker also ignored the constraints that predation, disease, or any of a host of other factors might exert on male searching behavior; his model simply assumed that the risk of predation and disease is equal in all zones. The decision not to contend with these variables seems justified by the results of his observations, which confirmed that males distributed themselves in the zones about dung pats as if their only goal was to maximize the opportunity to contact females (Fig. 1.4).

Had Parker's data contradicted his prediction, it would have been necessary to adjust the hypothesis to take into account the possible ef-

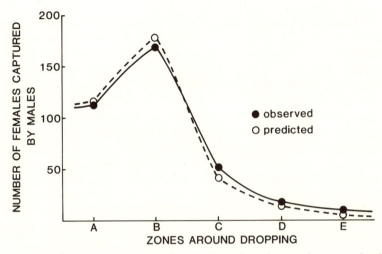

1.4   The predicted distribution of male dung flies about a dung pat closely matches the observed distribution. The assumption on which the expected distribution is based is that each individual male will behave in ways that enable it to reproduce as well as other males. (From G. A. Parker, 1974c.)

fect of other variables and to test new predictions. But in light of his results, he makes a satisfactory case that male searching behavior has the adaptive function of enabling males to contact females at as high a rate as possible. This does not mean that there is nothing else to say about male searching tactics in dung flies. Borgia (1981, 1982) has shown that male size, density of the male population, and female choice have to be considered in order to fully understand male searching behavior and its adaptive significance. The fundamental point is that despite the limitations and simplifying assumptions of an approach based on optimality theory, the use of this method to generate testable predictions seems justified by its record of success. Nevertheless, an optimality approach is (fortunately) not the only way in which to test adaptationist hypotheses. The evolutionary biologist can turn also to the comparative method to examine ideas about the adaptive value of a trait.

The Comparative Method and Evolutionary Predictions

In applying optimality theory to behavior, one matches what an animal actually does (and how it affects individual reproductive success) against a theoretical optimum. The comparative method takes a different route to determine adaptive function, by testing whether there is a correlation across species (or populations) between certain traits and particular ecological factors. The underlying premise is that similar environments—and thus similar selection pressures—should result in the evolution of analogous traits in unrelated species (convergent evolution). On the other side of the coin, divergent pressures acting on even closely related species that may have once been similar should lead to the evolution of distinctive adaptive characteristics (divergent evolution). If we find that for a variety of species the presence or absence of a special ecological condition is reliably associated with the presence or absence of a specific characteristic, then it is logical to conclude that the characteristic is an adaptation to that condition (Williams, 1975).

The comparative method has a refreshing directness to it and a freedom from many of the assumptions and limitations of optimality theory. Williams has asserted that "for answering questions on function in biology comparative evidence is more reliable than mathematical reasoning" (1975:7). The successes of the comparative method are legion. Darwin relied heavily on this approach and used it extensively in his analyses of the function of secondary sexual characteristics and facial expressions (1872, 1874). For especially good modern examples of its application to behavioral issues the reader is referred to E. Cullen (1957), Williams (1975), Clutton-Brock and Harvey (1977), Alexander et al. (1979), and Hoogland (1981).

Although we are not able to provide a particularly elegant comparative test of our hypothesis for precopulatory riding of females by males of *L. hirsuta*, we can at least show how one might go about constructing such a test. We begin by once again returning to the basic costs and benefits of male guarding behavior. On the benefit side, remaining with a gravid female may improve the male's chances of ultimately inseminating that female; but on the cost side, riding on the back of one female reduces the genetic success of a male hippoboscid, because waiting for the female to become receptive decreases the number of other females that the male could encounter and mate with during the waiting period. The ratio of costs to benefits will depend upon ecological variables that affect such things as the frequency of mating by females in their lifetimes, the number of nearly receptive females a male can encounter per unit of time searching, the ease with which a male can determine that a female is about to become receptive, and the intensity of competition for available females.

Because costs and benefits change as environmental parameters change, it is possible to formulate general predictions about the correlation of traits with ecological factors. If our hypothesis for precopulatory female riding in *L. hirsuta* is correct, then similar behavior should occur in unrelated species faced with similar ecological pressures. Other things being equal, female riding should be more likely in species in which the following are true:

(1) Males can identify females that are about to become receptive (the cost of waiting is thereby reduced).

(2) Riding on nonvirgin females is associated with species whose females mate more than once at intervals throughout their lives. (If only freshly emerged females are receptive, there can be no genetic gain from riding on the back of a previously inseminated, gravid female.)

(3) Precopulatory mate guarding is more prevalent in species whose nearly receptive females are rare and dispersed than in species with relatively dense populations of females. If there are few receptive females, the cost of remaining with a potential mate falls, because the probability of encountering another superior mate in the time spent riding on one female is slight.

The ecological conditions affecting *Lynchia hirsuta* make the hypothesis of adaptive mate guarding plausible for this species:

(1) Females that are about to become receptive probably can be identified by their swollen abdomens or some other clue.

(2) Females are rare; usually no more than a single individual is found per quail (Tarshis, 1958), so that the continuous mate-searching strategy is unlikely to be highly productive.

(3) When depositing the prepuparium, the female leaves her host and goes to a shaded place in the environment, of which there are probably many suitable for her offspring. Males that abandon quail to wait by a shady spot are not especially likely to encounter a number of mates and are without food, besides.

In the locust, *Schistocerca migratoria,* male behavior is very similar to that of the fly. Males locate gravid, ovipositing females and remain by or on them, repelling other males (Parker, Hayhurst, and Bradley, 1974; Parker and Smith, 1975). As in *L. hirsuta,* locust females mate at intervals. They become receptive immediately after depositing a batch of eggs. Potentially receptive females can be readily identified by their oviposition behavior, in which the abdomen is buried in the soil (Fig. 1.5). Moreover, it is known that locust females preferentially use the sperm of their most recent mate. Even though a female may not be a virgin, a male can benefit genetically by copulating with her.

In animals still more distantly related to hipoboscid flies, such as spider mites (Potter, Wrensch, and Johnston, 1976), various crustacea (J. T. Manning, 1975, 1980; Shuster, 1981), and a salticid spider (Jackson, 1980), precopulatory mate guarding also occurs. Potter (1981) has shown that in a spider mite males can identify females about to undergo a final moult and become receptive adults. Males fight more strenuously for possession of these females than for those that are farther from becoming receptive. In the isopod *Thermosphaeroma thermophilum* males can also determine when females are about to undergo the sexual moult that precedes a bout of sexual receptivity (Shuster, 1981). Males carry about-to-become-receptive females, which are very rare, underneath their bodies for periods of up to several days while rejecting females that are not near the sexual moult (Fig. 1.6). These cases of apparent convergent evolution in male behavior among unrelated animals support the contention that factors such as cues that identify nearly receptive individuals, and the rarity of receptive females, favor the evolution of precopulatory guarding of females, even if they are not virgins.

Examples of divergent evolution among related species can also be used as a source of evidence for a particular generalization about adaptation. Because related species share common ancestry, they can be expected to share behavioral features as well, unless they have been subjected to divergent selection pressures in the past. Relatively little is known about male behavior of hippoboscid flies (but see Graham and Taylor, 1941; Bequaert, 1953). In some species such as *Lipoptena*

1.5   A female locust ovipositing (her abdomen inserted in the ground), with a male in close attendance. He will try to impregnate her as soon as she finishes laying her eggs. (From Parker, Hayhurst, and Bradley, 1974.)

1.6   A male isopod, *Thermosphaeroma thermophilum,* that is carrying a female under his body in (unconscious) anticipation of the time that she will become receptive. (Photograph by S. Shuster.)

depressa, a parasite of deer, males appear to behave similarly to males of *Lynchia hirsuta* in that they discriminate between females with and without a mature larva. Whereas females in the early stages of pregnancy are not clasped by males, it is unusual to find a female with a mature larva in her uterus that does not have a male clinging to her abdomen (I. M. Cowan, 1943). In marked contrast, previously inseminated females of *Stilbometopa impressa* are not mounted and ridden by a guarding male. In this species, also a parasite of quail, females are believed to mate once shortly after achieving adulthood (Tarshis, 1958) and therefore males cannot gain genetically by guarding a nonvirgin.

Difficulties with the Comparative Method

The comparative method appears to provide a simple and direct way to test hypotheses about the adaptive function of a trait. Although it is *the* major source of such tests, it is not without its shortcomings. All interspecific comparisons are clouded by the fact that separate species have distinct ancestry and therefore any similarities (or differences) among them may be the result of historical accident rather than similar (or different) selection pressures. Moreover, different species are subject to a great variety of environmental forces, some of which may confuse or confound the usually simple correlation an observer is trying to discover between a single category of selection and a specific trait. Although males of two species may share the socioecological pressure of having females that exhibit a repeating cycle of receptivity, they may differ greatly in the food resources they exploit or in the kind of predator that attacks them. These factors in one species may have effects that outweigh selection in favor of mate guarding and thus the two species differ behaviorally, even though one common factor may be acting in the predicted direction. Because all things are rarely equal, one runs the risk of discarding a basically sound idea because of confounding variables. (Still, better this risk than the danger of too readily accepting a false hypothesis.)

A related criticism of the comparative method is that it lacks the rigor of the controlled experiment, in which the observer systematically manipulates the parameters of interest while controlling all the other variables (Stearns, 1976). In ideal applications of the comparative method, however, the confounding effects of phylogenetic constraints and contradictory ecological factors are perfectly controlled for by the selection of a sample of species that truly randomizes the effects of all but the key variables (Alexander, 1979). If a sufficiently large sample of unrelated species provide the data for the test, all irrelevant features should cancel one another out, leaving an unambiguous correlation between a trait and an ecological pressure. The ideal is admittedly never achieved in practice, but the same is true for the experimental approach to biological hypotheses. The degree to which an experiment is able to control confounding variables is usually a function of whether the experiment is conducted in the field or in the laboratory. The gain in control over variables in the lab is achieved at the expense of some relevance to natural conditions, so that there is always some doubt about the results of experimental tests as well as comparative ones.

The recognition that total certainty is not attainable in science does not free us from the search for the most effective means to analyze the possible adaptive value of behavioral traits. More thought is required by students of behavior on the problems of how to state null hypotheses precisely, how to collect a fair sample of data bearing on a hypothesis of interest, and how to employ statistical tests appropriate to the data (Clutton-Brock and Harvey, 1977; Maynard Smith, 1978a). But at least a beginning has been made and there are potentially effective ways to test evolutionary hypotheses.

Philosophy and Organization of Our Text

Having acknowledged some of the problems with using experimental, observational, and comparative techniques we trust our readers recognize that the ideas about adaptive function presented in this volume are tentative, sometimes highly speculative, and subject to revision or rejection in the light of new evidence. We survey a spectrum of topics related to reproductive competition, with the aim of showing how characteristics of insects may be interpreted from an adaptationist viewpoint. And we rely heavily on selectively collected examples of apparent convergent evolution as support for statements about the relation between evolved behavioral characters and environmental pressures. Much of what is presented here is far from the last word on the subject. Our hope is that the hypotheses reviewed in this book will stimulate others to look at the behavioral world of insects with a willingness to test selectionist predictions.

Our text begins with an examination of the diversity of reproductive modes in insects (Chapter 2) and the still unresolved issue of the adaptive value of sexual reproduction. We then review how sexual reproduction creates the selection pressures associated with sexual selection (Chapter 3). This sets the stage for a discussion of how one component of sexual selection—intrasexual selection—favors traits that enable members of one sex (usually the males) to outcompete others of the same sex for access to mates and fertilization opportunities (Chapters 4–11). We conclude with a survey of epigamic selection, which is sexual selection for traits that make members of one sex (usually the males) attractive to members of the other sex (Chapters 12–14). Our central theme will be that male-male competition and selective mate choice by females are at the heart of insect reproductive behavior, and that an appreciation of the role of sexual selection is central to an understanding of the sexual tactics of insects.

2

Modes of Reproduction

THE MODES OF REPRODUCTION of insects are diverse and remarkable and have stimulated the invention of some equally remarkable labels such as pedogenesis, polyembryony, thelytoky, gynogenesis, heterogony, facultative deuterotoky—and our favorite, arrhenotokous parthenogenesis. Faced with such formidable terms, one may be tempted to take up the study of mammals. Rather than do this, we shall examine the catalogue of reproductive techniques, sexual and asexual, from an evolutionary perspective. Our goal is to identify some costs and benefits of the different modes of reproduction by examining their genetic consequences in different ecological settings. In keeping with the working hypothesis established in the first chapter, we shall offer interpretations of the various reproductive methods consistent with the notion that they are evolved adaptations that promote the genetic success of individuals.

The natural history of an aphid provides an appropriate introduction to this topic. Aphids may be small and familiar insects, but their life cycles are highly complex and often involve an alternation of different methods of reproduction. Williams (1975) has argued that intraspecific variation in the manner of reproduction provides a particularly important source of information on the function of the different methods. If a correlation exists between a particular ecological variable and a particular reproductive system, then one can logically conclude that the trait is an adaptation for that condition. Aphids therefore hold the promise of helping us identify the adaptive value of sexual and asexual reproduction, a goal that is the central theme of the chapter.

The Life Cycle of *Pemphigus bursarius*

Temperate-zone aphids, like the British lettuce root aphid (*Pemphigus bursarius*) studied by Dunn (1959), are confronted with plant growth patterns that vary greatly from season to season and from species to species. The behavior and life history phenomena of aphids are beautifully adapted for the exploitation of their host plants, as the following account of *P. bursarius* makes clear. In the spring eggs that have overwintered in crevices of poplar trees hatch, giving rise to immature founder females (reproductive type 1) just as the leaf buds of the tree are opening in April. The founder crawls onto the stalk of a young leaf, where she causes the formation of a gall on the leaf petiole (Fig. 2.1). She does so by repeatedly inserting her piercing mouthparts into the petiole in a small circular area. The aphid evidently inoculates the plant tissues with growth inhibitors and stimulators, because the tattooed area experiences greatly reduced growth, whereas the region immediately adjacent undergoes rapid cell proliferation. Through her injections the female eventually creates a hollow gall around her in which she feeds and grows, passing through several immature stages of increasing size.

As the female develops into a nearly sessile adult without wings and with reduced legs and antennae, the gall grows too, reaching the size of about 10 × 15 mm at the time the aphid gives birth to the first of 100 to 250 daughters. The founder female reproduces *parthenogenetically* (she is not fertilized by a male); moreover, her offspring are born living, as nymphs rather than eggs. The nymphs develop inside the gall, feeding on plant fluids and excreting quantities of sticky honeydew. Their own waste might entrap and kill them, were it not for the fact that the founder female secretes many strands of wax from special glands on her body. When her daughters are born, they move about and reduce the waxy filaments to a powder that coats each drop of honeydew; this forms a safe, nonsticky pellet that eventually falls out of an opening in the gall.

The foundress' daughters, when mature, have fully developed legs,

2.1 (A) Gall made by a North American *Pemphigus* aphid that, like the lettuce-root aphid, exploits poplar leaves as a food resource and shelter. (B) Scanning electron micrograph of a founder female on a poplar leaf in the earliest phase of gall formation. (Photographs by T. Whitham.)

antennae, and wings, and soon leave their shelter. About 9 weeks have passed since their mother formed the brood chamber. After shaking their wings to remove wax particles, the daughters fly off in search of a new plant host—lettuce being a favorite. The winged females (reproductive type 2), like their mother, reproduce parthenogenetically and ovoviviparously and can do so soon after locating their new host because they carry embryos inside their bodies that were well developed even before they left the poplar gall. The young aphids initially feed on lettuce leaves, but later leave the plant and move on well-formed legs over the soil until they find a crevice or worm tunnel that leads them to a root of the lettuce plant. There they continue to feed and grow, developing into *wingless,* sedentary adults with reduced legs, which reproduce parthenogenetically and ovoviviparously (reproductive type 3). Their offspring initially are relatively mobile and find unexploited portions of the root system upon which they settle prior to producing another generation like themselves. In August, as the growing season for lettuce passes, a generation produced by root-feeding parthenogenetic females leaves the roots while still immature and crawls out onto the base of the plant or its lower leaves. They develop wings as adults and fly from lettuce to poplar trees, where they seek crevices in the bark. These individuals (reproductive type 4) parthenogenetically produce only sexual progeny, males and females, that lack mouthparts altogether (type 5 reproductives).

The sexual females remain in the bark cranny where they were born and await the arrival of a male, which walks about in search of a mate. The females are, according to Dunn, essentially one large egg with legs. After mating, each female lays a single giant, fertilized, greenish-white egg covered with a light dusting of wax. It is this egg that, after the passage of winter, gives rise to another founder female, which initiates a new annual cycle of lettuce root aphid generations.

The functional significance of this elaborate pattern of production of different kinds of aphids over time (Fig. 2.2) may be related to seasonal changes in the availability of food. Young poplar leaves and growing lettuce plants are transitory, patchily distributed, and in great demand by a host of other herbivores. The founder female's ability to produce a large number of offspring in a short time takes advantage of both the short burst of rapid growth in the poplar and the relatively low competition for food early in the season. Her ability to cause the poplar to produce additional plant cells for her own use and that of the first generation of daughters makes it possible for large numbers of her offspring to mature while remaining in the shelter of the gall.

The founder's daughters leave the poplar at a time when its produc-

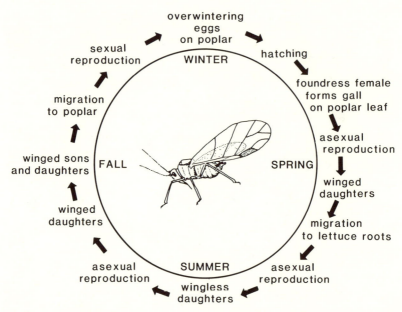

2.2 Annual cycle of the lettuce-root aphid showing the shift from asexual to sexual reproduction at the time when individuals produce overwintering offspring that will face new conditions in the spring. (Drawing by J. Smith.)

tivity is decreasing and search for a new host at the start of its growing season. Their fecundity is much less than the mother's, presumably because they must divert materials away from production of embryos and into the legs, antennae, wings, and muscles that make long-distance dispersal possible. Nevertheless, as soon as they find a suitable host, they begin at once to generate offspring that will tap the host at a time when competition for this food is relatively low. A single aphid alighting on lettuce can, over the course of the summer and through several generations, produce enormous numbers of descendants that use the resources she has discovered. But as populations grow and the lettuce season comes to an end, individual females may receive a variety of signals (tactile stimulation, decreasing photoperiod length) correlated with the optimal time to begin producing female forms that return to poplar, so that their descendants will be in a position the next spring to exploit the growing season of the tree (Lees, 1966; Hardie, 1980).

From the perspective of this chapter, a key question about aphid life history is why some females gain genetically by producing male and female progeny that must engage in sexual reproduction in order to contribute to the next generation. Before we can try to answer this question, we need to define precisely the difference between sexual and asexual reproduction.

The Definition of Sex

To many, the different modes of reproduction of aphids would seem to be classic examples of sexual (biparental) reproduction and asexual (uniparental) reproduction. Regrettably—or happily, depending on your point of view—nature is less clear-cut. What is one to do with a hermaphroditic individual that may fertilize itself, as is true for many species of plants, for example? How should one categorize species in which a male partner is required to provide sperm that activate development of the egg, but that do not actually enter and fertilize the egg? The modern consensus is that the essential ingredient in sexual reproduction is the creation of offspring that differ genetically from the parent (owing to mechanisms other than mutation). With the exception of some bacteria and a few other organisms, this means that sex is typically synonymous with meiosis (Ghiselin, 1974; Williams, 1975; Maynard Smith, 1978b). Meiosis helps guarantee that an individual's progeny will differ genetically from the parent because the gametes or sex cells produced meiotically have two important properties: (1) the chromosome number of the gamete is reduced by one-half from the diploid cell that gave rise to it; and (2) in the process of chromosome reduction there may be *recombination* or exchange of genetic material between homologous chromosomes. (There are of course two sets of chromosomes in each diploid cell, one donated by the mother and the other by the father.)

The first property, chromosome reduction to one set, is essential if the offspring (created by the union of two gametes) are to have the same total number of chromosomes as the parents. The second property, recombination, ensures that the organization of genes on the chromosomes of the gametes will not be identical to that found in the diploid cells of the parent. This is far from trivial, because interactions among genes along a chromosomal strand may affect the timing and rate of production of enzymes in cells. These epistatic interactions influence the metabolic capacities of cells and their course of development. Altering the physical arrangements of genes on chromosomes may alter the phenotype that results from the gene-environment interplay. Therefore, meiosis produces progeny that differ genetically from their parents.

Asexual reproduction generates offspring that are identical genotypically to their parent through mitosis. The cellular mechanism of mitosis permits *exact* duplication of the chromosomal contents of a diploid cell. If the duplicated cell is employed as an offspring, it will be genetically the same as its parent, and it will also be identical with any other mitotically produced offspring created by the parent.

Sexual Reproduction in Insects

The vast majority of insects reproduce sexually, employing meiosis to produce gametes. In most sexually reproducing insects, gender is well defined, males producing sperm that unite internally with eggs developed within females. This pattern is so familiar to us that we tend to consider all other modes of reproduction aberrant or even vaguely indecent. Yet other systems of sexual reproduction have their insect representatives. Hermaphrodites are exceptionally rare, known to occur only in one genus of coccid bugs (scale insects). These creatures produce sperm and eggs and then fertilize themselves (Engelmann, 1970). Equally rare are cases of gynogenesis, in which females use sperm only to activate egg development but not to provide genetic material for the nucleus of the egg. This pattern is practiced only by a few moths and by one beetle. The males (which come from closely related species) are victimized by the gynogenetic females, as they supply sperm, time, and energy for no genetic reward. Indeed J. E. Lloyd (1979a) suggests that females of this sort may have retained a "dependence" on copulation and sperm-triggered egg development in order to continue to secure sperm and other ejaculate for their *nutritional* value (see Chapter 12).

Gynogenesis is merely a special form of parthenogenesis (reproduction without fertilization). Sexual parthenogenesis (production of unfertilized offspring that differ genetically from the female parent) is relatively common among insects, having been recorded for at least some representatives of almost all insect orders (R. F. Chapman, 1971). It is characteristic of all female ants, bees, and wasps, which employ meiosis to generate gametes and are able to lay unfertilized eggs that become haploid males. Through recombination the chromosomes of these individuals have been reorganized so that the genome of a male not only is reduced relative to its mother but also is restructured.

In a very few other insects (such as some scale insects) meiosis is employed to generate diploid eggs; in the course of egg formation two haploid pregametes fuse to produce a diploid cell, which then develops without benefit of fertilization (Miller and Kosztarab, 1979). Again the genetic organization of the offspring's chromosomes need not be identical with that of the parent; the relatively slight differences between them are sufficient to label this a form of sexual parthenogenesis.

Still another mode of meiotic parthenogenesis involves the pairing of homologous chromosomes in a diploid pregamete and crossing over without the subsequent doubling of each chromosome and the two meiotic divisions required to produce a haploid gamete. The diploid cell's chromosome number is not altered in this abbreviated form of meiosis, but the arrangement of genes on the chromosomes may be modified by recombination. Cognetti (1961) argued that some aphids may employ this system. If they do (and there is contradictory evidence on this point—see Addicott, 1979), aphids engage in sexual reproduction.

Although aphids may or may not reproduce asexually, this mode of reproduction is perhaps widespread among the insects. Mitotic parthenogenesis without alteration of the parental genome probably occurs in at least 12 orders of insects and is well documented in some snout beetles, chironomid midges, phasmid walkingsticks, and book lice (Suomalainen, Saura, and Lokki, 1976). These insects reproduce mitotically, thereby removing the possibility of crossing over in the formation of a daughter cell with the parental genome.

A special form of asexual reproduction that involves *meiotic* parthenogenesis may occur in certain grasshoppers if the parental genome is exactly restored following meiosis (R. F. Chapman, 1971). In cases of this sort the offspring will be genetically the same as the parent.

A number of insects alternate between sexual and asexual reproduction at different stages of their life cycle. Not unexpectedly, this phenomenon has been labeled (heterogony). One of the most remarkable heterogonic life cycles is exhibited by a few species of cecidomyid midges (Fig. 2.3 and Engelmann, 1970). While there is much variation within the group, some species have the capacity to omit the adult phase altogether. Instead, the larvae become sexually mature and lay eggs, or alternatively they may produce cells that develop within their bodies into daughter larvae (pedogenesis) that will eventually kill the mother larva by consuming her tissues. The offspring in turn may repeat the sacrifice of their mother, or they may ultimately develop into adult males and females. The adult flies reproduce in the typical sexual pattern.

Several parasitic Hymenoptera combine meiotic and mitotic reproduction in a different way. Females produce eggs meiotically and then introduce one or a few eggs (which may be fertilized or not) into the

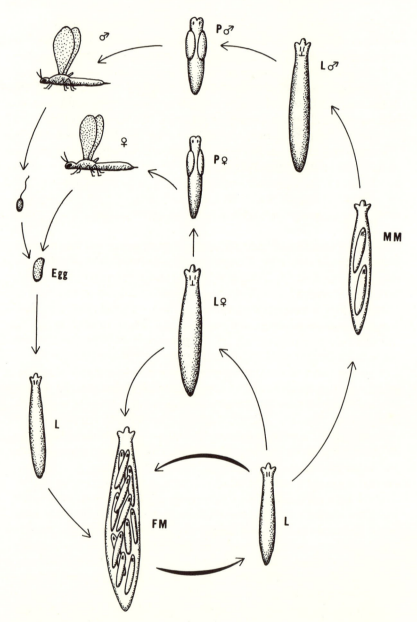

2.3 Life cycle of a cecidomyid midge that can reproduce asexually (as a larviform female that parthenogenetically produces offspring within her body) or sexually (as winged males and females that disperse to new habitats). *FM* = larva that produces only daughters, *L* = larva, *MM* = larva that produces only sons, *P* = pupa. (From Camenzind, 1962.)

larger egg of a host species (often a moth). The host egg hatches into a larva and matures. The parasite's egg gives rise to a mass of undifferentiated cells, each of which then develops into a separate larva (polyembryony). In this way a single egg can give rise to a thousand embryos, a clone of genetically identical siblings that emerge from the host larva only when the victim is finished growing and has been consumed entirely by its burden of parasites (Engelmann, 1970).

Table 2.1 summarizes the reproductive modes discussed in this section and indicates for each the degree to which the offspring will differ genetically from their parents.

The Adaptive Significance of Sexual Reproduction

We have established that varieties of both sexual and asexual reproduction exist in insects. This diversity creates a corresponding variety in the degree to which offspring resemble their parents genetically. Insects therefore provide the comparative material from which one might be able to establish correlations between ecological parameters and degree of genetic diversity in offspring. Information of this sort could help to elucidate one of the most difficult questions in evolutionary biology, that of the ultimate function of sexual reproduction.

At the heart of the matter is evidence which suggests that sexual reproduction is damaging to an individual's reproductive success. Recognition of the disadvantageous aspects of this mode of reproducing comes from the work of Maynard Smith (1971, 1978b) and Williams (1975, 1980). Although a number of arguments have been developed, one simple way to demonstrate the cost of sexual reproduction is to consider what a female loses by producing male offspring (Williams and Mitton, 1973).

Asexual parthenogens arising in a population with males and females will enjoy a huge reproductive advantage by producing daughters only (if certain conditions apply). Let the sexual female invest her resources equally in sons and daughters and assume that parthenogenetic females secure as many resources for offspring production as sexual females. Then the parthenogen will generate twice as many female offspring as the sexual female. Assuming that parthenogenetic daughters survive as well as those produced sexually, the proportion of the female population that is parthenogenetic will double from generation 1 to generation 2. The process will lead to the eventual elimination of the sexual females (and therefore of the sexual males) by the asexual parthenogenetic females (Maynard Smith, 1971).

Table 2.1   Methods of reproduction in insects and their consequences with respect to genetic similarity of parent and offspring.

Asexual reproduction: offspring share exactly the same genotype as the parent.

(1) *Mitotic parthenogenesis:* female produces eggs mitotically that develop independently of her, or germ cells that develop within her body (pedogenesis).

(2) *Meiotic parthenogenesis:* female produces eggs meiotically, but they do not differ genetically from the parent because of mechanisms that restore the parental genome following meiosis.

Sexual reproduction: parent(s) produce offspring that have different genotypes from the parent(s).

(1) *"Typical" sexual reproduction:* offspring are derived from the union of gametes produced by two separate individuals. Any allele present in the heterozygous condition in a parent has a 50-percent chance of not being represented in each offspring (if meiosis proceeds normally). Overall genotypic differences between parent and offspring depend upon the degree of inbreeding. In highly inbred populations that tend to be genetically uniform, a parent and its sexually produced offspring may share far more than 50 percent of their genes.

(2) *"Typical" sexual reproduction with polyembryony:* the same as above, except that a clone of offspring is produced mitotically from a single fertilized egg. All offspring differ from their parents to the same extent, but are genetically identical to one another.

(3) *Hermaphroditism (selfing):* offspring are derived from the union of egg and sperm produced meiotically by one individual. The degree of genetic variation between parent and offspring depends on the degree of heterozygosity in the parent, which determines the genetic differences among gametes. Usually the offspring of hermaphrodites are genetically quite similar to the parent (relative to parent-offspring similarity in biparental outbred species).

(4) *Sexual parthenogenesis:* a female produces eggs meiotically that can develop without uniting with a male gamete. Again, the degree of genetic variation from parent to offspring depends on the degree of heterozygosity in the parent. The genome of an offspring may nearly duplicate the parental genome in some cases, but still be different because of recombination and the resulting rearrangement of genetic material on the chromosomes. Gynogenesis is a special form of sexual parthenogenesis in which a meiotically produced egg is stimulated to develop by a sperm but is not actually fertilized by it.

The advantage enjoyed by parthenogenetic females applies far more broadly than originally thought (Williams, 1980). Sexual reproduction is an evolutionarily unstable condition even in species whose males make a paternal investment per offspring equal to that of the female (and therefore might double the production of offspring of their sexual mate). These populations are vulnerable to invasion by a mutant "cheater" female, which accepts the male's donation of nutrients or parental care but reproduces parthenogenetically via diploid eggs. The existence of gynogenetic females shows how female exploitation of males can evolve. The cheater female's genetic gain is twice that of a noncheater female, which uses sperm to fertilize her meiotically produced eggs.

Given that the cost of producing males is very large, can nonparthenogenetic sex confer reproductive benefits to individuals that more than compensate for these disadvantages? Or is the prevalence of sexual reproduction best explained as the outcome of group selection favoring sexual over asexual populations? There are strong arguments for both the group selection hypothesis (Maynard Smith, 1971, 1978b) and for the individual selection hypothesis (Williams and Mitton, 1973; Williams, 1975; Triesman, 1976; Glesener and Tilman, 1978).

Group Selection and Sexual Reproduction

In sexual reproduction we witness the widespread occurrence of something that theoretically should not be prevalent. One way to approach this paradox is to concede that asexual parthenogens enjoy a reproductive advantage in the short term, but that sexual reproduction is well represented among plants and animals because of the long-term processes of group extinction and survival. This is a classic group selectionist argument, which implies that evolution can result in a trait that reduces the relative reproductive success of individuals in the present in order to preserve the species as a whole in the future.

It is true that sexual reproduction has the effect of generating variation in offspring and combining favorable mutations, and thus it could be that sexually reproducing species survive longer than asexual ones. The lack of genetic flexibility of asexual species could lead to the accumulation of deleterious alleles and reduce the number of unusual variants available (although Mitter et al., 1979, found a high degree of genetic diversity among parthenogenetic populations of a moth). These features in turn could promote a high rate of extinction and therefore reduced representation of asexual forms in the pool of currently existing species.

Maynard Smith (1978b) reviews several lines of evidence in support of the group selection position. First, parthenogenetic strains appear infrequently, suggesting that viable mutants of this sort do not arise often. In the absence of an effective parthenogenetic competitor, sexual forms can persist indefinitely. Second, parthenogenetic varieties do appear to experience early extinction, as judged from an analysis of their taxonomic distribution. Third, parthenogenetic forms often coexist with but fail to replace the sexual ancestor from which they are derived, suggesting that their theoretical reproductive advantage is not realized in practice.

The familiar difficulty with the group selectionist view is that individual selection appears to be far more powerful than group selection in shaping the adaptations of a species. If viable asexual mutants do arise in sexual populations, it is hard to see why they would not very rapidly spread and eliminate the genes for sexuality from the population. In order to account for the persistence of sexual reproduction, is it possible that sex confers great benefits to individuals under certain ecological conditions?

Individual Selection and Sexual Reproduction

One can devise hypotheses that account for sexual reproduction on the basis of its short-term advantages to individuals rather than its long-term benefits to a population or a species. One approach, recently developed by Bernstein, Byers, and Michod (1981), is to propose that sexual reproduction originated when single haploid cells began to exchange DNA segments, an exchange that enabled individuals to acquire material to repair damaged regions of their own chromosomes. Cells that then spent an increasing portion of their life span in the diploid state might gain because deleterious mutations would not be expressed as frequently as in haploid cells.

A somewhat similar argument focuses on the ability of sexually reproducing individuals to create offspring with fewer mutations than the parents carry. Asexually reproducing females are condemned to pass on to their progeny all the mutations that their egg cells have acquired during the female's lifetime. Sexual females can avoid this if they possess mechanisms that reject defective eggs in favor of ones that through chromosome reduction and recombination happen to carry relatively few mutant alleles. Moreover, a female that mates with a male that provides a great excess of sperm is in a position to fertilize her eggs with sperm that are relatively mutation free. This could be achieved passively, simply by permitting the sperm to compete for access to the egg, for congenitally defective (mutation-laden) sperm would be weeded out. Alternatively, the female might employ active choice if she possessed physiological mechanisms along her reproductive tract (Kiefer, 1969) or on the surface of her eggs to discriminate against sperm that carried damaging mutations (Stanton, in press).

That sperm-screening devices have evolved is suggested by the finding that the frequency of recombination in the meiotic formation of sperm in various organisms is highly correlated with so-called gamete redundancy (the "extra" sperm donated by a male in each copulation above and beyond the number needed to fertilize the female's mature eggs). The greater the recombination frequency, the greater the chance for error in combining chromosomal segments and the higher the proportion of defective sperm. A high frequency of recombination therefore favors females with mechanisms that happen to eliminate recombination accidents; this in turn favors males that provide a sufficiently large number of sperm to ensure that some will pass the screening tests.

Under either an active or a passive selection process, the sperm that actually fertilize an egg should be far better than average in terms of freedom from mutations. The improved genetic quality of offspring produced through sexual means could therefore more than compensate a female for the disadvantage of producing males (Stanton, in press).

In addition to the mutation elimination hypothesis, an important (complementary) explanation of sexual reproduction states that under certain conditions individuals raise their genetic success by producing genetically variant progeny (Williams, 1975; Maynard Smith, 1978b). Individuals that generate genetically diverse offspring in an *unpredictably* and rapidly fluctuating environment may outreproduce individuals that generate offspring asexually. This converts the group selection hypothesis, based on supposed long-term advantages to the species as a whole, to one that focuses on the individual advantages of sexual reproduction based on *short-term* changes in the environment.

To make this individual selection argument clearer, Williams introduced the metaphor of the changing environment as a lottery in which the spaces available for colonization by offspring are the prizes. If the properties of these spaces cannot be predicted exactly in advance, the parent that produces variety in its offspring (tickets with different numbers) may succeed in placing more than twice as many progeny in the available spaces as a parthenogenetic female that produces a brood of daughters all having the same genotype (tickets with the same number). Maynard Smith has pointed out that implicit in this metaphor is the notion that competition among sexually produced siblings is reduced relative to competition among parthenogenetically produced ones. Individuals with different genotypes may be able to locate and exploit different kinds of vacant slots in their environment, whereas a gaggle of identical offspring will have the same ecological requirements and will compete directly for vacant slots of the same type.

The lottery analogy can be looked at in another way: diversity in tickets (genotypes) may improve the odds that some offspring will win a chance to exploit an open "niche" in an environment, or it may improve the chances that some offspring will not be exploited by enemies in the environment, such as predators, bacterial pathogens, or parasites. Jaenike (1978) first advanced this hypothesis, and W. D. Hamilton and his co-workers (1980, 1981) have provided sophisticated mathematical analyses of some genetic models that show how parasite pressure and environmental cycles can favor sexual over asexual reproduction. Viewed in this way, sexual reproduction permits an individual not to put all its genes in one basket. A pathogen or parasite may evolve rapidly, introducing unpredictable changes in the environment. If a new form could destroy one offspring, it might be able to destroy an entire brood if they were genetically identical (and therefore possessed identical physiological resistance to attack), whereas the genetically and physiologically diverse offspring of a sexual union might contain some especially resistant varieties (see Edmunds and Alstad, 1978, for possible evidence on this point). In this context if frequency-dependent selection by predators is widespread (and with it the rare phenotype advantage), then sexual reproduction becomes more advantageous. In this version of the lottery the reward of survival in an unpredictable environment of exploiters is more likely to be enjoyed by genotypically unusual individuals.

Mathematical treatments of various versions of the lottery model convinced Williams (1975) that the advantages of sexual reproduction could apply only under special conditions and only when large numbers of offspring were produced sexually by a female. But Hartung (1981) argues that in low-fecundity organisms in which males provide vast numbers of gametes with each attempt to fertilize a few eggs, females may be engaged in a kind of prelottery in which they test the gametes received and select the few with truly superior gene combinations. This hypothesis views the obstacles to fertilization imposed by the design and biochemistry of the egg membrane and the female's reproductive tract as a way of selecting not just mutation-free sperm but sperm with novel and compatible gene combinations. Unusual but developmentally competent zygotes may be able to outrace coevolutionary exploitative advances made by the animal's (sexually reproducing) enemies and to circumvent counteradaptations evolved by the animal's (sexually reproducing) prey (see also Triesman, 1976; Glesener and Tilman, 1978; W. D. Hamilton, 1980; Hamilton, Henderson, and Moran, 1981).

Tests of the Adaptive Value of Sexual Reproduction

A common thread running through most hypotheses on the genetic advantages to individual parents of sexual reproduction, including the lottery model, is that the benefits of sex apply to animals living in relatively unpredictable environments. If this view is correct, it should be possible to discover correlations between sexual versus asexual reproduction and the degree of stability of the environment. Williams (1975) noted that, as a rule, asexual propagules tend to remain close to the parent that produced them and to develop rapidly; they therefore will often reproduce in conditions similar to those experienced by their parent. Sexually produced offspring, on the other hand, typically disperse long distances from their parents or they begin growth and reproduction a considerable time after the death of their parents. Thus they are likely to encounter many environments that differ from those in which their parents lived. These correlations support the hypothesis that sex is an adaptation geared to the production of offspring that are likely to face changed and unpredictable conditions.

The comparative method has also been employed by Levin (1975) for plants and Glesener and Tilman (1978) for animals, to test the prediction that environmental predictability should favor asexual reproduction. They argue that biologically simple communities are characteristically found in geographic areas such as temperate regions, high altitudes and latitudes, disturbed habitats and islands, and island-like habitats. Individuals in these areas face relatively few competitors and predators from a limited range of species, so that the variability and unpredictability of interspecific interactions should be reduced. This makes the conditions favoring asexuality more likely to occur. There actually are relatively more asexual plants and animals in these geographic regions and habitats than in biotically complex environments, a finding that supports the prediction and its underlying hypothesis.

The general correlations between degree of unpredictability and mode of reproduction that have been established by interspecific comparisons can be supplemented by examination of the conditions under which members of the same species employ sexual as opposed to parthenogenetic reproduction. Animals such as aphids, whose life cycles feature alternating periods of different reproductive modes, provide the strongest support for the hypothesis that sexual reproduction can confer short-term genetic advantage on the individuals that practice it (Williams, 1975). The fact that sexual reproduction persists in these species, despite the occurrence of viable parthenogenetic alternatives, indicates that it must confer immediate individual advantage that more than compensates for the cost of sexual reproduction (but see Maynard Smith, 1978b).

Does the alternation of reproductive types correlate with changes in the predicted direction in the environment of the species? In aphids the foundress female that succeeds in finding a suitable food source promptly produces parthenogenetically. Her offspring develop rapidly (F. Taylor, 1981) and live near her (as do the lettuce root aphids, for instance) in an environment that may even be shaped by the mother (as is the poplar gall of the lettuce root aphid). This environment is stable for several generations, and identical copies of her genotype are likely to survive and reproduce within it. Typical sexual reproduction in the lettuce root aphid (and many others) is reserved for the production of offspring that will necessarily face a highly unpredictable environment. The sexual aphids mate in the late fall; their progeny, the foundress females, emerge in the spring on a plant almost certain to be different from the plants that supported their parents and the ancestors of their grandmothers.

Likewise in cecidomyid midges (Fig. 2.3), a female that locates an untapped fungus of the appropriate species has the capacity to reproduce parthenogenetically. But as the food supply is depleted, sexual forms are produced whose offspring will disperse in search of new patches of fungus with properties likely to be different from the (now consumed) fungus that supported the parent of the sexual males and females. Thus in aphids and fungus gnats, genetically diverse offspring are produced when the decline in productivity of a resource makes dispersal or winter diapause advantageous. This correlation is not invariant. For example, in the lettuce root aphid sexual forms are not produced at the time of movement from the primary host (poplar gall) to the secondary host (lettuce). If the lottery model is correct, one would anticipate that aphids should produce genotypically diverse offspring at times of dispersal—although this would depend on the nature of the environment provided by the lettuce roots. (For a different view on the timing of sexual reproduction and its significance see Meunchow, 1978.)

Many more tests of the relation between ecology and mode of reproduction are needed. One way to broaden their scope would be to venture beyond comparisons of the two extremes: sexual versus asexual organisms. As we described earlier, the different forms of sexual reproduction have different outcomes in terms of the degree of genetic diversity produced in offspring. If the function of sexual reproduction is to generate sufficient diversity in offspring to match the unpredictability in the environment, then there should be a correspondence between the degree of environmental unpredictability that the offspring will encounter and the kind of sexual reproduction employed by the parent.

The Significance of Inbreeding

One way to test this prediction would be to search for the ecological correlates of inbreeding. Inbreeding functions as a way in which a sexually reproducing individual may reduce, but not eliminate entirely, the genetic diversity in its progeny. If the environmental predictability hypothesis is correct, inbreeding should occur when an individual's offspring are likely to grow and reproduce under conditions that are only slightly different from those encountered by the parent. There have been no explicit tests of this prediction, to the best of our knowledge, and the issue is clouded by competing explanations for the occurrence of inbreeding, as the following case demonstrates.

In the eumenid wasp, *Euodynerus foraminatus,* the females often provision multicelled nests, from which eventually emerge several individuals of both sexes. In nests of this sort an emerging male may claim the area around the exit hole as his territory and if he can repel other males from the site, he will copulate with his sisters when they emerge (Fig. 2.4). D. P. Cowan (1979) estimates that about 40 percent of all matings are between siblings. Other eumenid wasps appear also to engage in a relatively high degree of inbreeding, and there is nothing about their foraging or nesting behavior that obviously sets them apart from other predatory wasps, which appear to be outbred for the most part. Cowan argues that sibling matings could be advantageous simply because parents become more closely related to their offspring if they happen to be brother and sister (see also W. M. Shields, 1982, forthcoming). If siblings mate, the resulting progeny share 75 percent of their genes with their (diploid) parents, whereas if they had outbred, each parent's genetic relatedness with its offspring would be the familiar 50 percent (the half-genome transferred from parent to offspring in the egg or sperm).

Donald Windsor (personal communication) and Dawkins (1979) point out, however, that inbreeding has a cost (above and beyond any reduction in fitness caused by the concentration of damaging recessive alleles in one's progeny) that may exactly balance the genetic benefit identified by Cowan. Assume (1) that sibling and nonsibling matings are equally productive in terms of surviving offspring, and (2) that an individual that mates with a sib loses a chance to mate with a nonsib. The second assumption is a reasonable one, especially

2.4   A marked male of the eumenid wasp, *Euodynerus foraminatus,* perched on a trap nest "greeting" a virgin receptive female that is about to exit. (Photograph by D. Cowan.)

for females whose production of eggs is often small and readily fertilized by one male's sperm donation, but also for males whose supply of time and energy to search for mates is limited. Given these conditions, an individual will have the same genetic success whether it mates with a sib or a nonsib. That is, if every individual produces four offspring, then each will donate the genes in four gametes to the next generation. If brothers and sisters mate, their genes will be combined in the same four offspring; if not, their genes will be distributed among eight offspring.

Another interpretation of this point is that an individual who mates with a sib prevents that sib from mating with a nonsibling (see Parker, 1978a, for a complete algebraic development of this argument). The individual therefore loses some nieces and nephews that otherwise would have been produced by its sib. If our second assumption holds, for every offspring produced by a diploid brother and diploid sister mating together, the brother and sister each lose one nephew or niece. The 25 percent gain in genetic relatedness with their inbred offspring is balanced by the loss of 25 percent of their genes by common descent, in the absence of nieces and nephews (Dawkins, 1979).

Naturally, if sister-brother matings (or any other inbred combination) are more productive on average than highly outbred matings, then the loss of distant relatives would be more than compensated by the increased number of highly related daughters and sons. In order to understand the heightened productivity of such matings, one would need to know what ecological factors enable relatives to produce more offspring by mating together rather than by seeking out nonrelatives. The low dispersal, stable environment hypothesis provides one possibility. One can imagine a situation in which some females are unusually able to pass on to their progeny some critical resource such as a dwelling place or even a superior supply of nutrients per egg. By providing these resources, a female would be controlling the environment her offspring experience, making it in effect more stable and predictable and therefore reducing the advantage of producing highly variable progeny. These resource controllers might benefit by inbreeding if they were a minority in the population as a whole. If an outbreeding male were likely to mate with average females, his offspring would be less likely to survive or reproduce than if he were to mate with his sister. In this case a female who mated with her brother would not lose so many nephews/nieces by utilizing her brother's reproductive activity and so could experience a net gain by inbreeding.

It is also possible that our second assumption of a 1:1 reduction in nonsibling matings for every sibling mating does not apply in every situation. If so, a female's gain of increased genetic relatedness with her offspring sired by a brother would not be completely offset by losses of nieces and nephews. In *E. foraminatus*, for example, territorial males wait at nests during the morning and mate with emerging females; in the afternoon they join nonterritorial males in patrolling nonaggressively for females at flowers. If there are more receptive females at flowers in the afternoon than in the morning, or if territorial males are better searchers than morning patrollers (the territorial individuals might be better rested), the time and energy spent acquiring sisters as mates need not lead to a 1:1 reduction in his reproductive success with nonsibs.

Cowan (1979) notes that a related advantage of sibling matings is the possibility of reducing the cost of sexually producing males. This cost to female parents is lowered if (1) they can control the sex of their offspring, (2) they can produce clusters of female progeny with one or a few male offspring, and (3) brothers succeed in mating with their sisters before nonsibs do. Hymenoptera females can presumably control the sex of their progeny by laying either unfertilized haploid eggs that become males or fertilized diploid eggs that become females. This has been unequivocally demonstrated for the solitary bee

Megachile pacifica (Gerber and Klostermeyer, 1970). A wasp female that produced a multicelled nest could potentially stock it with just enough males to ensure that her daughters would be fertilized. In many parasitic wasps it is common for a female to parasitize a host by laying a number of eggs on one victim. The sex ratio of her offspring typically is strongly biased toward females (W. D. Hamilton, 1967). Brothers reach adulthood before their sisters and wait nearby to mate with them as they emerge. Because one male can fertilize many females, the mother wasp need not produce as many males as females and so approaches the advantage of producing females enjoyed by parthenogenetic females, while at the same time retaining the potential benefits of sexual reproduction. Chapter 3 gives a full discussion of the evolution of sex ratios.

Inbreeding as an Incidental Effect

The spectrum of possible advantages and special conditions under which inbreeding may be adaptive should not blind us to the chance that in at least some species inbreeding is merely an incidental effect of some other adaptive aspect of the species' biology. For example, inbreeding by the eumenid wasp *E. foraminatus* may be simply a by-product of the efforts by males to monopolize access to receptive females. Because several virgin, receptive females may emerge from a single nest in this species, a male that controls a small area about the nest is able to secure a substantial number of matings. The genetic benefit to a male of copulating with several of his sisters may be great enough to outweigh the cost of inbreeding depression (primarily a decrease in his sisters' reproductive success through a reduction in the heterozygosity of the daughters produced).

Inbreeding depression has been demonstrated for a number of insects, and its consequences sometimes are severe. For example, an inbred line of the moth *Atteva punctella* exhibited more than twice the percentage of infertile matings as an outbred one (O. R. Taylor, 1967). Only 40 percent of the eggs laid after four brother-sister pairings in the moth *Hyalophora cecropia* hatched, as opposed to 90 percent from outbred crosses (Waldbauer and Sternburg, 1979). Inbreeding in the true bug *Oncopeltus fasciatus* leads to a sharp reduction (over 30 percent) in the number of eggs produced, and there are additional declines in the survival of the inbred progeny (Turner, 1960). In the wasps and other Hymenoptera, however, males are haploid and this has the effect of exposing damaging recessive alleles directly to selection, with the result that they tend to be eliminated from populations (Smith and Shaw, 1980). The proportion of polymorphic loci in the Hymenoptera ranges from 0.11 to 0.17, whereas in diplo-diploid insects the average is about 0.53 (Pamilo, Varvio-Aho, and Pekkarinen, 1978; see also Crozier, 1977, 1979). The genetic, as well as the external, environment can affect the costs of inbreeding. If fitness-decreasing recessives are relatively rare in the Hymenoptera (as these reports suggest), the disadvantageous aspects of inbreeding should be reduced (although not necessarily eliminated altogether, because inbreeding can reduce the reproductive success of females in some bees and wasps—see Mackensen, 1951; Kerr, 1974). To the extent that inbreeding is actually less costly for Hymenoptera, it should be more common in this group than in insects with typical diplo-diploid systems of sex determination.

To summarize, there are at least four major hypotheses to account for the evolution of inbreeding:

(1) Inbreeding leads to increased reproductive success in relatively stable or controlled environments in which a great deal of genetic diversity in offspring would not be beneficial.

(2) Inbreeding may, under certain circumstances, lead to a net genetic gain through increased representation of an individual's genes in offspring produced with a relative.

(3) Sibling matings enable a female parent to reduce the number of males she produces and so increase the productivity of her progeny.

(4) Inbreeding may be an incidental effect of a mating system in which some males are able to monopolize clusters of receptive females, which may happen to be their sisters or other close relatives.

Given the number of competing hypotheses on the function of inbreeding, careful thought is required to devise predictions and testing to help reject some of the possibilities. This remains to be done. It is appropriate to conclude on this note of incompleteness; so much must still be learned about the adaptive significance of the various modes of reproduction. But because insects exhibit such a broad range of reproductive methods, both interspecifically and sometimes intraspecifically, they potentially hold the key to testing hypotheses about the adaptive value of different patterns of reproduction. The basic problem is how to measure whether the advantage of producing genetically variable progeny is proportional to the degree of instability and unpredictability in the environments faced by the offspring. Much more work is needed to determine what aspects of the abiotic and biotic environment are unpredictable and how to measure this unpredictability—if not in absolute terms, at least in a relative way so as to permit careful comparisons between populations or species. If an accurate measure of environmental instability could be integrated with information on the degree of genotypic diversity among offspring and on the degree of short-term shifts in gene frequencies, the validity of competing evolutionary hypotheses might become more apparent.

3

Sexual Selection Theory

IN CHAPTER 2 we noted that sexual reproduction is typical of most insects, although it is achieved in more than one way. Despite the fact that the selective advantage of sex is not obvious, there is no doubt that its evolution has had great impact on the reproductive tactics of both males and females. The primary characteristic of males is their drive to secure mates, which leads to competition for access to females and the evolution of a host of traits associated with this struggle. Females, on the other hand, have the luxury of choosing among many potential partners; their preferences are expected to raise their genetic success and in turn exert pressure on males favoring traits considered desirable by females.

Darwin's interpretation of typical male and female behavior and his "discovery" of sexual selection is the starting point for this chapter. We examine a series of questions related to the foundation of sexual selection:

(1) Why are there just two sexes, which produce gametes of two different sizes?
(2) Why do females more often than males exhibit parental care and other investments in their progeny?
(3) Why are sex ratios near unity in most outbred species?

The answers to these questions support the argument that females usually are a limited (and selective) resource for which males compete. We shall discuss some case histories of insect reproductive tactics to illustrate the apparent effects of sexual selection on male and female behavior. The chapter concludes with an outline of the three basic mating systems of insect species.

Darwin and the Concept of Sexual Selection

The foundation for the theory of sexual selection, like the rest of evolutionary biology, was laid by Charles Darwin. In *On the Origin of Species* (1859:75) he wrote that sexual selection "depends, not on a struggle for existence, but on a struggle between the males for possession of the females; the result is not death to the unsuccessful competitors, but few or no offspring." The puzzle that Darwin sought to explain was the evolution of characters, particularly in males, that are detrimental to survival but that promote success in reproductive competition (Otte, 1979; West-Eberhard, 1979; Thornhill, 1980a). He recognized that the elaborate and conspicuous plumage of male birds, the complex courtship displays of certain animals, the heavy horns and antlers of sheep and deer, and many other so-called secondary sexual characteristics are costly in terms of survival. These traits require energy to produce and expose the individual to heightened risk of predatory attack, and yet they persist—presumably because they enable males to gain access to mates. Thus Darwin made a distinction between selection for traits that enhance survival (natural selection) and selection for traits that increase an individual's success in acquiring mates (sexual selection).

Most modern researchers no longer make the same distinction, because both natural and sexual selection have their evolutionary effects through the differential reproductive success of individuals. All traits have costs and benefits in terms of their impact on gene transmission. A character that promotes survival will persist only to the extent that survival is correlated with successful reproduction. A sexually selected trait may decrease the longevity of the individual but more than compensate for this cost by increasing the reproductive chances of that individual. Thus sexual selection is a category of natural selection. Both result in the evolution of phenotypic properties that tend to increase the reproductive success and gene propagation of individuals (Williams, 1966; E. O. Wilson, 1975; Dawkins, 1976; Alexander and Borgia, 1979).

There is general agreement, however, that sexual selection is a useful concept and that Darwin correctly identified the components of this evolutionary force, and many of their consequences. He proposed that sexual selection consists in two different pressures: (1) "a constantly recurrent struggle between males for possession of females," and (2) a process of mate choice, usually by females that "select those [males] which are vigorous and well-armed, and in other respects most attractive" (Darwin, 1874:109). By the "struggle between males" Darwin meant not only physical combat but an array of other modes of competition for females, such as searching skill or ability to attract females from a distance (Otte, 1979). These components of sexual selection have acquired labels—intrasexual selection and epigamic selection—that are now widely employed (Huxley, 1938).

Some workers have sought to clarify the distinction between sexual and other forms of selection by placing sexual selection in the broader context of social competition (see West-Eberhard, 1979). In this approach sexual selection becomes one of a variety of forms of social selection that arise whenever there is competition among members of a group for such things as food, dominance, reproductive rights, and access to mates. Social selection is contrasted with the pressures exerted on the members of a population by aspects of their physical environment such as climate, and by predators, parasites, and competitors of other species. The argument is that social selection is particularly likely to result in a coevolutionary spiral. When a mutant trait arises that confers an advantage in intraspecific social competition, it will spread through the population and create a new social climate. This in turn produces a new set of selection pressures, which set the stage for the success of new counteracting competitive responses (when the appropriate mutation occurs). Williams (1966:184) points to an example of the continuing coevolutionary process created by social (sexual) selection:

> Inevitably there is a kind of evolutionary battle of the sexes. If a male attempts to reproduce at all in a certain breeding season, it is to his advantage to pretend to be highly fit whether he is or not. If a weak and unresourceful male successfully coaxes a female to mate with him he has lost nothing, and may have successfully reproduced. It will be to the female's advantage, however, to be able to tell the males that are really fit from those that merely pretend to be. In such a population genic selection will foster a skillful salesmanship among the males and an equally well-developed sales resistance and discrimination among the females.

Thus the sexual preferences of females and the pressures of combat and competition among males may lead to the ever more refined (exaggerated) characters that aroused Darwin's curiosity in the first place. Much of this book is our attempt to explain characters bordering on the bizarre (the fantastic horns of certain male beetles, the complex penis structure of damselflies, the presentation of food gifts to females by male scorpionflies, and week-long copulations in walkingsticks) as the products of social selection for traits useful in an intensely competitive and coevolving sexual environment.

Sexual Selection and Bateman's Principle

Darwin and many other naturalists realized that it was usually the male of the species that fought to copulate and it was usually the female that appeared to exercise choice in selecting a copulatory partner. These observations lead to the deduction that males will usually vary more than females in the number of surviving offspring they produce. Almost all females will have a willing partner and so produce some fertilized eggs; in contrast, some males will fail to persuade any female to copulate with them or will be excluded from the mating process by more powerful rivals. The relative variance of the sexes in reproductive success proved to be an extremely important point in the development of sexual selection theory, but it was not *quantitatively* documented until Bateman did so in 1948. Between Darwin's work in the 1870s and that time there were only a few major developments in the field of sexual selection (R. A. Fisher, 1958; first edition, 1930) and little attention was paid to the evolutionary significance of social competition.

Bateman's study is now recognized as having started a new era of interest in sexual selection. In order to measure the variance in reproductive success of males versus females, Bateman placed three to five adult fruit flies, *Drosophila melanogaster,* of each sex in a number of laboratory containers for breeding. Thus each female could choose among a number of males, and each male had to contend with several rivals for the females. Genetic markers in the adults allowed Bateman to attribute individual offspring to specific parents and thereby obtain accurate data on the differential reproductive success of both sexes.

His findings were highly revealing. First, male reproductive success was in fact much more variable than female reproductive success (Fig. 3.1). The great majority of females copulated with only one or two of the males available to them. Despite the readiness of all males to court, many failed to secure a single mating. Of those that did

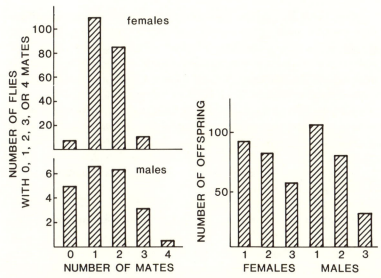

3.1 Bateman's study on the variance in reproductive success of male and female fruit flies. (*Left*) Many males have no mates, whereas almost all females
copulate at least once; quite a few males have three or four mates, whereas almost all females have just one or two. The data are from one experiment in
which three to five flies of each sex were caged together. (*Right*) The differences in the number of offspring produced by three males caged with three females far exceeds variation in reproductive success among the females. (From
Bateman, 1948.)

succeed in reproducing, most copulated with more than two females.
Statistical variance is a measure of the spread around the mean, which
is calculated by the formula $\sum (x_i-\bar{x})^2/N-1$. The average variance in
male mating success in Bateman's experiments was 2.5 times the female variance. Moreover, male variance in number of offspring produced exceeded that of the female population in every one of 64 experiments.

Second, male reproductive success was limited by the ability of individuals to coax the opposite sex to mate, whereas female fitness did
not depend on the capacity of females to "persuade" males to copulate
with them. All males exhibited courtship behavior and every female
was vigorously courted.

Third, the number of offspring produced by a male was positively
related to the number of his mates. In contrast, females that copulated
with two or three males had about the same number of progeny as
those that mated with only one male.

Bateman explained his results in terms of the energetic investment of each sex in their gametes. In *D. melanogaster,* sperm are very small relative to eggs. For the same allocation of resources a male can manufacture vastly more gametes than a female can. A male's reproductive success is therefore limited by his success at inseminating females (because one individual could potentially fertilize all the eggs of a great many females). A female's fitness is limited by her ability to produce gametes and not by her skill in coaxing males to inseminate her. Bateman argued that gametic differences created a limited resource for males (the females' eggs) and helped explain the evolution of the eagerness of the male flies to copulate in contrast to the sexual discrimination of females and their usual refusal to mate more than once. This principle of Bateman's applies broadly across the animal kingdom, because males produce smaller gametes than females in almost all sexually reproducing organisms.

The Evolution of the Sexes and Sexual Conflict

Why are there differences in the size of the gametes produced by males and females in most sexually reproducing species? Although sex occurs without the sexes in some isogamous unicellular organisms, the general rule is that this mode of reproduction is characterized by gametic differences. When gametes are not all of the same size, females are *defined* as those individuals that produce larger, relatively immobile gametes, which contain a considerable number of nutritive substances for early development of the zygote. Males are identified as the sex that produces smaller, motile gametes, which usually carry only genes. How did these differences come about? Why are there males and females?

Parker, Baker, and Smith (1972) provided an interesting model for the evolution of the sexes, one that has been discussed by Power (1976), Maynard Smith (1978b), and Alexander and Borgia (1979). Consider an ancestral diploid organism which exhibited sexual reproduction via the fusion of haploid gametes to form a zygote, but which had no distinct sexes. Assume that the zygote's survival probabilities were related to its size, and thus to the energetic contributions made by both gametes during fertilization. Further, imagine that most individuals produced gametes of intermediate size, but small and large gametes were made by a few mutant individuals; the size of a gamete produced by an individual was inherited by its offspring. As is the case for all biological beings, our hypothetical organism had only a finite amount of reproductive effort to expend. This led to a benefit-to-cost

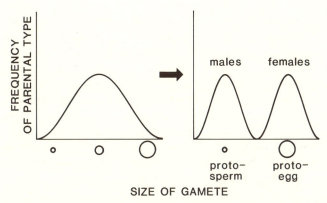

3.2 Disruptive selection and the evolution of male and female gametes. In the hypothetical initial population there was variation in gamete size, with most about average. Selection favoring the producers of very small and very large gametes could have led to evolution of the sexes. (Modified from Daly and Wilson, 1978.)

trade-off between the number and the size of gametes produced. Although individuals that invested in large gametes produced fewer, less motile gametes, their offspring had a high survival probability even when they were produced by union of a large and a small gamete. Individuals that created many small gametes did not improve their offspring's survival chances, but their mobile sex cells were more likely to find and fertilize other gametes. The occurrence of large gametes in the ancestral population led to selection favoring individuals capable of producing a greater number of small, motile gametes in order to exploit the abundant nutritive reserves of the large gametes. Intermediate-sized gametes would neither develop as well as large eggs nor fertilize as well as small sperm. Given these circumstances, disruptive selection (in which intermediate phenotypes are disfavored) acting on variation in gamete size in the primeval population would have led to the evolution of two distinct gamete types, a large development promoter (the egg) and a small skillful fertilizer (the sperm). Figure 3.2 illustrates this situation.

During the evolution of egg-producing females and sperm-producing males, conflict between individuals with different-sized gametes probably occurred. As pointed out by Williams (1975), sperm in one sense parasitize the parental resources of the egg. Perhaps selection originally favored large gametes able to avoid uniting with smaller ones in order to fuse with another large sex cell. But there would have been even stronger selection on "sperm" to overcome the defenses of "eggs." Small sex cells would only have survived if they had succeeded in uniting with a large gamete, because two fused "sperm" would have produced a zygote too small to survive. However, the zygotes formed by the union of a large and a small gamete could often have lived. Selection would have favored female gametes with traits maximizing the survival of zygotes formed by the union of egg with sperm. Selection on male gametes favored very different attributes associated with locating and fertilizing female gametes. As a result of disruptive selection, only two modes would be stable: one for male function and one for female function.

Sperm Design and the Competition to Fertilize Eggs

In addition to being small, sperm have other design features that reflect selection for success in fertilizing eggs (Fig. 3.3). Most sperm possess a long and powerful flagellum, large numbers of mitochrondria to power the flagellum, and an enzyme-filled cap that assists in penetration of the egg (Bacetti and Afzelius, 1976). These features clearly assist the male gamete in the competition with other sperm to reach and enter eggs.

Moreover, males usually donate large numbers of sperm with each ejaculate (although some mites and a few insects transfer less than two sperm per egg—see Cohen, 1971, 1975, 1977). Parker (1970c) suggested that sperm redundancy may result from selection created by sperm competition between male ejaculates (see also J. E. Lloyd, 1979a; Sivinski, 1980a). One way to cope with the possibility that a female has already received sperm from one's competitors is to flood her system with quantities of speedy gametes, some of which may reach her eggs before the rival sperm do. Some male insects include in their ejaculates both fertile and sterile sperm (those with an irregular number of chromosomes or no chromosomes at all). The infertile sperm are often larger than the fertile morph and so may serve to displace competing sperm placed within the female by other males. Moreover, the ejaculates of some other insects contain anucleated flagellar structures, derived from developing sperm, that may help their fertile companions reach locations where they are most likely to be used to fertilize eggs (but see also Chapter 12). In *Drosophila* fruit flies the male may manufacture as many as four sperm morphs that come to occupy different locations within the female's reproductive tract, perhaps to play different roles dealing with subsequent ejaculates received by the female from competitors of the male (Sivinski, 1980a).

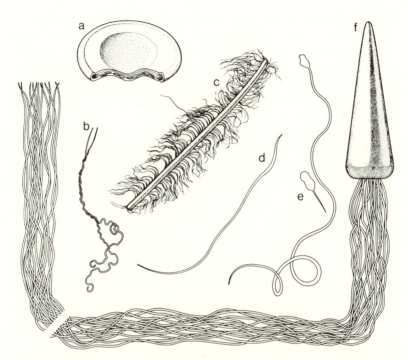

3.3   Diversity in arthropod sperm (not drawn to scale). (*a*) The immobile disc-like sperm of the proturan, *Eosentonon transitorium*. (*b*) Entwined sperm that swim in tandem from the thysanuran, *Thermobia domestica*. (*c*) A rod, the spermatostyle, carrying a vast number of spermatozoa of a gyrinid beetle, *Dineutus* sp. Both the spermatostyle and the sperm are transferred to the female during copulation. (*d*) The threadlike sperm of the firefly *Pyractomena barberi*. (*e*) Two sperm morphs from the symphylan *Symphylella vulgaris*. (*f*) A multiflagellate sperm from the Australian termite *Mastotermes darwinensis*. (From Sivinski, 1980a.)

Competition to fertilize eggs is complicated by the ability of females to store sperm within a storage organ known as the spermatheca. In many species the male's sperm must have the ability to reach the spermatheca and adopt a position within it such that the female is likely to select his sperm for use when fertilizing her eggs (see Chapter 11). The advantages of monopolizing the storage space may be responsible for the evolution of the bizarre giant sperm of the ptiliid or featherwing beetles (Dybas and Dybas, 1981). Males in the genus *Bambara* produce sperm that range from one-third to two-thirds the length of the adult itself. Sperm received by a female are placed within the spermatheca. There is a remarkable correspondence between the length of the sperm and the dimensions of the spermatheca (Fig. 3.4). Because relatively few sperm are enough to completely occupy the storage organ, other sperm received by the females may not be able to be stored. This explains why field-collected females are often found with sperm protruding from the vagina, apparently unable to enter the female completely and find space within the spermatheca.

Likewise the barbed sperm of certain grasshoppers, and the ability of some insect sperm to form an entangled mass, may help prevent their removal or displacement from the female, increasing the probability that the donor will succeed in fertilizing some of his mate's eggs (J. E. Lloyd, 1979a; Sivinski, 1980a, forthcoming).

The Significance of Maternal Care

Male reproductive behavior, like the morphology of their sperm, is related to the race to reach and fertilize eggs. The greater caloric investment per female sex cell is not the only reason why males compete for access to eggs. The sexual difference in the investment per offspring

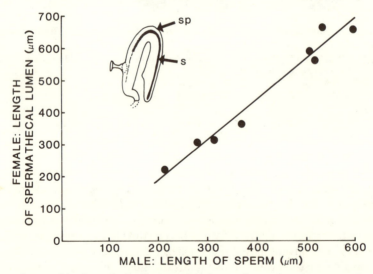

3.4   Sperm (s) of a male featherwing beetle as it lies within the spermatheca (sp) of a conspecific female. The close relation between sperm length and spermathecal length is shown for a number of different species of these beetles. Perhaps by fully occupying the spermatheca, the sperm acts as a block to rival sperm. (From Dybas and Dybas, 1981.)

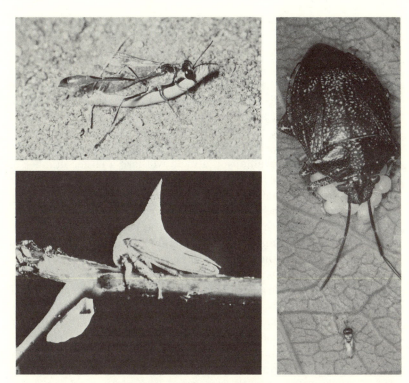

3.5   Maternal insects provide parental care in addition to large gametes. (*Top left*) Many wasps nourish their progeny with food they gather from their environment. This *Ammophila* wasp provisions her brood cells with moth caterpillars. (Photograph by J. Alcock.) (*Right*) There are species of pentatomid stink bugs whose females guard their eggs against small parasitic wasps (one is below the female) until the young emerge. (Photograph by W. G. Eberhard. By permission of the Smithsonian Institution Press, Washington, D.C.) (*Bottom left*) A maternal membracid bug guarding her young, which have recently hatched. (Photograph by T. Wood.)

is often magnified by the female's allocation of additional resources and time to the development of her fertilized eggs. Parental care is far more likely to be maternal than paternal (Fig. 3.5). What is given to one offspring cannot be used in other ways, notably to produce additional gametes. Thus in species with maternal care, eggs are even scarcer relative to sperm than they would be in the absence of parental behavior by females. With opportunities for egg fertilizations reduced, males compete ever more fiercely for the remaining opportunities. A male that fertilizes an egg may not only "win" its intrinsic developmental benefits, but also the mother's parental allocation.

Why is it that females, above and beyond their greater energetic investment per gamete, are more likely than males to devote additional time, energy, and risk to help their offspring survive? This question has yet to be fully resolved (Werren, Gross, and Shine, 1980; Gross and Shine, 1981) but there are some promising hypotheses, one of which might be labeled the "relative reduction of fecundity hypothesis." Remember that each unit of reproductive resources devoted to existing offspring results in the loss of a certain number of additional gametes. Because sperm are smaller than eggs, for each unit of resources diverted to parentalism, a male will lose a larger number of gametes than a female. Admittedly, males rarely donate one sperm for each egg to be fertilized; but the *potential* loss of future offspring produced will generally be greater for the paternal male than for the maternal female. Other things being equal, the benefit of parental care must be higher for a male than for a female if the benefit-to-cost ratio is to be favorable for the evolution of paternal behavior.

A second hypothesis is based on the reality that the reliability of paternity is generally less than the reliability of maternity. In other words, a male that copulates with a female is less likely to be the biological father of the young she eventually produces than the female is likely to be the biological mother. Females usually retain control of their gametes and unite their eggs with sperm of a mate of their choosing. Males relinquish control of their gametes and are rarely able to force their mates to use their sperm with absolute certainty. This is true of insects because they universally practice internal fertilization. While a male may donate his sperm to a female, he has no guarantee that she will ever use any of them to fertilize her eggs. This further reduces the benefit-to-cost ratio of paternal care relative to maternality, because of the persistent risk that a male would assist the progeny of an unknown rival. As a result, the evolution of investments by males in resource-rich sperm, the eggs of their mates, or the zygotes produced by their mates becomes less likely (Alexander and Borgia, 1979).

Because insect females are gamete acceptors rather than egg donors, they are physically closer to their progeny (Williams, 1975; Gross and Shine, 1981). In insects and other animals with internal fertilization a male may begin to exert some influence on the survival of his offspring only after the fertilized eggs are deposited or the young are born. By the time a male is likely to become an effective paternal investor, he may be separated from his young and unable to help them. Yet when the eggs are laid the mother is certain to be present and so can more easily provide care for her progeny, if ecological pressures such as predation or resource scarcity make this an adaptive response.

Parental Investment Theory

The amount of parental care offered by females varies a great deal among species. Moreover, paternal behavior occurs in insects and other animals and may even, albeit rarely, exceed the allocation of resources provided to the young by their mother. Trivers (1972) recognized that to understand the evolution of sexual behavior, one must know more than the relative investments per gamete made by males and females. He saw the need for a common currency for all the different ways in which individuals may assist in the development and survival of their young, from the nutrition provided in the gamete to the risks taken by a parent in driving predators from the young. To this end he proposed the concept of parental investment as "any investment by the parent in an individual offspring that increases the offspring's chance of surviving [and hence reproductive success] at the cost of the parent's ability to invest in other offspring" (1972:139). Thus if the female offers food to one of her existing offspring rather than consuming it herself, her donation represents a parental investment, because she will have fewer materials with which to produce additional eggs. Likewise, attempts to defend one's progeny is parental investment if the attempt exposes the protector to risk of dying from predatory attack—and with death, the loss of future opportunities to reproduce.

The relation between sexual selection and parental investment is based on the proposal that "the sex whose typical [average per offspring] parental investment is greater than that of the opposite sex will become a limiting resource for that sex" (Trivers, 1972:173). The argument represents an extension and refinement of Bateman's ideas, because parental investment embraces all the components of parental "sacrifice" for an offspring, not just the energetic investment in a gamete. Accordingly, hypotheses based on parental investment theory generate predictions that do not follow from Bateman's principle. For example, should males of a species exhibit greater parental investment per offspring than females, then the males should show less variance in reproductive success than females that compete for access to males—despite the fact that male sperm will always be smaller than the eggs of conspecific females.

Trivers provided evidence that the relative parental investment determines the evolution of sexual differences in behavior, by reviewing

3.6   Examples of mating effort by male insects. (*Above*) A male asilid fly, *Heteropogon stonei,* in the midst of elaborate and time-consuming courtship of a perched female. (Photograph by J. Alcock.) (*Below*) A copulating pair of katydids, *Orchelimum delicatum.* The male, on the left, transfers a large spermatophylax, which the female consumes after mating. To the extent that the nutritious spermatophylax induces the female to use her mate's sperm to fertilize her eggs, it constitutes mating effort by the male. (Photograph by D. Gwynne.)

observations on the behavior of species with parental males. As predicted, in monogamous species with paternal care, *both* sexes compete for mates and exhibit courtship and apparent discrimination in mate selection. In polyandrous bird species, in which females may mate with several males, complete sex-role reversals are sometimes seen; females may be larger, more competitive, and more colorful than males (Jenni, 1974; Emlen and Oring, 1977). Chapter 13 presents

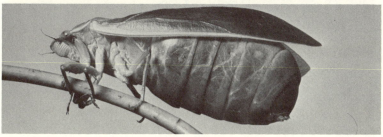

3.6 *(continued)* *(Above)* Male fireflies of the genus *Photinus* take risks in searching for mates. The dismembered male shown here has responded to the flash of a predatory *Photuris* female and has paid the ultimate price for his sexual endeavors. (Photograph by J. E. Lloyd.) *(Below)* A male bladder cicada *Cystosoma saundersii* has a greatly enlarged abdomen which serves as a resonating chamber. Males sing noisily, an activity that exposes them to predators. The energy expended by the male in singing and in development of the large abdomen also represents mating effort. (Photograph by D. Young.)

similar evidence on the effects of high paternal investment by male insects.

Reproductive Effort and Its Components

The proportion of an organism's total available energy that is used in reproduction can be labeled *reproductive effort* (Williams, 1966; Hirshfield and Tinkle, 1975) and can itself be subdivided into *mating effort* (the component of reproductive effort expended in attempts to acquire mates) and *parental effort* (the sum of parental investments in offspring made by the individual) (Low, 1978; Alexander and Borgia, 1979). We have seen that typically—but not invariably—males allocate most of their energies to mating effort, whereas females make a greater commitment to parental effort. The typical male provides nothing to a female and his progeny other than his sperm. These nutritionless gene carriers are themselves a form of mating effort, because the properties of sperm have evolved in the context of the race to fertilize eggs. The male's time and energy are devoted to searching for or calling in females, fighting with other males for possession of sites likely to contain females, and defending their actual or potential partners against their rivals (Fig. 3.6). Not only are these activities energetically demanding but they expose the actor to risk of death from exhaustion, predation, or injuries sustained in male-to-male combat. A prediction from Trivers' model is that the differential mortality of the sexes should be related to the relative parental investment of males and females. It is well known that in species with low male parental investment (in which some males acquire several mates) mortality rates of males exceed those of females (Trivers, 1972). We shall show in Chapter 5 that male insects put themselves in mortal danger far more frequently than females when going about the business of reproducing. A "typical" male insect appears to be trying to maximize the number of times he mates (or the number of eggs he fertilizes).

Males of some species, however, do supply females with material benefits prior to, during, or shortly after mating. These resources may be in the form of nuptial gifts of food collected by the male, or glandular secretions, or nutritious spermatophores—or in extreme cases, the male himself may be consumed by the female. Mating presents have been viewed as male parental investment (Trivers, 1972; Thornhill, 1976a; Boggs and Gilbert, 1979; G. K. Morris, 1979, 1980) in that they potentially or actually enhance the survivorship of developing gametes carried by a male's mate. In some cases nourishment provided by a male goes into the eggs he fertilizes. Alexander and Borgia (1979) have suggested, however, that these male activities are usually best considered as mating effort, not parental effort, because they have evidently evolved to help the male secure matings, or to increase the likelihood that a female will use his sperm after it is transferred. That the female uses male-contributed nourishment for her own nutrition or for that of her eggs may be incidental to the male's primary goal, which is to fertilize eggs.

Because nuptial gifts need not promote the survival or development of a male's progeny, Gwynne (forthcoming a) suggests that we categorize them as a form of mating effort—nonpromiscuous mating effort. Gift giving according to this approach represents an allocation of resources from the male that reduces his ability to increase the number of his mates. By contrast, promiscuous mating effort includes all those activities whereby a male attempts to find mates but does not offer his partners costly inducements, which require time, energy, and risk taking to acquire.

Nonpromiscuous mating effort is similar to parental effort in its effect on the operation of sexual selection. When males contribute benefits to females in exchange for matings, they may become the sex that is the limiting resource. Because females benefit from receipt of the presents, whether or not their mate's progeny also gain, they may compete for access to males in order to receive the resources associated with copulation. Males on the other hand may become somewhat discriminating, in an attempt to secure the greatest genetic return for their gifts (for example, by mating with relatively fecund females). Therefore one can argue that the limiting sex will be a function of how the sexes differ with respect to three factors: (1) the resources invested in construction of a gamete, (2) any additional postzygotic parental investment expended per offspring, and (3) any nonpromiscuous mating effort offered in exchange for a mating. The degree to which the sexes diverge in the total resources secured per copulation should determine the intensity of sexual selection acting on the males and females of a species.

Evolution of Sex Ratio

Relative gametic investments and other material benefits available to a mate are not the only factors that will influence sexual selection. Females and their eggs are usually a limited resource for males *in part* because the sex ratio of most species is 1:1. Theoretically the differences in reproductive success of males would be lessened if adult sex ratios were strongly female biased (Fig. 3.7), or if sex ratios were adjusted in relation to the extent of sexual competition among males (the greater the potential competition, the more females in the population). But in most outbreeding populations the ratio of males to females tends not to deviate from unity.

The sex ratio of a population can be explained in proximate terms by the operation of meiosis and the various mechanisms of sex determination (White, 1973). We shall consider the ultimate factors in terms of the selective forces that have produced 1:1 sex ratios, whatever the underlying mechanism for creating males and females. It is entirely possible to interpret the typical sex ratio as an adaptation im-

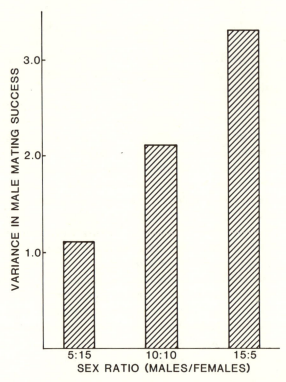

3.7   Variance in mating success (number of copulations) of male *Panorpa*, as affected by the ratio of males to females in a cage experiment. When females are relatively few, many males are prevented from mating. (From Thornhill, unpublished.)

portant for individual reproductive success. (See Williams, 1966, for a critique of group selectionist hypotheses on the evolution of the sex ratio.) R. A. Fisher (1958) supplied the key argument. He pointed out that it is advantageous for an individual to produce a 1:1 sex ratio of offspring, because every individual in a bisexual species has a mother and father, each of which supplies half the genome of each offspring. When the sex ratio is unity both sexes are, on the average, of equal value for the genetic perpetuation of the parent.

Imagine a population of animals in which males and females occur in the ratio 1:10. Because the sexes contribute equally in a genetic sense to each successive generation, the average male in this population has ten times the reproductive success of the average female. Thus, selection will favor genes that bias the sex ratio of an individual's offspring toward the uncommon sex (male). The disparity in the numbers of males and females will decrease as a result of such selection. Should males become more abundant than females, selection will favor individuals that produce more females. The result will be a stable equilibrium around 1:1.

Fisher's explanation for a numerical sex ratio of unity applies only when males and females are equally costly for parents to produce (R. A. Fisher, 1958; Trivers and Hare, 1976). It is possible that one sex may be cheaper than the other because fewer resources are needed to rear it successfully, or because of reduced mortality of that sex during the period of parental care. Consider a hypothetical situation where sons and daughters show similar mortality schedules during rearing, but sons require only one-half the resources of daughters. For example, adult males might be half the size of adult females, and thus half as expensive to produce. Therefore a set of parents could create two sons for the cost of one daughter. In a population in which the male-to-female sex ratio was 2:1, the average reproductive success of one son would be half that of a daughter. But if the parental cost of a male were also one-half that of a female, the average genetic return *per unit of parental resources* would be the same for a son or a daughter. Therefore the 2:1 sex ratio would be evolutionarily stable, given the conditions of this example. In a population where a male required one-third the allocation of parental resources needed for a female offspring, the evolutionarily stable sex ratio would be 3 males to 1 female.

Consideration of differential benefits and costs to parents of male versus female offspring led Fisher to his general theory of sex ratio evolution: selection will favor parents that invest equal amounts in each sex. This will result in a 1:1 sex ratio only when a son and a daughter are equally expensive to rear, which is the common case.

Sex Ratio Bias under Inbreeding

The idea that parents should evolve the capacity to equalize their investments in sons and daughters assumes outbreeding (R. A. Fisher, 1958). As first pointed out by W. D. Hamilton (1967), sex ratio theory must be adjusted to apply to organisms that inbreed (Trivers and Hare, 1976; Alexander and Sherman, 1977; Charnov, 1978). Because brothers mate with sisters in certain ants, bees, and wasps (among other insects), a female can maximize her genetic propagation by investing more in daughters than in sons. A female need only produce enough sons to fertilize her daughters' eggs in order to maximize the reproductive value of her brood.

In the Hymenoptera, unfertilized (haploid) eggs develop into males and fertilized (diploid) eggs develop into females, which allows the mother to manipulate the sex ratio precisely. She need only fertilize an egg to produce a daughter, fail to fertilize an egg to produce a son. In those species that exhibit brother-sister matings, brothers compete with one another for sexual access to their sisters (on local mate competition see W. D. Hamilton, 1967; Alexander and Sherman, 1977). One prediction is that under these conditions a female parent should produce relatively more females than males because one son can fertilize many daughters (additional sons would be wasted parental effort).

The prediction has been tested by Werren (1980) in a species of wasp, *Nasonia vitripennis*, that exhibits varying degrees of local mate competition. The wasp parasitizes pupae of a wide range of flies. She first immobilizes the host pupa by stinging it, then lays a number of eggs upon it. The sex ratio per host typically is highly biased toward daughters; about 3 males emerge among an average of 29 offspring per pupa. Males emerge first and compete for the single exit hole through which their sisters will emerge. Males mate with females immediately after they crawl through the exit hole. Females copulate only once and then disperse. Males are capable of inseminating many females; they cannot fly off, since they possess only rudimentary wings.

Not all matings in *Nasonia* occur between sibs. This species shows a degree of superparasitism, whereby two females may lay eggs upon the same host. The second female can detect previous parasitism. The two broods emerge synchronously. This situation is ideal for determining the relationship between the extent of inbreeding and the sex ratio bias toward daughters. The first female to lay on a pupa should produce only enough sons to inseminate her daughters; the second female should produce more males than the first because her sons, if sufficiently numerous, may gain opportunities to fertilize the many daughters of another female.

If an already parasitized host is small, the second wasp will lay only a few eggs; but if it is large, the second female may lay as many eggs on it as the first wasp did. On the basis of a simple model Werren predicted that in instances when only a few eggs were laid, they should all be sons, whereas when the second female laid many eggs, as few as 25 percent should be males. When the second female lays a small number of eggs, her offspring will emerge from the site with a relatively large number of females deposited by the first wasp. If the second female's offspring are males, they will on average have several mates and enjoy high reproductive success. But if conditions permitted the

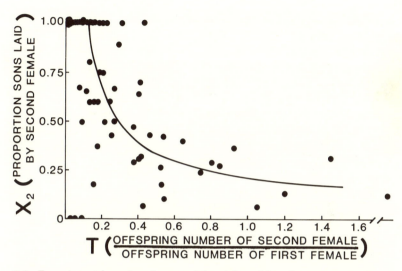

3.8 Experimental results of a test of the ability of female wasps, *Nasonia vitripennis,* to control the sex ratio of their progeny in an adaptive manner. As predicted, when a female lays only a few eggs in a host containing many female progeny of another wasp, she produces almost all sons. But when the second female lays many eggs, she produces only about enough sons to fertilize her daughters. (From Werren, 1980. Copyright 1980 by the American Association for the Advancement of Science.)

second wasp to lay many eggs on a parasitized host, and a large proportion were male, her sons would be likely to engage in profitless competition with their brothers for sisters. This favors an increasing bias toward females in the second wasp's brood. Werren's predictions were met with incredible precision (Fig. 3.8). The second wasp to lay in a host pupa is following an optimal sex ratio strategy by manipulating the sex ratio of her brood in relation to the extent of local mate competition.

Werren's findings provide strong support for Fisher's theory by showing that when the assumption of outbreeding is violated, sex ratio bias moves in the predicted direction—toward females. There are, nevertheless, some alternative explanations for female-biased sex ratios (Colwell, 1981; D. S. Wilson and Colwell, 1981) and there is room for additional work on this problem.

Parental Ability to Vary Sex Ratio in Outbreeding Species

Nasonia vitripennis is an example of a species whose females are able to adjust the sex ratio of their progeny in order to cope with variation in the degree of inbreeding their offspring are likely to experience. In purely outbred populations, the same ability to invest to a greater or lesser degree in males may also be adaptive (see Wittenberger, 1981a). As discussed later in this chapter, most species are polygynous in the sense that males show more variation in reproductive success than females owing to sexual competition. Trivers and Willard (1973) reasoned that in polygynous mammals females in good condition should bias their sex ratio toward sons. A healthy, well-fed female should be better able to produce healthy, large sons than a less well-nourished female. Healthy, strong males are more likely to acquire a harem and enjoy exceptional reproductive success in a polygynous species than smaller, weaker males. On the other hand, a mother in poor condition would tend to produce daughters rather than competitively disadvantaged males, because female mates are always in demand and therefore are likely to have some reproductive success. A small or sickly son would usually fail to reproduce at all.

Although Trivers and Willard applied their hypothesis to mammals, it also has relevance to the relatively few insects that are polygynous and also show parental care. In those insects in which there is no parental aid to offspring beyond the nutrition placed in eggs, sex ratio manipulation might be adaptive if a female could vary egg size based on her nutritional condition and at the same time determine the sex of the zygote arising from each egg. Egg size does vary in certain insects and appears to be related to offspring fitness (Capinera, 1979).

Another context in which females might be expected to bias investment toward one sex or the other in polygynous populations involves male genetic quality (Thornhill, 1979a). Because of high variance in male reproductive success in polygynous populations, a female mating with a genetically superior male, regardless of her nutritional state, might experience greater genetic success by biasing the sex ratio of her progeny toward sons, inasmuch as her male offspring should be especially attractive to females. But a female mating with a lesser male would pass on more genes to the next generation by allocating her resources to female offspring.

Under the nutrition hypothesis and the male genetic quality hypothesis, individual parents may bias their investment toward sons or daughters. But again, if males become more numerous than females in the overall population, their average reproductive value to a parent will fall, favoring individuals able to produce daughters. This will shift the populational sex ratio back toward unity and create a sex ratio highly conducive to male-male competition.

Types of Sexual Selection

Given that sex ratios are usually 1:1, that eggs are always larger than sperm, and that average parental investment per offspring is almost never equal for males and females, the stage is set for sexual selection. Members of one sex, usually the males, will be under selection to compete with one another for possession of individuals of the other sex, usually the females, that invest more in each progeny (or offer more of a critical resource in return for mating). This form of sexual competition generates *intrasexual* selection.

The limiting sex (typically females) may choose among potential mates, exerting *epigamic* or *intersexual* selection pressure for characteristics attractive to them. We shall discuss the evolutionary products of each type of sexual selection briefly and provide illustrative examples.

Intrasexual selection is responsible for the evolution of a broad spectrum of attributes related to competition for mates and their gametes (Table 3.1). These traits can be categorized as precopulatory, postcopulatory, or postfertilization depending on when they have their effect in the reproductive process. Before copulation can occur, the sex competing for mates must locate potential partners either by searching them out or by attracting them. This favors individuals skillful in selecting the optimal times to find or attract mates and with the ability to select the habitat in which receptive partners are most likely to be found. Moreover, in many species (but not all) mate searchers may engage in direct combat for possession of partners, or for territories containing resources attractive to mates, or for high dominance status within a group if this correlates with access to mates. The result has been the widespread evolution of aggressive skills and structural attributes, such as large size and fighting equipment, that enable individuals to win violent contests (see Fig. 5.12). At the same time, intrasexual selection in combative species has favored individuals that can avoid damaging competition (by withdrawing from aggressive interactions they cannot win) in order to live to fight another day.

Sexual competition may continue even after copulation has taken place. In general, this expresses itself in the attempts of males to copulate with females that have already received sperm from other individuals. On the one hand, this favors males that hide their mates from rivals or prevent takeovers from occurring. On the other hand, if there are egg fertilization gains to be derived from mate stealing, sexual selection will favor males powerful enough or stealthy enough to take

Table 3.1   The products of sexual selection.

Intrasexual selection
   Precopulatory competition for access to potential mates
      (1) Skill in mate location
      (2) Production of effective mate-attracting signals
      (3) Aggressive competence in the defense of mates and territories
      (4) Capacity to avoid damaging interactions with rivals
   Postcopulatory competition for access to eggs
      (1) Mate concealment
      (2) Mate guarding
      (3) Ability to find and take protected mate from original partner
      (4) Ability of sperm to displace competitor ejaculates
   Postfertilization destruction of rival zygotes
      (1) Ability to induce abortion of fertilized eggs
      (2) Infanticide

Epigamic (intersexual) selection
   Mate discrimination by choosy sex
      (1) Rejection of members of wrong species
      (2) Selection of genetically superior conspecific partner
      (3) Selection of partner with useful resources or services
   Attributes that make opposite sex attractive to discriminating sex
      (1) Attractive courtship behavior
      (2) Morphological characters considered attractive by opposite sex
      (3) Material benefit attractive to opposite sex

---

females from their original males (Parker, 1970a,c). If females do accept sperm from more than one male, there will be competition among the ejaculates within her favoring sperm capable of outracing or displacing rival gametes in the battle for egg fertilizations.

   Even after fertilization has taken place, it is still possible for male-male competition to exercise itself if an individual is able to induce a female to abort, resorb, or otherwise destroy the zygotes fertilized by the opponent and to accept the new male as her partner (Trivers, 1972; Schwagmeyer, 1979). A male may also reduce a competitor's fitness by destroying the rival's offspring, especially if the mother, once diverted from parental care of the deceased brood, becomes receptive to the infanticidal male (Hrdy, 1979). Intrasexual competition at the postfertilization level does not appear to have been discovered in the insects, but attributes related to precopulatory and postinsemination competition are widespread among the group. As an example consider the bibionid fly, *Plecia nearctica,* a species whose common name, the lovebug, suggests that its behavior should be an appropriate introduction to the world of sexual competition among the insects.

Intrasexual Selection in the Lovebug

The lovebug is a typical outbred insect in that it has an adult sex ratio of about 1:1 and its males engage primarily in mating effort while the females invest largely in parental effort (Thornhill, 1976b, 1980b). In the Gulf states of the United States and in some coastal areas of Mexico, lovebug males in the early morning fly up from the grass litter of the pastures and fields where they have spent the night and begin a period of hovering flight facing into the wind. Large numbers of males may gather above locations from which other flies are emerging (from pupal cases in the ground). A swarm of males may form, extending from 0.3 to 7 m above the ground, with the highest male density in the lower section of the aggregation. A hovering fly will pursue and often contact any insect that passes close to it. If the object is not a female or a copulating pair, the male returns to his hovering spot. Because moving objects close to a male are often other males (which may hover a few centimeters from each other), intraspecific interactions are almost continuous.

Females do not hover. Instead, during the morning and afternoon emergence periods, they crawl up vegetation in the field and take flight through the swarm. Sometimes a female is grasped even before she can begin to fly; if captured in flight, usually after flying only a few centimeters, the female will sail down with her partner to the ground cover where copulation begins (Fig. 3.9). A male that captures a female in flight must often wrestle with other males for possession of her. As many as ten males have been observed clustered about a single female, each attempting to copulate.

The swarm persists until midmorning, at which time the aggregation breaks up until the late afternoon; then it forms again for another few hours of flight. During the middle of the day most pairs and single individuals have dispersed from the emergence area to feed on pollen and nectar from flowers in the surrounding region. Males remain with their mates while they feed. Copulation may last for as much as 3 days in the field and it averages 56 hours in the lab (thus the title of lovebug). After completion of mating, females typically lay their eggs in the soil or ground litter and die without copulating again.

Almost every aspect of lovebug behavior reflects the impact of sexual competition among males. Swarms form at the times of day when

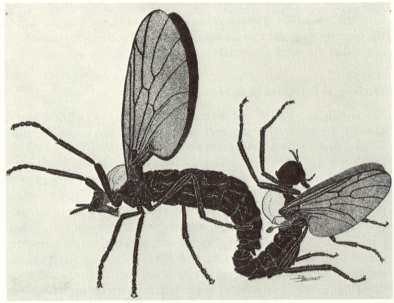

3.9   A pair of lovebugs, *Plecia nearctica,* in copula, which they may maintain for a period in excess of 2 days. The female is on the left. Note the sexual dimorphism in eye size, which is related to mate detection by males. (Drawing by D. Bennett.)

females are eclosing and are available for mating. Males hover over places from which females are most likely to emerge. They are highly motivated to copulate and will pursue any object close in size to a conspecific female that passes near them. Each male attempts to defend his hovering point within the swarm by bumping into nearby rivals. Large males are apparently able to displace smaller flies upward; the mean thoracic length of males swarming in the half-meter zone close to the ground is 2.1 mm, whereas males collected from the top of the swarm average only 1.5 mm. Because large males are in the best position to capture emerging females when they fly up from the ground, they enjoy disproportionate mating success. The average thoracic length of copulating males is 2.2 mm, as opposed to 1.9 mm in the swarming population as a whole.

Once a female has been captured, sexual competition continues, with other males attempting to take her from her original discoverer. This presumably favors males that attempt to remove a captured female from an area with a high concentration of competitors by drifting to the ground to mate there. The male that achieves a copulation remains with his partner for a very long time, perhaps using copulation as a form of postinsemination guarding. By staying with a female until she is ready to oviposit, the male may prevent her from mating again and so will fertilize all her eggs. Because at any one moment there are many more waiting males than receptive females, the long time spent guarding one female is not unusually costly to the copulating male in terms of missed opportunities to capture additional females. His mate searching and mate defending behavior assist him in transmitting his genes in a social environment of intense intrasexual competition for mates.

Epigamic Selection

It is highly likely that because large males enjoy a competitive advantage in the swarm, some males mate while others fail to do so. In contrast, all females are eagerly pursued by males and probably all succeed in copulating (if not taken by a predator). Thus the variance in male reproductive success probably exceeds that in the female population. This has not been quantitatively established, however, nor has the contribution of female choice, if any, to variance in male reproductive success been demonstrated in the lovebug.

In theory, females of the lovebug and other insects should favor (1) males that are of the same species and possess sufficient quantities of sperm, (2) males of superior genetic quality, and (3) males that provide exceptional material benefits for the female or parental investment for the female's progeny (Table 3.1). Receptive lovebug females are able to identify suitable members of their own species and thereby avoid hybrid matings, with the production of sterile or low-viability offspring. Whether female lovebugs also evaluate males on the basis of criteria (2) and (3) is more problematical, but certainly within the realm of possibility.

For example, female lovebugs might make discriminating choices among males by behaving in ways that stimulated competition among males for access to them. An early emerging female might wait to take flight until the swarm of males above her had reached some minimum density. Or a female that had been grasped by a small male might slow the rate of descent to the ground to permit other males to reach the pair and displace the original male. By "encouraging" competition females could improve the chance of mating with a male of superior perceptual ability, flight speed, strength, or size. These traits, if heritable to some extent, might raise the fitness of their male (and perhaps female) offspring. Even if the variation among males were the result of environmental factors (such as the nutritional history of the larvae) rather than genetic differences, a female could conceivably gain by mating with a relatively powerful flier because a large, strong male might increase the female's mobility during the long copulatory and foraging period. Sharp and colleagues (1974) have documented that the male plays an important role in the flight efficiency of the copulating pair.

As we shall discuss in detail in Chapters 12 and 13, convincing evidence on female choice is limited. This is particularly true with respect to the genetic superiority criterion. The case for female choice in relation to the material benefits that she will receive from a partner is more substantial. The example of *Hylobittacus* scorpionflies (Thornhill, 1976c, 1979a, 1980c,d), to be presented in detail in the chapters ahead, will show that females do not accept or use sperm from males unless they also receive a nuptial present of sufficient size. Here we shall briefly illustrate the topic of female choice by returning to Bateman's subjects, *Drosophila* fruit flies.

There is now a large literature on the "rare male advantage" in *Drosophila* (see reviews in Petit and Ehrman, 1969; Ehrman and Parsons, 1976; Ehrman and Probber, 1978). In experimentally manipulated populations female fruit flies are more likely to mate with males of the less common genotype. The outcome of a rare male experiment is shown in Fig. 3.10. Note that in each case the rare genotype mated significantly more often than expected. What seems to happen is that females "sample" a number of courtships and permit males with a rare odor (correlated with genotype) to copulate with them (see, for example, Spiess and Carson, 1981). A mechanism of this sort often results in the production of offspring with greater heterozygosity than otherwise. If most males in a population carry a particular genotype, most females will also possess it. Therefore by employing the rule "mate with an unusual male," a female will tend to select a partner genetically different from herself. Heterozygote offspring may enjoy a variety of fitness advantages, such as avoidance of the effects of deleterious homozygous recessive alleles and promotion of a greater range of enzyme variation, which may help buffer the individuals against variation in their environments. Thus a female fly could gain by choosing a rare male as a partner, because of the complementary genetic contribution of such a male.

There have been some suggestions, however, that the rare male advantage is an artifact of the experimental procedures followed (Markow, 1980; Kence, 1981). For example, in choosing the group of males to be placed together with a population of females, a few males

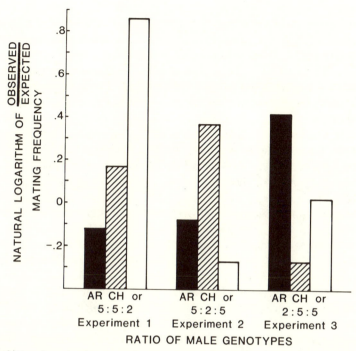

NATURAL LOGARITHM OF OBSERVED/EXPECTED MATING FREQUENCY

AR CH or
5 : 5 : 2
Experiment 1

AR CH or
5 : 2 : 5
Experiment 2

AR CH or
2 : 5 : 5
Experiment 3

RATIO OF MALE GENOTYPES

3.10 Results of experimental tests in which groups composed of different combinations of male fruit flies of three genotypes—Arrowhead (*AR*), Chiricahua (*CH*), or orange-eyed (*OR*)—were presented to 12 AR and 12 CH females. The relative reproductive success of each genotype was a function of its rarity. (From Leonard and Ehrman, 1976.)

of one type (those selected to be the minority males) would often be taken from the top of their holding vial (because they were most accessible). The larger number of males of the majority type would be removed from positions throughout their holding vial (in order to secure a sufficient number of individuals). Males taken from the top of vials have been shown to be more active than those resting lower down. Thus the minority and majority samples may not be comparable with respect to male vigor, and the advantage of the rare type may reflect the more active searching or courtship behavior of males taken from the tops of vials rather than female preference for an unusual phenotype/genotype (Markow, 1980).

Although there is some reason for caution in accepting the rare male advantage for *Drosophila*, there is little doubt that female fruit flies can make extremely subtle discriminations among potential mates. Markow, Quaid, and Kerr (1978) have shown that if a female is given a choice between a male that has mated twice recently and a virgin male, she is more likely to accept the virgin male (Table 3.2). This is not because the recently mated males are reluctant to court or mate; the speed with which experienced males initiate courtship, the duration of courtship, and the length of copulation (when successful) match that of virgin individuals. Females will mate with a twice-mated male when they have no choice and will produce about 200 progeny. But if they copulate with a virgin male (or with a twice-mated male that has had 24 hours to recover from his previous matings) offspring production rises to about 300. Thus female choice improves female reproductive success by 50 percent.

The reduction in fertility of *Drosophila* males that have copulated several times in a short period is due not to a depletion of the sperm supply but to depletion of the ejaculatory fluid produced by the male accessory glands (Lefevre and Jonsson, 1962). Males fail to transmit additional sperm when they do not have an appropriate amount of added glandular secretion to provide the female. It has been shown that implants of the accessory glands in females reduce their receptivity (Burnet et al., 1973). If females partially adjust the duration of nonreceptivity following copulation to the quantity of accessory fluids they have received, a male that tried to transfer sperm in the absence of these secretions might gain little or nothing from the transfer (when female *Drosophila* mate again, they predominantly use the sperm of their most recent partner; Boorman and Parker, 1976). One wonders if the accessory secretions might not serve a nutritional function (T. A. Markow, personal communication); if so, females might be discriminating among males on the basis of the material benefits provided during copulation (see Chapter 12.)

Table 3.2   Relative mating success of virgin male fruit flies versus males mated twice previously when one of each type was placed with a receptive female. (From Markow, Quaid, and Kerr, 1978.)

Interval between previous matings by twice-mated male and the choice experiment	Male chosen		Females mating with virgin	x^2	P
	Virgin	Twice mated			
≤ 1 hr	219	139	61%	17.88	< 0.01
24 hr	84	91	48%	0.28	NS

Our main point is that female choice is a reality in certain insects. The ability to make sophisticated discriminations among males may raise female reproductive success and exert pressure on males to exhibit those traits that females consider attractive. We now turn to an examination of why there is diversity in the mating systems of insects.

Sexual Selection and Mating Systems

Rivalry among males and the operation of female choice combine to determine the nature of the sexual association between the males and females of a species. These associations vary considerably, creating diversity in mating systems. Our first task is to define the different types. This is not simple, because although there is general agreement that the terms polygyny, monogamy, and polyandry should be used to label different mating systems, there is no universally accepted set of definitions for these words. Various classification schemes differ, first, on whether the presence or absence of a pair bond or other lengthy association between copulatory partners is a key criterion in defining different mating systems (see for example R. K. Selander, 1972; Wittenberger, 1981a, versus Emlen and Oring, 1977). Second, there is a difference of focus, with some authors applying their definitions to individual males and females, while others discuss mating systems as they apply to entire species.

For our part we shall dispense with the requirement that certain mating systems be identified on the basis of pair bonding between male and female. This potential feature of the sexual relationship can (and will) be treated as a separate issue, distinct from the question of how many mating partners an individual acquires. Thus polygyny will be defined here as the mating system that results when some males copulate with more than one female in a breeding season. By our definition, a male that is pair bonded with one female but also copulates even once with another female with whom he has only a brief association is a polygynous male, not a monogamous one. Monogamy occurs when a male and female have only a single partner per breeding season. Polyandry refers to one female's having more than one male as a mate during a breeding season.

This classification has several advantages. It focuses the distinction between mating systems on a single criterion, the number of copulatory partners per individual. It eliminates the need to define what is meant by a "pair bond" or "lengthy" association, an often sticky issue (as Wittenberger, 1981a, acknowledges). Pair bonding in the avian sense is rarely observed in insects, in any case. Moreover, we need not introduce a fourth general category, "promiscuity," to accommodate those instances in which pair bonds (however defined) are absent. The term "promiscuity," although widely used to describe certain bird and mammal mating systems, in common parlance implies an absence of selectivity in the choice of mates, an implication that is often untrue for insects and other so-called promiscuous animals.

Moreover, a classification of mating systems based on numbers of mates per individual can be used at either the individual level or the populational level, whichever is convenient. An individual male can be categorized as polygynous or monogamous, depending only on the number of mates he secures. At the population level one can characterize an entire species as polygynous if a reasonable percentage (5 to 10 percent) of the males acquire several mates. In such a species many males may actually secure only one mate and therefore may be treated as monogamous individuals; a good many others may experience enforced celibacy because their more successful rivals have monopolized the receptive females. A monogamous species is one in which all or almost all males and females never acquire more than one mate per breeding period. In a polyandrous population a small to moderate proportion of the females acquire more than a single mate.

Usually the three options are considered mutually exlusive at the population level, but this need not be so. Even in some vertebrates (and often in insects) a significant fraction of both the males and females mate with more than one member of the opposite sex, in which case the population can be characterized as polygynous/polyandrous.

One can measure quantitatively the degree to which a species (or population) is polygynous, monogamous, or polyandrous. This requires data on the statistical variance in the reproductive success of males and females, data of a sort similar to those gathered by Bateman (1948) for *Drosophila melanogaster*. The degree of polygyny could then be given as the degree of variance in male reproductive success (Daly and Wilson, 1978; Wade and Arnold, 1980; but see Payne and Payne, 1977); the degree of polyandry could be measured in the same way, but using data on the variation in reproductive success within a sample of females. This is not to say, however, that in truly monogamous species there would be zero variance in reproductive success because of an absence of competition for mates. Differences in the number of surviving offspring produced by monogamous individuals could occur as a result of differences in the abilities of males to acquire a mate early, rather than late, in the breeding season. It is conceivable that pairs which copulate sooner have more time in which to produce and, in some cases, care for their progeny. In addition, members of the same sex may compete for superior partners (that is, more fecund females or more highly parental males) whose "possession" translates into increased fitness. Nevertheless, the variance in reproductive success in monogamous species should generally be *relatively* low.

Securing the necessary data on the reproductive success of individuals is not an easy task, particularly for males in species in which there is uncertainty about the effectiveness of some copulations. One cannot assume that all matings result in egg fertilizations for a male, nor can one assume that the number of eggs fertilized per mating is identical. Still, the value of having even a crude estimator of the degree of polygyny or polyandry is potentially great. Such a measure would permit a more quantitative approach to tests of the relationship between the kind of mating system employed by a species and the ecological factors to which it was subjected. For example, a prediction derived from sexual selection theory is that the variance in male mating success should be related to the distribution of food resources in species whose females make copulation contingent upon receipt of food under a male's control (Emlen and Oring, 1977). If the key resources are limited and clumped, some males may be able to monopolize them and as a result secure a large proportion of all copulations.

This prediction has been tested experimentally with *Panorpa* scorpionflies (Mecoptera) (Thornhill, 1981, unpublished). In these species males compete for possession of dead arthropods which attract females (Fig. 3.11). In exchange for sexual access, males permit their mates to feed at the carrion they defend. The abundance of dead crickets available for competing males was manipulated in large field enclosures into which ten individually marked males and ten females were placed. Two, four, or six carrion items were introduced into each of 28 enclosures. The males were observed intensively and the number of copulations per individual was recorded for the period of a week. As predicted, the statistical variance in mating success declined significantly (from 6.4 to 4.0 to 2.5) as the density of defendable resource items increased. When more dead crickets were available, the intensity of sexual selection declined; more males were able to hold a food resource and the copulations were more evenly distributed.

It would be useful and interesting to gather comparable information about the intensity of sexual selection occurring in natural populations, but to do this is obviously more difficult than to do so in controlled laboratory conditions or in field enclosures. One has to discriminate between the effects of natural selection and of sexual se-

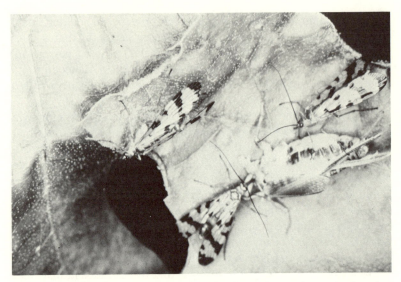

3.11   A male *Panorpa mirabilis* scorpionfly (*bottom*) guarding a dead cricket upon which a female is also feeding. The third individual is a rival male excluded from the prey. (Photograph by N. Thornhill.)

lection on the variation among individuals in reproductive success. This requires a way to identify the specific impact of sexual competition and mate choice on reproductive success. Wade (1979; Wade and Arnold, 1980) argues that the variance in the number of copulations gained by different males is a reasonable estimate of the intensity of sexual selection. Thornhill (1977) has been able to translate data on copulation frequency by males into rate of egg gain, thereby permitting a comparison of the number of fertilized eggs attributable to individual males and females in a natural population of another scorpionfly, *Hylobittacus apicalis*. (The reproductive behavior of this species is discussed in detail in Chapter 12.) Individually marked males were followed as they mated and hunted prey to feed their mates, and females were tracked as they copulated and laid eggs between copulations. The technique of following individuals is possible in this species because mecopterans fly weakly among low herbs, matings are frequent, and females lay large, visible eggs one at a time by dropping them among the leaf litter while hanging from a leaf or twig of a low herb. If some reasonable assumptions are made about paternity in this species (Thornhill, 1977), the variation in the number of eggs fertilized by males per day ranged from 0 to 58. Some males

(23 percent of the total) that copulated were unable to mate long enough to transfer sperm and so gained nothing from their copulation(s). (Not included in this analysis were males that failed to mate because they could not obtain prey or because they died in the attempt to do so.) The variation in female reproductive success was much lower, 3 to 14 eggs laid per day, with most females laying an average of 12.

These data gathered in a field study reinforce Bateman's laboratory work. The variance in male reproductive success (as estimated by egg fertilization rate) was much greater than the variance in female success. One could use data of this sort to produce a ratio of the two variances. The result would be a single figure that could be used to characterize the mating system of the species. This ratio (male variance/female variance) should be nearly unity for a monogamous species, greater for a polygynous species, and less for a definitively polyandrous species.

The use of quantitative descriptors of mating systems for the purpose of testing comparative patterns is in its infancy. Despite the difficulties of the approach, it seems likely that insects will provide excellent material for work of this nature. The adult phase of the life history of most insects is short, eliminating the need to follow individuals over several years in order to get even a crude measure of relative reproductive success. Moreover, large samples are needed for the variance approach, and for the most part these can be more readily gathered with insects than with vertebrates. Even without precise quantitative information on the degree of polygyny or polyandry exhibited by a species, it is clear that some mating systems are more common than others. Sexual selection theory helps us understand the reasons.

Sexual Selection and the Prevalence of Polygyny

By far the most common form of mating system in all animals, including insects, is polygyny. Given our discussion of the operation of sexual selection, it is relatively easy to see why this might be the case. The two conflicting, extreme options for males are to allocate limited supplies of time, energy, and risk taking to mating effort or to parental effort (Trivers, 1972; Low, 1978; Alexander and Borgia, 1979). The question is, which allocation yields the greatest return for a male in terms of genes contributed to the next generation? The equation is strongly tilted toward mate acquisition rather than paternal behavior, simply because of the potential males have (given their vast supply of sperm) for fertilizing many females (each of which has a relatively small number of eggs). Because females typically make a large parental investment per offspring, a male can rely on his mates' investments (in nutrients for the eggs and other parental activities) to produce viable offspring. Copulatory success, then, is likely to be an accurate measure of genetic success for a male.

So our question becomes, what enables some males to express their potential for polygyny, while others fail to do so? Emlen and Oring (1977) and Bradbury and Vehrencamp (1977) have argued that certain socioecological factors make it *economically* feasible for some males to monopolize access to females, either directly or indirectly. In some animals, for example, predator pressure favors the formation of more or less permanent groups of the prey species, which defend themselves in various ways against their enemies. Some males may be able to control these groups, expelling rival males and patrolling their ready-made harems of females, because the area to be defended is relatively small. Expenditures associated with defense of the group are manageable, whereas in other species the direct defense of a number of females would be extremely costly because of the generally dispersed distribution of potential mates.

The *indirect* monopolization of females can occur even if receptive individuals do not congregate in permanent groups, provided that the resources used by females occur in discrete, small clumps that some males can defend. For example, in the *Panorpa* scorpionflies, females feed on a patchily distributed resource, dead arthropods. These small patches may attract numerous females over time, and the male that can defend the area containing the resource monopolizes mating opportunities (assuming that those females that visit the area are receptive).

If the possibilities for direct and indirect defense of mates are poor because of the dispersed nature of females and the resources they exploit, extreme polygyny can still evolve—but along a very different route. If males do not make a parental contribution to their offspring, female choice may revolve purely around attributes that indicate the genetic quality of the sperm she will receive from a male. In certain species of this sort males behave in ways that appear to be self-advertisements of genetic quality. For example, in some Hawaiian fruit flies males gather at prominent tree ferns and compete intensely with one another for "symbolic" territories that do not contain groups of females or useful resources (Fig. 3.12). Nonetheless, females visit the arenas to acquire a mate whose behavior or location within the group indicates his social dominance over other males and, perhaps, his genetic superiority. The variance in male reproductive success may be very great in species of this sort (Borgia, 1979).

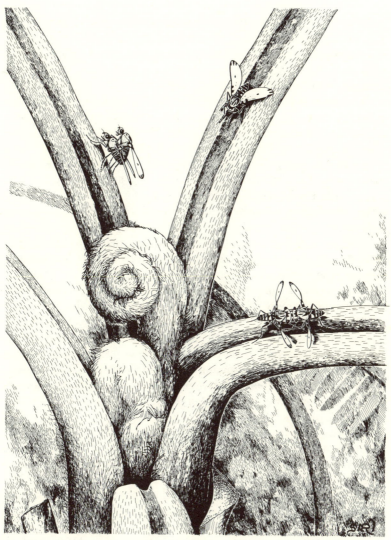

3.12 A group of male Hawaiian fruit flies, *Drosophila heteroneura,* at their lek, a collection of closely spaced territories on a tree fern. The fern does not contain emerging, ovipositing, or food gathering females. Males battle for control of perch territories (*foreground*). Females visit the fern solely to select a mate from the males present; a copulating pair appears on the left. (Drawing by L. S. Kimsey.)

Thus in a variety of different ecological settings a few males are able to achieve some (small) fraction of their polygyny potential at reasonable cost. But not all environments favor male polygyny. Monogamy is not surprising from the female's perspective, because her limited number of eggs can often be fertilized by sperm received from a single male. It is puzzling (and unusual) for a male to advance his genetic interests by mating with only one female rather than attempting to donate his physiologically inexpensive sperm to many females (Orians, 1969). Nevertheless, as we shall document more fully later on, there are ecological circumstances that combine to decrease the mate monopolization potential of males (Emlen and Oring, 1977) and also to raise the benefits for males of engaging in parental effort. In birds, for example, females commonly make it advantageous for a male to be parental by requiring a prolonged courtship and by synchronizing their breeding activities with those of their neighbors. The possibility that a male could succeed in courting two or more females and inducing them both to copulate under these conditions is very low. In general, synchrony of mating by the female population reduces the possibility of polygyny (Knowlton, 1979). If males are able to provide useful materials to the female or her offspring, then monogamy stemming from a male's commitment to one female and her progeny can theoretically yield greater reproductive success for the male than attempts at polygyny. Again, mating effort and parental effort are essentially incompatible; the male that allocates effort to futile mating attempts cannot devote these energies to improving the survival chances of his mate's offspring.

If monogamy is puzzling at first glance, polyandry is even more so. We have just noted that from the female's perspective, there is generally little or nothing to be gained from mating with more than one partner. Males rarely offer anything other than sperm in return for the copulation. In species with "typical males," therefore, females mate simply to acquire sufficient gametes to fertilize their eggs, and one mating usually suffices. Only when copulation gives a female special access to resources or parental assistance is polyandry likely to result in reproductive gains for her. Here the conflict of the sexes should be most clearly developed, because efforts by a female to mate multiply may be opposed by the female's first mate as he attempts to monopolize fertilization of her eggs. Particularly if mating commits a male to donate anything of value above and beyond his genes, he should be reluctant to share his mate with another male. Although in many species a female would probably gain if she had a polyandrous arrangement (a "harem" of paternal helpers), the members of the harem would almost always be at a disadvantage to rival males that avoided sharing one female (and her eggs) with other males (Orians, 1969).

In light of sexual selection, the extreme rarity of polyandry (particularly in pair bonding animals) and the moderate rarity of monogamy become more explicable. The relative frequency of the three basic mating systems reflects the action of sexual selection on the range of possible ecological conditions; it is usually possible for some males to monopolize several females, or for females to gain by choosing those few males that will have the greatest positive effect on their fitness. Polygyny is the common outcome.

4

Timing of Mate Locating

THE USUAL RESULT of sexual selection is to favor males that are skillful in the competition with others for the chance to fertilize the large eggs of their females. The next eight chapters explore the remarkable variety of consequences of intrasexual selection on the evolution of male reproductive behavior in the insects. We begin by describing the behavior of the anthophorid bee, *Centris pallida*, to which we shall refer repeatedly to illustrate the multiple effects of sexual selection on male behavior (Alcock, Jones, and Buchmann, 1976, 1977). In this chapter the particular focus is on how males time their mate locating activities to achieve maximum advantage. Males of *C. pallida* show how decisions on when to become a sexually active adult and when to search for mates on a daily basis are related to reproductive competition.

Male Reproductive Tactics in a Solitary Bee

In May in the Sonoran Desert of Arizona, the palo verdes begin to bloom, each green-barked tree producing an extravagant profusion of yellow flowers in an otherwise austere environment. These flowers contain a pollen and nectar resource that provides the energy base on which the bee *Centris pallida* depends. Just as the first palo verdes are coming into flower, males begin to emerge from underground cells in which they have spent an entire year (first as an egg, then as a larva, then as a pupa, and finally as an adult). They gnaw away the mud cap that has sealed their brood chamber and burrow upward through the soil to the surface. In an open sandy clearing surrounded by mesquite and palo verde trees, many males may emerge over the span of a few mornings from an aggregation of nests their mothers had constructed the previous year. Individual bees probably remain for several days in the area where they emerge, visiting nearby flowering shrubs and trees for nourishment.

During the early morning, however, adult males are not likely to be feeding. Beginning at about 0700 the first males begin their patrol flights, cruising at high speeds a few inches from the ground over an area as large as 40 sq m. By 0900 in a dense emergence site, hundreds of these large, gray, nearly bumble-bee-sized males may be patrolling broadly overlapping home ranges, filling the air with a resonant humming. Individuals make no effort to keep other males out of their searching areas and zoom about as if oblivious to one another's presence. By 1030 or 1100, sites that were once alive with cruising males are largely abandoned and have become quiet once again. Newly adult bees, male and female, generally emerge from the soil between 0730 and 1000.

Occasionally a male will emerge with some defect that prevents it from flying, and it soon expires on the superheated sand. The dying male (and later its corpse) will be pounced upon dozens or hundreds of times by patrolling males, many of whom make prolonged and frenzied efforts to copulate with their fellow male.

Patrolling males commonly alight on the ground even when not attracted by the body of a conspecific. These males usually walk about for a short while with their antennae held close to the surface. They may then fly up and resume their rapid patrolling or they may remain earthbound and begin to dig, using their jaws to loosen the soil and their legs to remove it from the excavation pit.

Other males frequently alight near the digger and approach him. As soon as he is touched by an intruder, the digger will respond— usually by kicking out with his hind legs. If the newcomer persists, the digger will stop his work and whirl to face his opponent. Although the two bees may have ignored each other when flying in their home ranges, on the ground they do not; fights for control of the digging pit are common.

Eventually, despite frequent interference from his fellows, one of the males removes 1 or 2 cm of topsoil, exposing an about-to-emerge bee in its emergence tunnel (Fig. 4.1). This individual may be a male,

4.1　A paint-marked male *Centris pallida* digging down to reach a female (*top photos*), then meeting her (*lower left*), and mating with her (*lower right*). (Photographs by J. Alcock.)

in which case the digger grasps the emerging bee as it exits only to release it after a brief struggle. If the uncovered bee is a female, the digger slips onto her back and immediately copulates. He then withdraws his penis and begins to stroke the female with complex synchronous movements of his legs, antennae, and abdomen. He may be interrupted by intruders that seek to pull him from his mate, sometimes successfully, but usually the male flies with his partner to a nearby tree to continue the stroking interaction for a few minutes more and to copulate again. The male then releases his mate and searches for additional virgins.

As the spring progresses, the number of searching males declines rapidly, whereas the number of nesting females increases. The female bees, once mated, do not copulate again. Unlike males, they tend to avoid areas from which they have emerged and choose other locations in which to nest. Many females may come to adopt the same area, forming a new nesting aggregation. Each female constructs her own burrow(s) and provisions the beautifully formed brood cell (one per nest) with a paste of palo verde pollen and nectar, topping this with a more soupy mixture of the same substances (Alcock, 1979d). When sufficient supplies have been placed in the cell, the female lays her egg on the surface of the provisions, constructs a firm mud cap for the cell, and then fills the burrow by kicking sand down the tunnel. Her off-spring, if it survives, will emerge a year later and will enter a new round of reproductive competition.

Note that males of this bee behave in much the same way as males of the unrelated lovebug (Chapter 3). The reproductive effort of male digger bees and lovebugs is devoted entirely to mating, for they engage in no parental activities. Their lives revolve around attempts to copulate. Males emerge as adults just prior to peak periods of availability of receptive females. When females are emerging, males compete intensely for strategic locations where they will be able to contact potential mates. All males are expert at detecting receptive females (so much so that they can find such females even before they emerge from the soil). Although there is no detailed quantitative evidence on the variance in copulatory success in populations of the bee, the imprint of sexual selection can be seen in almost every attribute of male reproductive behavior. We begin our analysis by examining how males time their sexual endeavors in order to secure as many copulations as possible.

The Timing of Metamorphosis to Adulthood

The factors that determine when female insects make the transition to active adulthood (and therefore are available for copulation) are numerous and complex. In seasonal environments variation in temperature, humidity, resource productivity, and other parameters may singly or in concert affect the probability of reproductive success for females. In C. pallida, as in most other insects, receptive females are not equally abundant year round. If a male is to maximize his mating opportunities, it is to his advantage to time his reproductive efforts to coincide with the period when receptive females are most likely to be encountered. His ability to do this will depend upon "decisions" that his physiological system makes on (1) when to reach adulthood, (2) when to begin searching for mates after achieving adulthood, and (3) when during each day to search for mating partners. We shall consider each of these temporal aspects of reproductive development and mate locating strategy in turn.

In *C. pallida* and many other insects that depend on a few species of food plants or prey, the timing of adult metamorphosis is critical inasmuch as their food can be collected only during a restricted season. Females of *C. pallida* must emerge from diapause during a 4- to 6-week period in the late spring if they are to gather brood cell provisions from flowering palo verdes. Males must "anticipate" female emergence if they are to copulate often. Because their lives are so entwined with the flowering period of palo verdes (and ironwoods), one wonders if the bee has evolved the ability to detect cues correlated with conditions that will promote flowering in these trees. If they could do this (while buried beneath 15 cm of soil and sealed within a tough brood cell capsule), they could be assured of becoming active adults at the optimal time for resource gathering (females) and for mate locating (males). Judging from the widespread occurrence in solitary bees of adult emergence in synchrony with the flowering of preferred food plants, bees such as *C. pallida* must have some mechanism to detect when food plants are producing food. Perhaps the most dramatic example of this phenomenon is deferred emergence, in which failure of the host plant to flower coincides with failure of female and male bees to emerge—as reported for *Halictus aberrans,* a specialist on *Oenothera latifolia.* This bee evidently can postpone emergence for an entire

4.2   Mayflies, *Hexagenia bilineata,* attracted to automobile headlights on the night of their mass emergence. (From Fremling, 1970.)

year, if necessary, to gain a chance to forage (and mate) successfully (Stephen, Bohart, and Torchio, 1969).

Another selection pressure that affects the life history properties of many insects is predation. Female mayflies and some other aquatic insects (Tokunaga, 1935) may gain a survival advantage by emerging en masse (Hynes, 1970; Edmunds and Edmunds, 1979; Sweeney and Vannote, 1982) over a period of a few hours on a relatively few days each year (Fig. 4.2). Inasmuch as they are highly edible and almost defenseless, mayflies are avidly consumed by a host of animals. Yet because some species occur in large populations, all of whose members emerge more or less simultaneously, the feeding capacity of the local predators may be swamped. If so, synchronous emergence improves the chance that a particular individual will escape capture; this favors males that have a maturation schedule coincident with females, both because of increased survival chances and because of improved mating opportunities.

The extreme example of the predator-swamping tactic is probably represented by the periodical cicadas, which possess one of the most extraordinary life histories of any insect. There are a number of species of cicadas in the eastern United States that require either 13 or 17 years to complete development. The current view is that there are three species that emerge together in each cycle; that is, three species of cicadas with 13-year development cycles and three others with 17-year cycles (Alexander and Moore, 1962). In any given region populations of all three species appear in the same year. Thus one brood has emerged every 17 years along the Atlantic seaboard from Virginia to Connecticut, with records beginning in 1724; another has been recorded from Iowa since 1844, with the emergence taking place one year after the coastal brood. As many as 100,000 adult cicadas may exit from their underground chambers in a single acre during a 2-week period in the right year (Fig. 4.3).

As in the mayflies, synchronized emergence may enable an individual to increase its chance of avoiding predatory attack simply because there are too many victims for the predators in the area to consume (Alexander and Moore, 1962; Lloyd and Dybas, 1966), despite the fact that cicadas evidently are attractive prey for many animals.[1] In addition, there is the possibility that the noisy calling of thousands upon thousands of cicadas may produce intense sound levels that confuse

1. Marlatt (1898), in his monograph on periodical cicadas, presented the reports of a number of entomologists/chefs on the palatability of cicadas to humans. The consensus was that the superior method of preparation involved frying the insects after they had been dipped in batter. Marlatt evidently was not tempted to experiment personally, as he merely offered the culinary opinions of others.

or repel predators capable of dealing with small isolated groups of calling males. Smith and Langley (1978) have shown that the stress calls
(which may reach 110 db) of a single cicada are aversive to grasshopper mice and induce errors in predatory attack by the mice. Presumably the noise produced by hundreds or thousands of chorusing cicadas also has a deterrent effect on some of the cicada's enemies.

Thus both predator pressure and seasonal variation in resource
availability can influence the optimal time for females to reach adulthood. This in turn creates selection on males to time their emergence
so as to be sexually active during the most propitious period to encounter mates. As a first approximation, one might expect male and female
emergence patterns to be more or less the same. Yet this does not
seem to be the usual case, for reasons that we shall now explore.

Protandry

The bee, *Centris pallida,* and the lovebug fly, *Plecia nearctica,* are examples of insects in which the males tend to emerge somewhat before
the females. This pattern, known as protandry, is standard for solitary
bees and wasps (see Fig. 4.4; Stephen, Bohart, and Torchio, 1969;
Evans and West-Eberhard, 1970) as well as for some other groups
such as butterflies (Shapiro, 1970; Scott, 1972; Wiklund and Fagerström, 1977), mayflies (O. W. Richards, 1927), and mosquitoes
(Nielsen and Nielsen, 1953). A variety of hypotheses has been advanced to account for the relatively early emergence of males, including (1) prevention of inbreeding, (2) enhancement of the selective
process through removal of unfit males during the prereproductive period, (3) reduction of prereproductive death of females through facili-
tation of their rapid fertilization after eclosion, and (4) improvement
of male reproductive success because early-emerging individuals
would have access to more females than late emergers.

Wiklund and Fagerström (1977) have reviewed this set of hypotheses from an individual selectionist perspective. They argue that if the
first proposal were correct there should be as many cases in which females emerge somewhat before males (protogyny) as species in
which protandry is the rule, and yet examples of protandry far exceed
those of protogyny. Furthermore, they point out that hypotheses (2)
and (3) are group selectionist, hypothesis (2) blatantly so and hypothesis (3) in a more subtle manner. In both arguments the

4.3 The cicada *Fidicina pronoe* emerges in numbers during a restricted period of the dry season in Costa Rica. (Photograph by A. M. Young.)

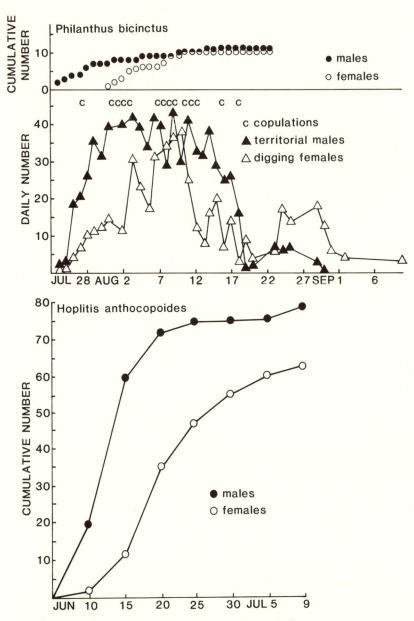

4.4 Protandry in solitary Hymenoptera. (*Top*) The cumulative number of males and females of the sphecid wasp *Philanthus bicinctus* that emerged within a cage over a period of days. (*Middle*) Males of *P. bicinctus* generally emerge several days before their females in order to prepare for the competition for mating territories. Territorial males secure copulations with females early in their nesting cycle. (*Bottom*) The cumulative number of males and females of the megachilid bee *Hoplitis anthocopoides* captured and marked in one population over its flight season. (From Gwynne, 1980, and Eickwort, 1972.)

presumption is that males put themselves at risk *solely* for the benefit of females (or the species as a whole). If males really sacrificed their reproductive interests for females or the species, then a mutant male that avoided the sacrifice should gain a reproductive advantage and selection would eventually eliminate the risky prereproductive emergence of males. In the fourth hypothesis, the risk of prereproductive death is viewed as one cost of a trait that may yield a net reproductive gain for a male. Wiklund and Fagerström provide a sophisticated mathematical treatment that shows how males can gain access to more females through early rather than delayed emergence. We shall employ a much simpler numerical example to make the same point.

The model of Wiklund and Fagerström was based on a butterfly species whose females mate once, shortly after eclosion. We also select for our example a species of this sort and assume further that (1) the life span of all males is the same, and (2) all males have equal searching or competitive abilities vis-à-vis mate acquisition. Under these conditions a male's reproductive success is the sum of the number of emerging females divided by the total number of searching males per day for each day of his life. If each male emerges in such a way as to achieve at least as many encounters with females as any other male in the population, then the sum should be the same for all males. This would occur if the daily ratio of emerging (receptive) females to surviving (searching) males were the same throughout the breeding season.

In our example in Table 4.1 the numbers of emerging females rise to a peak and then fall over a period of 8 days. If all males survive 8 days or more, then clearly they should all emerge shortly before the day on which the first female emerges. If, however, we set the life span of males equal to some fraction of the female emergence period, then males will achieve equal reproductive success only if they emerge gradually over a number of days early in the female emergence period. (If all males begin searching on day 1, they would have equal reproductive success, but the situation would be evolutionarily unstable inasmuch as a late-emerging mutant male would enjoy much higher reproductive success, having all the late-emerging females to himself after the death of his competitors.)

We conclude that given the assumptions on which the model rests (single-mating females that copulate soon after reaching adulthood), males should emerge on average somewhat sooner than females. This prediction is supported by observations of the lovebug and the digger bee. In fact, the correlation between single-mating females and protandry is strong throughout the solitary Hymenoptera (Stephen, Bohart, and Torchio, 1969). For example, in bees and wasps that nest in

Table 4.1   Numerical example of the optimal pattern of male emergence, given the staggered emergence of females that mate only once, early in their adult lives.

Population parameters	Day							
	1	2	3	4	5	6	7	8
Number of emerging females[a]	1	2	3	4	4	3	2	1
Number of emerging males[b]	4	4	4	4	4			
Cumulative number of living males if life span equals 4 days	4	8	12	16	16	12	8	4
Mean reproductive success of each male per day[c]	0.25	0.25	0.25	0.25	0.25	0.25	0.25	0.25
Cumulative reproductive success of each male	$0.25 \times 4 = 1.0$ for all males							

[a] Mean date of emergence for females is day 4.5.
[b] Mean date of emergence for males is day 3.
[c] Assumes equal searching abilities of all males.

burrows in woods, Krombein (1967) collected data on more than a hundred species and found that males were almost always placed in the outer cells of linear nests, females in the inner cells. This arrangement requires protandry because if the females were to emerge before the males, they would destroy their brothers as they pushed and gnawed their way down the burrow tube to the exit.

A somewhat more subtle and less well supported prediction from the model is that as male life span increases, an ever larger proportion of the total male population should emerge on the first few days that females are available. Moreover, there would seem to be no advantage to emerging on days before the first day of female emergence (unless there were some maturational advantages to be gained before beginning to search for mates, or unless early emergence gave a male a competitive advantage of some sort, as in acquisition of a territory or of a nuptial gift to present to a female). Michener and Rettenmeyer (1956) note that in the solitary bee *Andrena erythronii* the onset of male and female emergence is rarely more than a day or two out of phase, although the male emergence period is more compressed than that of the female. In contrast, males of *Philanthus bicinctus* may emerge a substantial number of days before the first female (Fig. 4.4; Gwynne, 1980). Males of this wasp, unlike males of *A. erythronii,* are highly territorial; the early days of the male's life are spent feeding at flowers in preparation for attempts to acquire and defend a territory. Perhaps individuals that emerge some days before the first females are better able to achieve the physiological condition that promotes success in the intense competition for prime territories.

A key assumption in all these arguments is that females mate immediately or quite soon after achieving adulthood. If they do not, then male emergence should be tailored to the temporal pattern of availability of *receptive* rather than recently emerged females. Figure 4.5 presents data on the emergence of males and females of two species of muscid flies of the genus *Fannia* (Tauber, 1968). *Fannia femoralis* is a species with single-mating females that copulate within a day or two

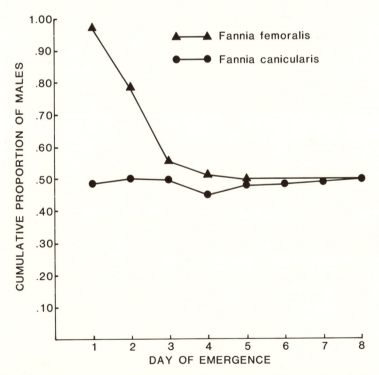

4.5 Comparative emergence patterns of two species of *Fannia* flies; *F. femoralis* exhibits protandry, while *F. canicularis* does not. (Data from Tauber, 1968.)

of eclosion and, as predicted, protandry has evolved. In its close relative *F. cannicularis,* however, males and females emerge on an identical schedule. But there are differences in the maturation of sexual receptivity in males and females; males become sexually competent 2 to 3 days after eclosion, whereas females require a 4- to 5-day period. As a result the maturational pattern is effectively protandrous, as the time by which half the males will have achieved sexual readiness is 1 or 2 days before the date by which half the females will have become receptive.

We can also ask what would be the effect of discarding the assumption that all males had the same expected life span or mate-acquiring ability. It is possible that selection can favor developmental schedules that take into account the body size or metabolic condition of the larval male. For example, males with a short expected life span but an exceptional capacity to monopolize receptive females would gain an advantage if they were to emerge just before the peak abundance of receptive females. The corollary prediction is that males of lower-than-average ability to compete with rivals for territories or access to females should emerge at times when the intensity of competition is relatively low. There are some data in support of this hypothesis from studies of scarab beetles (described in Chapter 9), which show that smaller males, less able to defend a burrow resource attractive to females, emerge sooner than larger individuals.

Multiple Mating by Females and Male Emergence

In some species protandry has evolved even though the females mate several times during their lives. The females of most butterflies actually mate more than once (Burns, 1968; Ehrlich and Ehrlich, 1978), yet protandry is the rule in this group. The same is true for some grasshoppers (Weissman and French, 1980), some beetles (see, for instance, Linsley, 1959; A. Milne, 1960), and *Pogonomyrmex* ants (Hölldobler, 1976). The key question is what effect does multiple mating by the female have on the genetic gains enjoyed by her first mating partner. If these gains are substantial, there can still be an advantage to males that emerge in such a way as to maximize their contacts with *virgin* receptive females. For example, even though female butterflies may mate a number of times, they alternate matings with relatively long intervals of non-receptivity during which a number of eggs are laid. Moreover, the reproductive potential of a virgin may greatly exceed that of an older receptive female (Suzuki, 1978). Thus a male that mates with a virgin is reasonably assured of fertilizing a substantial fraction of her eggs at a time when she is likely to lay a relatively large proportion of her lifetime supply. This favors males that time their emergence early to encounter relatively more virgins than older, previously mated individuals.

The same argument may apply to beetles like *Phyllopertha horticola*, whose males emerge a day or two before the first females, which also mate at intervals alternating copulations with bouts of underground oviposition (A. Milne, 1960). Females often lay all or the vast majority of their lifetime output of eggs after a single copulation, sometimes after the first mating. This may make it advantageous for males to anticipate the emergence of females, despite the fact that their mates may later copulate with other males.

There is, however, an alternative hypothesis for early male emergence that is not related to the competition for mates (L. D. Marshall, personal communication). If large size is achieved at the expense of prolonging the larval stage, then insects that emerge sooner will be smaller than later-emerging ones. In many insects relatively large females enjoy greatly increased fecundity, whereas relatively large males do not secure disproportionate reproductive success. In such cases females are often larger than males. In these species protandry might occur if mortality risks are greater for the immature stages than for adults. Males would achieve adulthood sooner at a smaller size and so escape the predators and parasites that attack the larvae; females would prolong development, become larger, and gain the fecundity benefits that are great enough to outweigh the risk factors. The "developmental constraints" hypothesis predicts that in species with no size dimorphism, male and female emergence should be synchronous, and in species in which males are larger than females, later emergence of males (protogyny) should be the rule.

Protogyny

The "sexual competition" hypothesis for delayed male adulthood argues that if a female's most recent partner is able to displace or supplant sperm donated by earlier mates (through *sperm precedence*), the evolution of protandry becomes less likely. When sperm precedence occurs, males should emerge to secure at least as many preoviposition copulations as any other males, rather than timing emergence in relation to the availability of fresh virgins. Because there is often a lag between female emergence and egg laying, males may begin to appear well after the peak of female emergence has passed. The degree to which male emergence is clumped at the onset of oviposition should be a function of the expected longevity of males; the more long-lived the males, the greater the proportion that should begin adult life in the first few days of the female oviposition period.

It is difficult to test these ideas, because examples of protogyny are relatively rare. But in the tephritid fly *Rhagoletis completa* females do become receptive after each oviposition event and mate at intervals throughout their lives. Four years of emergence data collected by Boyce (1934) rule out protandry in this species and show a slight biasing toward females early in the flight season and toward males later.

The same pattern may apply to the mosquito *Aedes taeniorhynchus.* Although most culicine mosquitoes are protandrous, this species is not. The females are unusual in having a repeating 5-day cycle of feeding followed by oviposition beginning on day 2 of their lives, with egg laying usually on the seventh, twelfth, and seventeenth days (Nielsen and Nielsen, 1953). It is conceivable that after each bout of oviposition females regain their receptivity, enabling later-emerging males to enjoy greater reproductive success than in most mosquitoes.

A more clear-cut case of protogyny comes from the solitary bees. As noted previously, wood-dwelling Hymenoptera are almost universally protandrous, but males of *Anthidium* spp. (megachilid bees) are usually not seen until some days after the first females appear (Jaycox, 1967; Stephen, Bohart and Torchio, 1969). In these bees inner cells are occupied by male progeny and females are placed at the front near the exit to the linear nest, so that females must leave the natal nest before their brothers (Krombein, 1967). Females of *Anthidium* mate throughout their lives with whatever male succeeds in grasping them. If sperm precedence occurs in this species, there is no genetic advantage for males that maximize their contacts with virgin females. Instead, preferred mates are those on the verge of oviposition. Thus males may wait to complete development until some females have finished nest construction and have provisioned a cell or two (with *unfertilized* eggs that will become males). Only then are copulations likely to yield fertilization opportunities for a male.

Another group in which multiple mating by females and protogyny are correlated is the Odonata (Fig. 4.6). For two species of aeshnid dragonflies, males have been reported to be scarcer than females early in the flight season only to become relatively abundant later (Corbet, 1957; Trottier, 1966). This example is misleading, however, because in the aeshnid *Anax imperator* it is also known that males have a shorter period of prereproductive adulthood than females. Therefore males tend to arrive at the oviposition and mating site several days be-

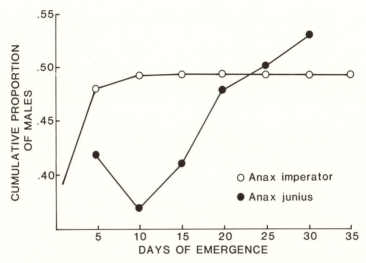

4.6 Emergence data from two species of dragonflies. In *Anax imperator* and *A. junius* males tend to emerge after females, but in both species males become sexually active before their females. (Data from Corbet, 1957, and Trottier, 1966.)

fore the first females make an appearance. Through early arrival males may gain a competitive advantage over their rivals and enjoy superior opportunities to monopolize the first females to oviposit at the site. Although there is suggestive evidence for sperm competition in multiple-mated females, individuals probably lay their eggs at intervals of several days and copulate only once per oviposition bout. (Aeshnid females are generally able to avoid males, if they so desire, and mate just once before laying each clutch of eggs—Waage, forthcoming.) Thus in *Anax*, as with the flies of the genus *Fannia*, differences in the speed with which males and females attain sexual maturity after becoming adults must be considered in analyzing the strategic aspects of male and female developmental schedules.

The Onset of Mating Behavior after Reaching Adulthood

Why do individuals of some insects pass through a nonreproductive adult phase? The cost of delay in reaching sexual maturity is clear. The longer the prereproductive interval, the more likely the insect is to fall victim to a predator or accident of some sort before it has a chance to begin reproducing. Males of many insects, such as the digger bee *C. pallida* and the lovebug *P. nearctica*, begin searching for mates promptly, and many females accept copulatory partners shortly after adult metamorphosis or eclosion. Thus it is possible for an adult insect not to engage in prereproductive activities before beginning to reproduce. Yet there are many insects that defer copulation.

The duration of the prereproductive interval, if it exists at all, is highly variable. In a number of Diptera and Lepidoptera 2 to 14 days is common for the time required to become fully "sexually mature" after eclosion (Boyce, 1934; Bequaert, 1953; Harris et al., 1966; Teskey, 1969; Scott, 1972; Fowler, 1973; Piper, 1976; G. Dodson, 1982). In some species of mosquitoes the proximate basis for the delay in mating by males has been attributed to the time required for the male's genitalia to become properly oriented. When the male exits from his pupal case, his sex organs are "wrong way round" (Gillett, 1971); it takes a day or so for the terminal segments of his abdomen to rotate 180 degrees, after which copulation becomes possible.

In some Odonata, Orthoptera, and Coleoptera, a prereproductive adult phase of 2 weeks to several months has been reported (Haskell, 1958; Halffter and Matthews, 1966; Kirk, 1971; Campanella and Wolf, 1974; Zelazny, 1975; Corbet, 1980). In the grasshopper *Trimerotropis occidentalis* females do not mate until 9 to 14 weeks after becoming adult (Weissman, 1979). Another species with an unusually long lag between eclosion and copulation is the carpenter bee *Xylocopa virginica*, whose males and females emerge in June and do not begin to mate until March or April of the next year (Gerling and Hermann, 1978). Some females of the face fly *Musca autumnalis* also overwinter as virgins to mate in the following spring (Teskey, 1969).

There is a variety of possible explanations for the delay in copulation; most apply more plausibly to short delays rather than to the exceptionally prolonged cases just discussed. In the first place, prereproductive phases can occur for reasons other than those related to sexual selection. For example, an insect in the preadult stage may not be able to detect with complete precision the cues that signal the onset of the optimal reproductive season. If so, it may be advantageous to emerge in anticipation of the suitable time, but to copulate only when the optimal moment actually occurs (Weissman and French, 1980). A special example is provided by many species of social insects, especially ants and termites. A new generation of reproductive forms can be maintained in the safety of the colony nest until workers determine that external conditions are best suited for the establishment of new

4.7   A group of winged reproductive males and females of *Pogonomyrmex* ants gathered at the exit of their nest prior to beginning their nuptial flight. (From Hölldobler, 1976.)

colonies. The reproductives are then ushered to the surface (Fig. 4.7) and allowed to disperse when factors such as rainfall, relative humidity, temperature, and wind speed give the males and females the best chance to locate mates, escape from waiting predators, and find a site in which to found a new colony (E. O. Wilson, 1975; Hölldobler, 1976). Although sexual selection appears to be unimportant in the evolution of this strategy, it may play a minor role nonetheless. While a colony's reproductives wait to emerge from the nest, the pool of adult reproductives in neighboring colonies increases as new individuals are added to the cohort waiting to be released. When they all leave their nests in synchrony, a queen from any given colony will have a wider range of possible mates than she would have had if she had been permitted to depart earlier. This may improve the female's chances of mating with a physiologically competent male or males that were able to outrace or otherwise outcompete many other males for access to her.

Nevertheless, for the social insects the timing of nuptial flights is probably linked to special nesting conditions that occur without predictable pattern. Similarly, for adult scavengers, parasites, and predators whose food resources are unpredictably distributed and widely dispersed, there may be advantages in waiting to copulate until the food source has been located and consumed. From a female's perspective, postponing copulation may be safer than announcing her location (to predators) in order to attract searching males, especially if males will probably be at the food resource when she discovers it. Moreover, the transport and maintenance of sperm may carry a physiological cost that can be reduced if she avoids mating until she has a mature batch of eggs ready to be fertilized, or until she reaches a prime reproductive habitat. Perhaps these factors have influenced the evolution of female receptivity patterns in the carrion-feeding calliphorids and blood-feeding pulicid fleas. In many blow flies, of which *Lucilia cuprina* is a prime example, a protein meal is required for maturation of the ovaries of the adult female and the onset of receptivity (Barton-Browne, 1958a,b; Brome et al., 1976). The same is true for some blood-feeding flies (Ross, 1961; Gillett, 1971). In the pulicid fleas, of which the rabbit flea is the best studied, the conditions for reproductive maturation of the adult female are even more restrictive: the female will develop mature eggs only if she consumes blood with a relatively high estrogen level (such as that found in the blood of pregnant does or newborn young) and she will copulate only in the presence of baby rabbits (Mead-Briggs and Vaughan, 1969; Rothschild, 1975). Thus in these species females only acquire sperm when they have encountered the unusual conditions favorable for the production of offspring.

In both the calliphorids and fleas, however, there are species whose females copulate *before* beginning to feed (Kamal, 1958; Mead-Briggs and Vaughan, 1969). An analysis of the ecological differences between species with and without a precopulatory feeding requirement would perhaps increase our understanding of the evolutionary significance of the deferral of mating by adult females.

There are some other speculative reproductive benefits that females might secure by refusing to mate immediately upon reaching adulthood. Young adult females of some insects are soft-winged and soft-bodied and vulnerable to damage from courting males (Loher and Gordon, 1968; L. E. Gilbert, 1976). If they were able to postpone copulation, females of these species might avoid the risk of injury and be better able to test and reject inferior suitors (those incapable of rapid pursuit and capture). In addition, by waiting to mate until they were able to move away from the point of emergence, females of some species might reduce the chance of insemination by brothers. If inbreeding reduces reproductive success, selection might favor females that rejected males until they left previously emerged brothers behind.

Benefits to Males of Deferred Mating Attempts

Adult males, as well as females, may be better able to exploit certain resources as adults than as immature individuals; if so, this would promote the transition to adulthood at a stage when reproductive mechanisms might be incompletely developed. At some point the insect may be able to complete its maturation more rapidly as an adult than by prolongation of the larval or nymphal stages. Sexual selection may influence the evolution of this life-history pattern if deferred mating attempts improve a male's chances of successfully attracting or courting a female, or improve his ability to defeat rivals in the competition for mates.

Perhaps males that provide nuptial gifts, including such things as salivary deposits or difficult-to-capture prey, benefit by postponing mating attempts until the necessary materials have been gathered or manufactured. If females make acceptance of a male contingent upon receipt of the nuptial gift, males without a suitable present that tried to copulate would receive little or no genetic return for their efforts. Furthermore, if the number of young resulting from a copulation is a function of the materials (other than sperm) received from the male (such as spermatophores or accessory gland products), it could be adaptive for the male to ignore receptive females until he had gathered the goods needed for an effective mating (one that would result in offspring bearing his genes). It may be relevant that males of the cockroach *Diploptera punctata* do not court females until they are at least 4 days old, whereas females are receptive on emergence (Stay and Roth, 1958). Males of this species transfer to their mates a substantial spermatophore. There is some evidence that the spermatophore is costly to produce (for instance, a male that attempts to mate twice in quick succession takes many hours instead of about 30 minutes to produce and pass on the second spermatophore). Moreover, the size of spermatophore seems to be related to the fecundity of the female that receives it (indicating that materials other than sperm are donated to the female). Females that mate with relatively young males that can only offer a small spermatophore have far fewer progeny than those that pair with older males whose spermatophores are clearly larger (Stay and Roth, 1958). The implication is that males must feed for some time as adults to acquire the key nutrients for spermatophore production and that premature attempts to reproduce will yield few, if any, offspring.

Intrasexual competition, as well as epigamic selection, can favor postponed copulation by males. If fighting among males for mates is

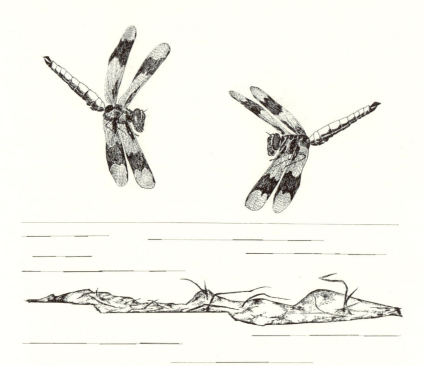

4.8 Two males of *Plathemis lydia* engaged in an aggressive display interaction over a patch of oviposition substrate attractive to females. (Drawing by J. Krispyn; from Matthews and Matthews, 1979.)

severe and costly, a male may gain by spending his early life feeding and acquiring energy reserves before entering the competition. This explanation is favored by Campanella and Wolf (1974) for the long interval between attainment of adulthood and attempts to secure mates by males of the dragonfly *Plathemis lydia* (Fig. 4.8). Male fitness in this insect depends heavily on the ability to acquire a territory at a female oviposition area in the face of extremely stiff competition for the relatively limited number of desirable sites. While a male is on a territory, he does not feed. An individual that enters the fray too early is either unlikely to displace older, more experienced males with previously acquired energy reserves or is unable to maintain its advantage for a sufficient period to encounter receptive females.

Daily Patterns of Mating Behavior

Having made "decisions" about when in its life cycle to become an adult and when to initiate copulation (or attempts to find mates), individual insects are still faced with the problem of how to allocate their time on a daily basis during the breeding phase of their lives. Reproductive attempts are probably never distributed evenly over a day, but instead peak or are restricted to a few hours (or even minutes) during the day. There is no "typical" pattern. Diversity is the rule even within closely related species (Hardeland, 1972). In the armyworm moth *Pseudaletia unipuncta* (Guppy, 1961) matings are most frequent 5 to 6 hours after sunset; in the oriental fruit moth *Grapholita modesta* the primary mating period begins 2 to 3 hours before sunset (Baker and Cardé, 1979), and in the noctuid *Trichoplusia ni* the peak is from midnight to dawn (Shorey and Gaston, 1964, 1965). In the Queensland fruit fly, *Dacus tryoni,* copulations are limited to a 30-minute interval around dusk (Tychsen, 1977), whereas in its close relative *D. neohumeralis* mating occurs in the middle of the day (Fletcher and Giannakakis, 1973).

Among the solitary bees a similar broad spectrum of daily patterns is represented. Copulations in *Anthidium maculosum* can occur from 0800 to 1800, although there is a modest midday peak (Alcock, Eickwort, and Eickwort, 1977); but in *Ptiloglossa guinnae* males are sexually active for only a brief period shortly before sunrise (Roberts, 1971). In a number of species of carpenter bees (*Xylocopa* spp.) mating is limited to the late afternoon or evening (Velthuis and de Camargo, 1975a,b; Marshall and Alcock, 1981). The anthophorid bee *Caupolicana yarrowi* has two mate-locating periods in the day, one very early in the morning around dawn and the other in the early evening (Hurd and Linsley, 1975).

To understand why there are interspecific differences in daily mating periods, one would have to analyze why females limit their receptivity to some portion of the day. In some species such as *C. pallida*, females mate promptly after emerging, and their emergence is limited to a few hours in the morning (probably because soil temperatures become unacceptably high as the day progresses in the Sonoran Desert). Variables such as temperature and humidity that follow a daily cycle and influence the success of metamorphosis can generate a cyclic pulse of fresh adult females during some hours of the day. If these females are receptive, then males can be expected to concentrate their searching activities during these times.

Other conditions may affect when receptive females are most likely to be contacted. Foraging may yield greater gains for females during some fraction of the day, either because of a daily cycle of prey activity or flowering, or because of the pattern of activity of competitors for the resource. If foraging females are receptive, males may time their mate locating behavior to coincide with the resource gathering behavior of their mates.

It is also possible that females of some species may be nonreceptive while foraging or ovipositing and will devote time for mating only during that period of the day when resource collecting or egg laying is least likely to be productive. If all females adhered to the standard limited mating interval, there would be no advantage to males that inspected or harassed foraging or ovipositing females outside the mating hours (if forced copulation is not a possibility). This in itself could improve the behavioral efficiency of females. A system of this sort, however, is unlikely to be evolutionarily stable; if even a few females accept mates outside the limited mating interval, selection will tend to favor males that inspect foraging or egg laying females (provided inspection costs are not high) in order to find the rare receptive individuals and so boost their reproductive success.

To what extent can female mate choice promote the evolution of a restricted mating period? If some ecological pressures force females to mate during a limited period, then selection will favor males that search and compete for mates during the appropriate interval. This should tend to increase direct and indirect competitive interactions among males (particularly if rendezvous sites for mating are also limited), because all or a large proportion of the sexually active males in the population would be engaged in mate locating behavior at the time. Provided the female had the option of rejecting potential suitors and selecting one from many, there could theoretically be gains to females whose daily pattern of receptivity further stimulated male-male competition.

It is suggestive that many of the insects that have relatively long breeding seasons and whose males compete for symbolic territories (those that contain neither emerging receptive females nor resources attractive to nesting or foraging females) have a daily mating period of only 2 to 3 hours (Table 4.2). A prime example is *Bittacus strigosus,* the only member of its family that does not engage in nuptial feeding. Males battle for perching territories for an hour or so, and females appear to choose males on the basis of their dominance status (Thornhill, 1977, unpublished). If females are sufficiently long-lived to have the time to make a careful choice, they may stimulate competitive interactions among males by synchronizing and limiting the daily period

Table 4.2   Examples of insects that have prolonged breeding seasons, symbolic territoriality by males, and an abbreviated daily period of mate attraction.

Species	Breeding season	Daily mating period	Reference
Papilio zelicaon (papilionid butterfly)	4 mo	2–3 hr	O. Shields, 1967
Atlides halesus (lycaenid butterfly)	3 mo +	2–4 hr	Alcock, unpublished
Cuterebra spp. (bot fly)	3–5 mo	2 hr	Catts, 1964; Alcock and Schaefer, 1983
Euglossa imperialis (orchid bee)	3–4 mo	2½ hr	Kimsey, 1980
Centris adani (anthophorid bee)	3–4 mo	2–4 hr	Frankie, Vinson, and Colville, 1980
Xylocopa hirsutissima (carpenter bee)	2 mo +?	2 hr	Velthuis and de Camargo, 1975a
Bittacus strigosus (scorpionfly)	3 mo +	1 hr	Thornhill, 1977

of receptivity. This increases the probability that a territory owner has been thoroughly tested by rivals and perhaps enables a female to select a truly dominant male as a mate, thus raising the fitness of her progeny.

Sexual Selection and the Daily Searching Pattern of Males

Given that there is a more or less limited daily period when receptive females may be available, the question then becomes how males should behave with respect to this interval. Should they search over the entire period or only during the peak phase of female availability? The answer is dependent to a large extent on such variables as the number and relative skillfulness of competitor males, the mating system of the female (monogamy or polyandry), and the mortality risks associated with mate location behavior.

When we consider species whose receptive, single-mating females are available in a roughly normal distribution over the daily period of several hours and whose males have equal searching ability, then the situation parallels the one described earlier in connection with the timing of male emergence relative to female adult metamorphosis. If the probability is uniformly high that males will survive the entire mating session that day, then all males should arrive at the searching site shortly before the first receptive female is likely to appear. In this way no male would permit a rival to enjoy greater reproductive gain than his own (assuming that males do not vary in the degree to which one day's mate searching reduces life expectancy). If, however, the probability of mortality is high during the mating hours, such that the expected searching time per male is some fraction of the daily breeding period, then males should stagger their arrival times over the early portion of the mating hours so that an individual is not outcompeted for mating opportunities.

Conditions favoring anticipation of female arrival by at least some males need not be limited to those described above. For example, males of the eumenid wasp *Epsilon* sp. perch on brood cells contain-

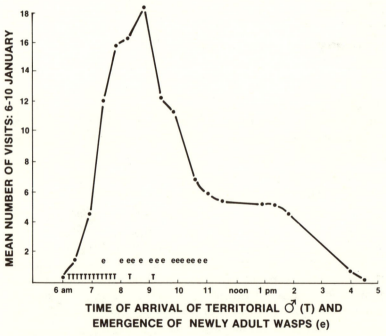

TIME OF ARRIVAL OF TERRITORIAL ♂ (T) AND
EMERGENCE OF NEWLY ADULT WASPS (e)

4.9   The time of first arrival (*T*) of territorial males of an Australian eumenid wasp that defends a cluster of cells from which receptive females may emerge. Female emergence (*e*) generally occurs after a territorial male has arrived. The peak number of visits to the territory site by nonterritorial intruders (*line graph*) occurs during the time when receptive females are most likely to be emerging. (From Smith and Alcock, 1980.)

ing potential mates (virgin females that have not yet emerged but may do so during midmorning). A male typically arrives to defend his territory and await female emergence an hour or so before a female is likely to exit from her brood cell and copulate with the male (Fig. 4.9; Smith and Alcock, 1980). By monopolizing a patch of cells, a territory owner gains access to most or all of the emerging females, which become nonreceptive after mating. There are nonterritorial males as well, which appear to patrol a route that takes them from the vicinity of one cell cluster to another. Like the territorial males, they appear to anticipate the peak period of female emergence, focusing their visits to the vicinity of the cluster when the first females are likely to emerge. (Some females escape the attentions of the territory owner and are probably captured and mated by nonterritorial patrollers.)

In a number of damselflies and dragonflies females mate multiply; still, a male can potentially gain a substantial number of progeny if he is the first to capture and inseminate a receptive female. In some species like *Tanypteryx hageni* females are unlikely to mate a second time on any given oviposition bout and therefore the male that grasps a female just before she oviposits presumably fertilizes most or all of her eggs released during the bout. Males consistently come to their territories near oviposition areas somewhat before the females arrive to mate and lay their eggs (Clement and Meyer, 1980).

There are, however, species whose males may not begin to search for mates until after some receptive females have appeared at a mating area. In the dragonfly *Plathemis lydia* females come to the pond to mate and oviposit from midmorning to late afternoon, but there is a strong peak in the middle of the day (Fig. 4.10). Males of this species compete so strongly for oviposition sites favored by females that there is a regular turnover in owners during the course of a day. By marking individuals, Campanella and Wolf (1974) showed that a reproductively active male first inserted himself into a territory as a submissive, nonterritorial patroller. On subsequent days the male might become a dominant territory owner, but initially only very early in the morning or late in the afternoon, outside the most productive time for securing mates. Each day thereafter the male would shift the time he spent on the territory as the top male closer and closer to the optimum period (Fig. 4.11). Thus the daily search pattern of males is a function of their relative maturity and the intensity of competition for the limited number of territories that attract females. On any one day a particular individual will search for females for only a limited fraction of the total mating period, with his lifetime strategy geared to acquiring a territory during prime time on at least one day in his life.

Thus far we have operated under the assumption that males should

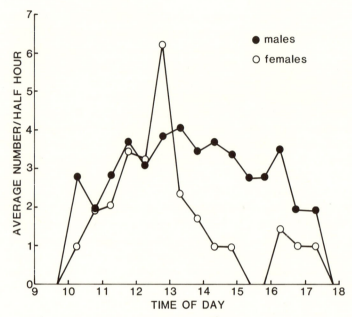

4.10 Males of the dragonfly *Plathemis lydia* arrive on their territories some-what before the first females come to the water to mate and lay eggs. (From Campanella and Wolf, 1974.)

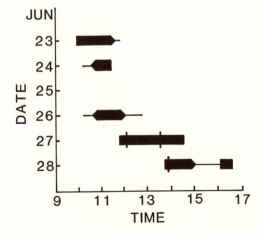

4.11 A typical record of territorial activity by one male dragonfly *Plathemis lydia*. The thick bars represent times when the male behaves in an aggressive territorial manner; the thin vertical bars represent copulations. Note that the male was territorial during the prime hours on only one day, June 27. (From Campanella and Wolf, 1974.)

attempt to maximize the number of copulations per day. But once again we must consider species in which females mate frequently and then oviposit, using the sperm of their last mate to fertilize all or most of their eggs. Here the critical factor becomes the daily distribution of oviposition events, rather than the daily distribution of receptive females. The goal of the male in this instance may be to maximize the number of matings with females on the verge of ovipositing.

A possible example is provided by the bee *Nomadopsis puellae* (Rozen, 1958; Rutowski and Alcock, 1980). Females of this species are believed to lay an egg at the end of the morning foraging period (individuals are able to collect sufficient provisions from their food plants for one brood ball over the course of the morning, after which the flowers close). Females make several foraging excursions from their nest burrow during the morning and will mate at any time with a male that grasps them. However, males that are active early in the mating period have a low probability of genetic gain per copulation (given the assumption of sperm precedence and the near-certainty that their mates will copulate again at a time closer to oviposition).

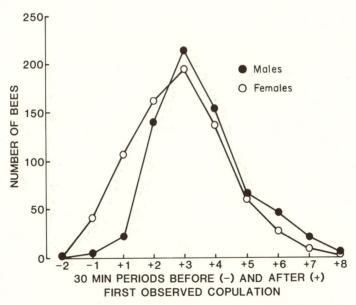

4.12 Differences in the total number of males and females of the bee *Nomadopsis puellae* counted along a transect during two censuses taken at half-hour intervals before (−) and after (+) the first copulation observed on a given day. (Data from Rutowski and Alcock, 1980.)

Males may therefore benefit by conserving their time and energy until the foraging period draws to its close. In fact, when females begin foraging, no males are present searching for them and it is only later in the morning that males begin to fly from flower to flower in search of a partner (Fig. 4.12).

We have seen that whether on a daily or an annual cycle, males time their mate locating activity to improve their chances of fertilizing eggs. This requires developmental programs that enable males to become adults and to begin hunting for mates at certain times, not others, as well as a sensitivity to the daily pattern of availability of profitable mates. The great diversity among species in the exact schedule of male development and mate searching can be understood in part as the consequence of sexual selection in different social environments. For example, the timing of mate locating appears to be strongly affected by whether females mate once or multiply. In monogamous species, males search when virgin females can be found most often; when females mate multiply, males are more likely to search during the periods when egg laying is about to occur.

5

Motivation to Copulate

In CHAPTER 4 we established that male insects time their sexual activities to coincide with the peak availability of genetically valuable females, within the constraints imposed by their male competitors. In the examples used in that chapter—including solitary bees, cicadas, butterflies, and dragonflies—it was always the male and not the female that searched or signaled conspicuously for mates. Long ago Darwin (1874:222) noted this important difference between the sexes.

> We are naturally led to enquire why the male, in so many and such distinct cases, has become more eager than the female, so that he searches for her and plays the more active role in courtship. It would be no advantage and some loss of power if each sex searched for the other; but why should the male almost always be the searcher?

Males generally are more strongly motivated than females to copulate, because a male's genetic contribution often increases with each female he mates whereas a female's entire number of eggs can usually be fertilized by sperm received from one or a few matings. Thus, as a rule, males actively search for or signal for females in competition with other males and are prepared to copulate instantly should the opportunity arise. The differences between males and females in sexual motivation are the primary focus of this chapter.

Male Attempts to Locate Mates

Males of the digger bee and the lovebug are typical male insects in their eagerness to copulate. In both species males, not females, expend hours of time and considerable energy in flight searching for potential mates. The males of C. *pallida* energetically dig down to buried females after they discover them. Having uncovered or captured a virgin, the male wastes no time in grasping her to begin courtship and copulation. After a single mating females are no longer receptive, but male bees resume the search for additional mates as soon as one copulation is completed.

These differences between the sexes can be attributed to differences in the genetic gains associated with multiple copulations for males and females of C. *pallida* and other similar insects. A female digger bee may live a month or so, during which time she can probably build no more than ten nests, each with a single cell and egg. The number of sperm she requires to fertilize the eggs she will lay is therefore small, and she is likely to receive more than enough from a single mating. The female pattern of a single mating followed by continuous nonreceptivity enables the bee to devote her energies to nest building and cell provisioning, rather than to superfluous copulations.

By contrast, males of C. *pallida* have everything to gain and little to lose by trying to mate repeatedly. Their only contribution to a female is the presumably inexpensive ejaculate that they transfer during mating. Each copulation may produce as many as five or so female offspring, all of which will carry the male's genes (female Hymenoptera are diploid, whereas males develop from unfertilized eggs). Thus there is every reason to expect that in C. *pallida* (and other species whose males provide little parental investment per offspring) males will run greater risks and expend more time and energy to acquire receptive partners than will females.

In the next sections we review some evidence that male insects are generally more likely than females (1) to detect and pursue mates, (2) to initiate courtship and attempt to copulate, and (3) to fight for the chance to copulate. Particularly useful data come from cases of sexual dimorphism in the structural features of insects that contribute to an individual's ability to detect, court, and fight for partners.

Sexual Dimorphism in Mate Acquiring Mechanisms

The degree to which males and females attempt to search for and detect mates is reflected in the more elaborate locomotory equipment or sensory mechanisms of many male insects. For example, when there is sexual dimorphism with respect to flight capacity, the winged flying forms are generally males that seek out their wingless females to copulate with them (Fig. 5.1; Bennett, 1961). The exceptions are often inbred species in which a few wingless brothers inseminate their many sisters, as in a variety of wood-dwelling insects (W. D. Hamilton, 1978).

In addition, males of many insects possess the larger (or more elaborate) antennae and eyes (D. Schneider, 1964; McAlpine and Munroe, 1968; Tozer, Resh, and Solem, 1981). Differences between the sexes in the design of these features need not always be related to mate detection. For example, the unusual antennae of some male chalcid wasps and some fleas are used to grasp the female during courtship and copulation (Withycombe, 1922; Buckell, 1928). But often the function of the large male eyes or antennae is to locate fe-

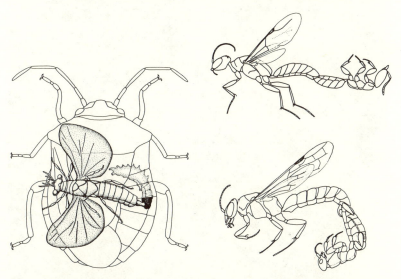

5.1 Sexual dimorphism in insects that is related to mate location: winged males and wingless females. (*Left*) In a strepsipteran insect, the wingless adult female lives within the body of a stink bug host. The winged male flies to her. While perching on the back of the stink bug, the male couples with the portion of her body that protrudes from the host. (From Kirkpatrick, 1937.)

Winged males of bethylid (*above right*) and thynnine (*below right*) wasps locate their wingless females and often carry them about in copulatory flight. (From Evans, 1960.)

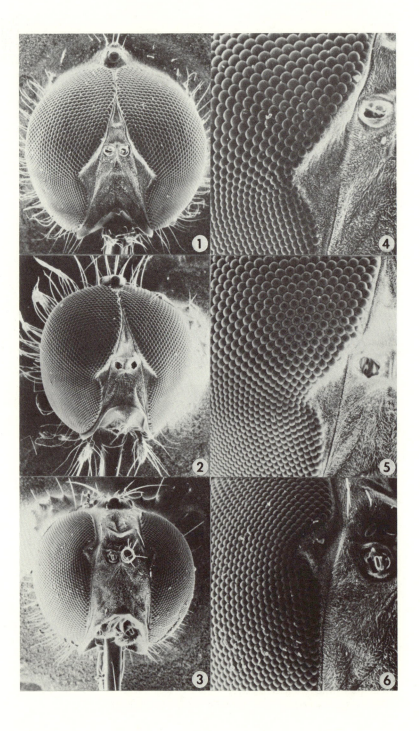

5.2   Sexual dimorphism in the eyes of insects. (1–2) The eyes of some male empidid flies (*Rhamphomyia*) that detect females flying overhead cover more of the head than the eyes of the female (3). Moreover, the male eye (4–5) is structurally distinctive with larger facets on the upper half of the eye than in the female (6). (From Downes, 1970.) The male bumble bee *Bombus nevadensis* (*above*) has enlarged eyes for the visual detection of females flying past his perch. (Photograph by J. Alcock.)

males, as in some mayflies, caddisflies, lepidopterans, Hymenoptera, and flies (see review by McAlpine and Munroe, 1968). Among the flies (Fig. 5.2) it is common for the eyes of males to cover more of the head than the eyes of females. In addition, the eyes of some swarming empidids (Downes, 1970) exhibit a peculiarity that may be related to spotting females: the dorsal surface of the eyes have enlarged facets (perhaps to achieve increased sensitivity to movement, facilitating detection of a female arriving at the swarm or a female passing near the perched or hovering male). Eyes of this sort are also possessed by males of some bumble bees that hover at landmark stations and dart after females passing overhead (Saunders, 1909; Alford, 1975).

5.3   Sexual dimorphism in the antennae of insects. (*Left*) Male moths typically have more elaborate antennae than their females. (Photograph by A. P. Smith.) (*Right*) A copulating pair of Australian crane flies. The male has the larger antennae. (Photograph by N. Thornhill.)

The rule that males allocate more to mate detection equipment applies also to antennae (Fig. 5.3). For example, males of the bumble bee, *Bombus ruderarius,* possess relatively large antennae and probably employ them to detect the odors associated with nests containing virgin gynes (Alford, 1975). Perhaps by having large antennae, males can accommodate more receptor cells on the surface of the antennae and so better sense nest odors. The better the nest finder, the more mates a male can be expected to acquire.

In every case of sexual dimorphism in antennal structure mentioned by Borror and White (1970) in their field guide to North American insects (see Table 5.1), it is the male that has the more elaborate antennae (see also Schneider, 1964). For the mosquitoes and chironomid midges, the plumose antennae of the males are known to serve as mechanoreceptors for the detection of airborne vibrations from the wings of mature conspecific females (R. F. Chapman, 1971). In other groups the males' plumose, pectinate, or flabellate antennae presumably bear large numbers of pheromone detectors.

Not only may competition among males for access to females promote a relatively large investment in receptors, it may also favor indi-

Table 5.1   Dimorphism in antennal structure in insects: cases taken from Borror and White (1970).[a]

Order and family	Male antennae	Female antennae
Strepsiptera	Complex, often with lateral processes	Usually absent
Lepidoptera (moths)	Often plumose	Usually threadlike
Diptera		
Culicidae	Plumose	Antennae with a few short hairs
Chironomidae	Plumose	Simpler
Hymenoptera		
Diprionid sawflies	Pectinate	Serrate
Eulophid chalcidoids	Pectinate	Simpler
Tanaostigmatid chalcidoids	Four branches on terminal segments	Simpler
Neuroptera		
Dilarid lacewings	Pectinate	Simpler
Coleoptera		
Rhipiphoridae	Pectinate or flabellate	Serrate
Pyrochroidae	Plumose in some males	Serrate
Phengodidae	Plumose	Larviform
Cerophytidae	Pectinate	Serrate

[a] This list is far from exhaustive. For example, the antennae of some male cerambycid beetles are longer than female antennae (Linsley, 1959), and male antennae are more elaborate in some other beetles such as certain elaterids and meloids (Britton, 1970).

viduals that possess unusually sensitive receptors. In the silk moth *Bombyx mori*, the male's highly branched antennae carry a huge battery of receptor cells, the majority of which respond only to bombykol, the female sex pheromone. According to D. Schneider (1969), these males have evolved the ultimate in sensitivity to this substance, as they are able to detect a single molecule of it. This case is evidence for the extreme effects of competition among males in mate detection.

Exceptions to the Rule

The examples of convergent evolution in elaborate and specialized sensory equipment support the hypothesis that males are more strongly motivated to copulate than females. Even though females should theoretically be able to take advantage of the fact that they have a limited resource (fertilizable eggs) that makes them attractive to males, there are cases in which females engage in considerable mating effort, searching for and traveling to males with which they copulate. How can we account for these exceptions to the rule?

In some cases the female apparently searches for mates only in the unusual circumstance that males are temporarily in short supply. Such facultative mate searching by females has, for example, been reported in certain Orthoptera (Renner, 1952; Haskell, 1958; Cade, 1979a) and a cockroach (Engelmann, 1970) when females have been deprived of mates for some time. A female that has not been promptly fertilized may well gain if she can find a willing male and thereby avoid a still more prolonged prereproductive adult phase, with its attendant risk of mortality.

Second, in many of the exceptions females appear to travel to males in order to acquire the resources that they control. Male scorpionflies, for instance, defend carrion or insect prey or salivary secretions as an offering to females (see Fig. 9.2 later). Here the costs of female travel are outweighed by the benefits of consuming or otherwise using the valuable resources received from males. In cases of this kind females are not traveling solely to secure a copulation; therefore their motivation is different from the typical mate searching male, who engages in costly activities primarily, if not exclusively, for the opportunity to transfer his gametes to females.

A third category of exceptions to the rule that males, not females, search for mates is provided by those insects whose females travel to conspicuously signaling males. In some of the more notable cases, such as most singing crickets, katydids, and many acridid grasshoppers (Otte, 1977), males offer useful resources, such as nuptial gifts of various sorts or oviposition burrows, that more than repay females for their travel "expenses." But in other instances, such as the cicadas, males appear to offer sperm only. Partial compensation for female travel in all these insects may be freedom from the special risks of advertisement. The songs of noisy orthopterans and cicadas appear to enable certain enemies, both vertebrate (T. J. Walker, 1964; P. D. Bell, 1979a; Doolan and Mac Nally, 1981) and invertebrate (Cade, 1975, 1979b, 1981a; Soper, Shewell, and Tyrell, 1976; Mangold, 1978; Burk, 1982), to track their prey. Crickets, katydids, and cicadas have specialized sarcophagid or tachinid fly parasites that are drawn to their victim's signals. (The flies can even be attracted to loudspeakers that play recordings of the song.) The sarcophagid female sprays her larval offspring on the singer; these small maggots proceed to consume the male in a matter of a few days (Fig. 5.4). It may therefore be less dangerous for a female to travel to a male than to employ acoustical signals that might attract lethal flies, not mates, to her.

The observation that when female insects produce long-distance attractants they generally employ pheromonal signals (Jacobson, 1972) may support this argument. Typically, pheromones are liberated in such small quantities that they may be difficult for a signal exploiter to detect because of the requirement for highly specialized olfactory receptors (Alexander and Borgia, 1979; Thornhill, 1979a). Yet there is no guarantee that a predator will not evolve this capability (Sternlicht, 1973). For example, there is a clerid beetle able to detect frontalin, the female-released sex pheromone of certain *Dendroctonus* bark beetles (Vité and Williamson, 1970). In this case, however, the males that alight on the bark of a tree infested with burrowing females are at greater risk than the signaling females themselves, which are in their burrows (K. Raffa, personal communication).

Male-released sex pheromones that attract mates from a distance are relatively rare in insects (Jacobson, 1972; Thornhill, 1979a; Greenfield, 1981). The males provide either a degree of parental care (some bark beetles) or nuptial gifts ranging from prey items (some Mecoptera and some carrion beetles) to salivary or other glandular se-

5.4   Females of the parasitic fly *Euphasiopteryx ochracea* are able to track singing male crickets by their songs. A fly is shown larvipositing on its victim. (From Cade, 1979b.)

cretions (for example, other mecopterans, certain cockroaches, tephritid flies, and malachiid beetles). In one case in which males employ a sex pheromone but do not appear to offer copulatory presents, release of the scent is unusually dangerous. Males of the southern green stink bug attract not only females with their sex pheromones, but also a tachinid fly parasite; males are parasitized substantially more than females (Mitchell and Mau, 1971; Harris and Todd, 1980a). This exemplifies the prediction that when releasing an attractant signal exposes the signaler to unusual risk, females will travel and males will provide the signal.

The final category of traveling females constitutes those that go to a swarm or other aggregation of males to select a mate. The benefit that might exceed the costs of travel in this instance is the selection of a mate with superior genes. Males in groups generally compete for positions within the aggregation, for access to a particular perch, hovering location, or territory; females that mate with dominant males, then, secure sperm from individuals of tested competence, which should be genetically sound as well.

Swarming males, as well as noisy signalers and resource gatherers, often put themselves at special risk to make themselves attractive to females (Burk, 1982). There are numerous reports in the literature of insect predators that attack swarming flies (O. W. Richards, 1927; Downes, 1955, 1970; Blickle, 1959). For example, sphecid wasps of the genus *Oxybelus* often specialize on male flies; in one study all 84 prey of *O. subcornutus* were males of a single species of syrphid (Fig. 5.5; Peckham and Hook, 1980). It seems probable that these often conspicuous aggregations also provide tempting targets for vertebrate predators. The genetic benefits of securing copulations are so great for males that the high costs of swarming can be overridden if aggregating is a prerequisite for acceptance by females.

Thus there are several plausible hypotheses to account for why females sometimes travel to males. Even in these exceptional cases, the attempts by males to advertise their location or to secure resources attractive to females represent a greater allocation to mating effort than that made by their partners.

Male Initiation of Courtship and Copulation

Our attention now turns from the question of which sex undertakes to locate the other to the issue of which sex initiates courtship or copulatory contact. Again, the general prediction is that males are more

5.5   An aggregation of males of the fly *Hylemya alcathoe* waiting for females. These aggregations attract sphecid wasp predators, which capture males to provision their brood cells. (Photograph by J. Alcock.)

ready to court and copulate than females in species in which the cost to the male of mating consists solely in time and sperm donated. A male that meets a female has little to lose by attempting to assess her receptivity, provided that this requires minimal time and energy. Cautious females, on the other hand, gain if they can better select a partner of the appropriate species or choose a superior conspecific male. If a female has already mated and has sufficient sperm on hand, she "should" be reluctant to copulate again if her suitor has only sperm to offer (provided that it is not too costly to reject a male).

Females are generally selective in picking a mate. The contrast between the typical copulatory eagerness of males and the typical sexual reluctance of females is perhaps best illustrated by the almost total absence of specific mate rejection behaviors of males, and their abundance and diversity in females (Chapter 13). The rejection responses of females range from failure to assume the copulatory position and flight from the male to actual aggressive attacks on persistent suitors. In some blister beetles males will attempt to mate with a female for many minutes. If the female already has sufficient sperm, the courtship will yield her nothing and may interfere with her pursuit of other goals such as feeding and oviposition. In *Pleuropompha tricostata* the courting male stands directly behind the female, fanning his antennae over the end of her abdomen. A nonreceptive female that wishes to terminate the courtship slams her abdomen and hind legs into the underside of the male's head (Fig. 5.6; Pinto, 1973). As might be expected, this action abruptly arrests male display and permits the female to depart without her harasser.

Males behave differently. Not only do they rarely reject a chance to mate, no matter how many times they have already copulated, but they often court or attempt to copulate with females that are not responsive and that provide rejection signals (Table 5.2). In his field study of the asilid *Ceratainia albipilosa*, Scarbrough (1978) observed 2,299 courtships in which the male hovered in front of a perched female, oscillating back and forth and spreading his legs on the forward movements, for periods of as much as 3 hr. Despite the elaborate nature of the courtship, females generally rejected their suitors, preventing them from mounting by kicking and wing vibration or by decamping. Only 38 (2 percent) of the courtships resulted in copulation.

Likewise, the percentage of successful courtships among butterflies (Wiklund, 1977) and meloid beetles (Pinto, 1972) is very low. It is common to see dozens of cases where male solitary bees and wasps pursue and grasp nonreceptive females for every instance of a successful courtship (see also Table 5.2).

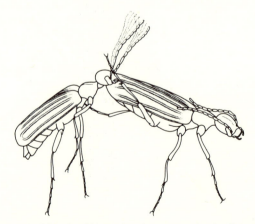

5.6　A female meloid beetle, *Pleuropompha tricostata*, in the act of rejecting an unwanted suitor by swinging her abdomen up under the head of the courting male. (From Pinto, 1973.)

Table 5.2   Examples of males that court or make copulatory attempts with nonreceptive females.

Species	Number of courtships or copulatory attempts	Percent that result in a mating	Reference
Muscidifurax zaraptor (pteromalid wasp)			
Virgin females	48	52	van den Assem and Povel, 1973
Mated females	17	0	
Ceratitis capitata (tephritid fly)			
Virgin females	58	72	Prokopy and Hendrichs, 1979
Mated females	26	15	
Physiphora demandata (otidid fly)	53	40	Alcock and Pyle, 1979
Odontoloxozus longicornis (neriid fly)	23[a] 25[b]	100 40	Mangan, 1979
Calopteryx maculata (calopterygid damselfly)	85[a] 77[b]	55 35	Waage, 1973
Philanthus bicinctus (sphecid wasp)	131	0	Gwynne, 1978
Trialeurodes vaporarium (whitefly—Homoptera)	243	40	Las, 1980
Danaus gilippus (nymphalid butterfly)	532	29	Brower, Brower, and Cranston, 1965
Tegrodera erosa (meloid beetle)	205	2	Pinto, 1975

[a] Females courted by territorial males.
[b] Females courted by nonterritorial males.

Male Mistakes in Courtship

The copulatory readiness of males expresses itself not only in their persistence in courting unwilling females but also in other "mistakes," including the courtship of other males (Table 5.3). Although it is possible that some "homosexual" interactions are really aggres-

Table 5.3   Examples of male insects that engage in homosexual courtship and copulation attempts.

Group	Reference
Field crickets	Alexander, 1961
Cockroach	Barth, 1964
Flea	Humphries, 1967
Cimicid bug	Lee, 1955
Pentatomid bug	Harris and Todd, 1980a,b
Corixid bug	Aiken, 1981
Naucorid bug	Constantz, 1973
Coreid bug	Loher and Gordon, 1968
Ichneumonid wasp	Heatwole, Davis, and Wenner, 1962
Nymphalid butterfly	Brower, Brower, and Cranston, 1965
Chironomid fly	Syrajämäki and Ulmanen, 1970
Calliphorid fly	Parker, 1968
Drosophilid flies	Spieth, 1952; Bennet-Clark, Leroy, and Tsacas, 1980; Tompkins, Hall, and Hall, 1980
Tephritid flies	Piper, 1976; Prokopy and Hendrichs, 1979
Muscid fly	Tauber, 1968
Syrphid fly	Maier and Waldbauer, 1979
Sacrophagid fly	Thomas, 1950
Anthophorid bee	Alcock and Buchmann, manuscript
Chrysopid lacewing	Henry, 1979
Tortricid moth	Sanders, 1975

sive in nature, in most cases males appear to be so primed to mate that they fail to discriminate initially between conspecific males and females. In *Drosophila melanogaster* the scent of freshly eclosed males is the same as that of receptive females; not surprisingly, young males sometimes are courted by older males (Tompkins, Hall, and Hall, 1980). Another instructive example is that of *Palmacorixa nana* (Aiken, 1981). This corixid water bug lives in aggregations, and homosexual mountings are common. But the males are not completely indiscriminate, because they tend to mount individuals larger than they are. Females are larger than males, so this strategy sometimes results in the capture of a female by a male. Totally misdirected courtship occurs when male insects display to members of other species or to completely inappropriate objects, such as twigs (Scarbrough, 1978), bananas and aluminum cans (Silberglied and Taylor, 1978), and beer bottles (Fig. 5.7).

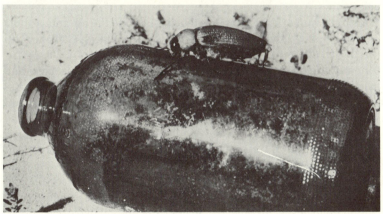

5.7 Misdirected copulatory attempts by male insects. (*Above*) Two male reduviid bugs (*Arilus cristatus*), one grasping the female on the bottom, the
other holding a fellow male. (Photograph by J. E. Lloyd.) (*Below*) Male of an
Australian buprestid beetle trying (unsuccessfully) to copulate with a dis-
carded beer bottle, which evidently possesses some of the key visual properties
of conspecific females, presumably in exaggerated form. Note the male's aedeagus, which is probing the bottle. (Photograph by D. T. Gwynne.)

In the digger bee, *C. pallida,* the intense sexual motivation of males sometimes leads them to copulate with dead males and females (see also Hurd and Powell, 1958) and to excavate preemergent males. (When this happens, the digger usually releases the uncovered male quickly but occasionally will mount and even attempt to insert his aedeagus in his inappropriate partner.) An olfactory cue appears to be involved in the location of underground conspecifics, judging from experiments in which males dug up portions of the bodies of male and female bees (Alcock, Jones, and Buchmann, 1977). The possibility that emerging males mimic the odor released by females in order to divert rivals or waste their time and energy is unlikely; diggers will uncover experimentally buried honey bees, wasps, and flies. Both visually and olfactorily, males appear to be responding to a general cue rather than a highly specific one limited solely to females. Although this sometimes leads to mistakes, it may be better to err on the side of reaching too many potential mates than too few. Because males of *C. pallida* tend to emerge before females, a digger male stands a better than 50-50 chance of uncovering a female when excavating an emerging bee. A male that hesitates when confronted with a potential mating may lose an opportunity to copulate to a less cautious, more active male.

A simple numerical example will help illustrate the advantages to a male of being relatively unselective in sexual contacts. Consider two genetically different phenotypes—males that practice a "leave-no-stone-unturned" strategy at the cost of making some mistakes, and males that exercise a "sure-thing-only" strategy at the cost of overlooking some ambiguous females. Let each digging attempt that uncovers a male cost the digger 5 minutes' lost search time. Let the "sure-thing-only" males average one female found and excavated per 500 minutes' search time (probably a reasonable estimate for *C. pallida*). Then the "leave-no-stone unturned" males need average no better than one female found per 100 digging attempts in order to enjoy a reproductive advantage. If we assume that the main cost of digging up males is the time lost, males can afford to make a very high proportion of mistaken contacts. As the time cost of unsuccessful contacts falls, the proportion of acceptable errors rises. Many insect males simply pounce on any passing object, immediately releasing inappropriate partners and copulating with the rare receptive female. There may be hundreds of unsuccessful contacts for each one that results in a mating.

The almost ridiculous ease with which sexual attempts can be elicited in male insects by researchers studying the sensory cues that control male sexual behavior attests to the low male threshold of sexual excitation. Dramatic experimental alteration of some aspect of the female usually has no effect on her attractiveness to males. In his study of the queen butterfly, *Danaus gillipus*, Brower (1963) painted a sample of females, using a variety of colors to create novel, bizarre patterns on the wings. Males courted and mated the painted females nearly as often as the unpainted ones.

Further evidence that male insects are sexually indiscriminate comes from studies in which they have been induced to respond sexually to very imperfect artificial mimics of conspecific females. Male butterflies are notoriously unselective in their early approaches to potential mates. In classic studies of a European fritillary (*Argynnis paphia*) and the grayling (*Eumenis semele*), Tinbergen and coworkers (1942) and Magnus (1958) were able to elicit sexual approaches by males to pieces of paper the general color of females. Their research showed that males of the two species do not attend to the fine details of color pattern when initiating a sexual pursuit; in fact, these details may actually detract from the attractiveness of a passing female. The experiments of Magnus with artificial models showed that males were drawn to the color orange. Yet females of this butterfly typically have wings with a complex pattern of orange and black patches, and one form has whitish-green wings. Although the wings of female butterflies have signal value, the message is far less refined than a human observer might initially suspect when looking at the remarkably intricate wing patterns of most butterflies.

The pursuit response to simple visual cues may help male butterflies avoid missing chances to court females. Similarly, copulatory responses can often be induced by simple cues. A classic case is the male mealworm beetle's response to the odor of females; Tschinkel (1970) showed that males could be aroused to try to mate with the tip of a glass rod dipped in an extract of the bodies of conspecific females (Fig. 5.8). Likewise, Tengö (1979) has shown that males of certain andrenid bees drawn to an area by an extract of glandular secretions taken from females will pounce upon leaves in the vicinity of the odor. In nature, the ease with which odor cues can stimulate copulatory behavior in bees and wasps has been exploited by a spectrum of plants, especially various orchids in Europe, Australia, and South America (van der Pijl and Dodson, 1966; Wickler, 1968). These plants typically provide an odor that mimics the sex pheromone of females of one species of anthophorid or andrenid bee or ichneumonid or thynnine wasp. Males are attracted to the plant and grasp a portion of the flower that incorporates some, but often far from all, of the visual characteristics of

5.8   A male of the tenebrionid beetle, *Tenebrio molitor*, sexually aroused by a glass rod that carries the scent of conspecific females. (From Tschinkel, Wilson, and Bern, 1967.)

the female (Fig. 5.9). In their efforts to copulate with the pseudofemale they acquire pollen from the plant, which they may transfer to another orchid of the same species if they are deceived again in the course of their mating patrols (Wickler, 1968; Stoutamire, 1974, 1975; C. H. Dodson, 1975).

We do not mean to create the impression that male insects are *totally* indiscriminate in their sexual behavior. A mistake in sexual recognition has costs for a male, ranging from the risk of dying to the loss of time spent in nonproductive pursuits or courtships that cannot be used for the location of fecund, receptive females. While males may make some errors in female identification, they generally do not persist long with an inappropriate partner. The mistaken courtships of twigs by males of the asilid *C. albopilosa* last just a few seconds, whereas courtships of female conspecifics may last many minutes (Scarbrough, 1978). Males of an halictid bee, *Lasioglossum zephyrum*, are also more discriminating than they appear at first glance (Barrows, 1975a,b; Barrows, Bell, and Michener, 1975). They can be tricked into pouncing on black ink dots placed on filter paper, provided the paper has been impregnated with the odor of a female (the female is placed in a vial with two pieces of filter paper for a day or two). But after males have pounced on a dot, they apparently remember the specific scent associated with it. If given a fresh piece of filter paper endowed with the odor of the same female (as much as an hour later),

5.9 Male of the Australian ichneumonid *Lissopimpla excelsa* on a flower of the orchid *Cryptostylis erecta*, with which it has recently attempted to copulate. The orchid produces a fragrance that mimics the sex pheromone released by female *L. excelsa* to attract male wasps as pollinators—note pollinia attached to the male's abdomen. (Photograph by A. J. Nicholson.)

they will dart at the dot far less often than a marked paper from a vial containing the odor of a different, new female (Fig. 5.10). When tethered live bees are used, unfamiliar females elicit a far stronger sexual response than previously contacted mates. Sexual selectivity based on recognition of past partners improves a male's chances of mating with a large number of different females rather than repeatedly with the same individual. This advances the male's genetic success. (Chapter 13 contains additional cases of adaptive mate selection by males.)

Male Fights for Mating Opportunities

A final line of evidence that male insects expend more mating effort than females comes from the observation that almost always males, not females, fight for copulations. We discuss how male insects employ aggression productively in Chapter 7, and there the reader will note that the number of insect species with fighting males is very

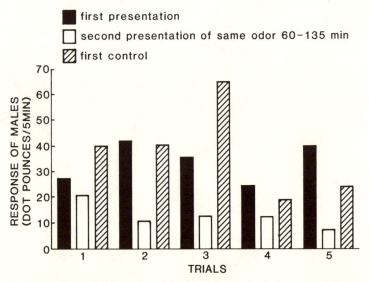

5.10   Mate recognition by males of the halictid bee *Lasioglossum rohweri*. Initially a high proportion of males in a group will pounce on a black dot on filter paper impregnated with the odor of one female (*solid black bar*). Later the males tend to avoid a dot on a fresh paper with the odor of this female (*open bar*) but respond heavily to a dot on a paper with the odor of a new, unfamiliar female (*striped bar*). (From Barrows, 1975a.)

great. Here our emphasis is on the sexual dimorphism in readiness to fight and in the structural adaptations for fighting. The disadvantageous aspects of fighting include not only the time and energy lost when struggling with an opponent or defending a territory, but also the heightened risk of injury and death and the construction costs of producing combat structures. Fighting costs are borne by males, not females, presumably because the payoff for successful male combatants (multiple copulations) may be extremely high.

The costs of aggression for males are far from trivial. Territorial males of the digger wasp *Philanthus bicinctus* expose themselves to predatory attack from robber flies when the wasps fly out to inspect a territorial intruder. Although such actions help maintain the integrity of the territory, they also result in a higher mortality for males than females at the hands of the intruding robber flies (Gwynne and O'Neill, 1980).

The risk factor includes not only predators but also conspecific rivals, which may be entirely capable and altogether willing to dismember and kill an opponent (Fig. 5.11). Males of some cerambycid beetles amputate the legs and antennae of their rivals (Linsley, 1959) and the galleries of wood-burrowing platypodid beetles "are often strewn with fragments of the vanquished" males who fought for (and lost) opportunities to mate with females in the nest chambers (O. W. Richards, 1927:305). Corbet (1957) states that a major cause of mortality among male dragonflies of *Anax imperator* is death by drowning, when one male forces his opponent into the water during an aerial clash over possession of an oviposition site. Tiny chalcid and agaonid wasps, although far less imposing than the emperor dragonfly, are no less disputatious (W. D. Hamilton, 1979). The males emerge first from their host (for example, a fly pupa in the case of some *Melittobia*) and then fight to the death to be the sole male present when the (receptive) females begin to emerge. The intensity of these fights is reflected in Buckell's (1928:19) observation that "a dead male, or even a small piece of one, will be fiercely pounced upon by another male, and dragged around and thrown about with a great show of anger, like a terrier with a rat." Even male butterflies, despite their gentle appearance, sometimes are potent warriors. In the swallowtail *Papilio indra*, males apparently defend territories around the larval food plant of their species, presumably either to contact females that are about to oviposit or to capture freshly emerged females. Males fight in the air, using their wings as weapons to damage their opponent's wings. After one aerial duel between two males that were fresh

5.11   Fighting male insects. (*This page*) A male of the bumble bee *Bombus fervidus* attempting to sever the wing of a rival by twisting it off in a fight for control of a territorial perch site near a nest with virgin gynes.

and whole at the outset, one individual had lost half a hindwing, one leg, and one antenna; both forewings had broken tips, and one forewing was torn down its entire length (Eff, 1962).

Finally, death can come about as a more or less accidental result of competition among males. In the beetle *Pleocoma oregonensis* as many as nine males fight their way into a female's burrow in the

5.11 (*continued*) (*Above*) Two male Japanese beetles jousting for possession of a female beneath them. (Photographs by J. E. Lloyd.) (*Below*) Two
male scorpionflies (*Harpobittacus nigriceps*) struggling for possession of a
large fly recently captured by the male on the right. The other male located the
prey-carrying scorpionfly by tracking the sex pheromone he was releasing in
an attempt to attract a female mate. (Photograph by N. Thornhill.)

ground. In the frenzied struggle for access to the mate, some males
are crushed to death in the narrow burrow (Ellertson, 1956). A similar sort of danger affects males of the elephant dung beetle, which
possesses elaborate horns rather like the antlers of deer or elk. Just as
in combat between deer bucks, the antlers of the beetle can become
interlocked and the two combatants are doomed to die (Arrow, 1951).

A test of the hypothesis that males more often than females engage
in costly disputes for access to mates (because they have more to gain
from multiple copulations) comes from examination of cases of sexual
dimorphism in fighting structures (Fig. 5.12). As predicted, it is almost always the male that possesses the enlarged crushing, slicing, or
grappling mandibles, the larger horns, the more powerful hind legs, or
the more pronounced abdominal forceps (Table 5.4).

The widespread occurrence of sexual dimorphism in structure and
behavior is testimony to the different selection pressures operating on
males and females. Because females are generally assured of a steady
supply of receptive males, they exhibit a much reduced tendency to
search for and court males and rarely prevent other females from interacting with a potential partner. Thus when sexual dimorphisms do

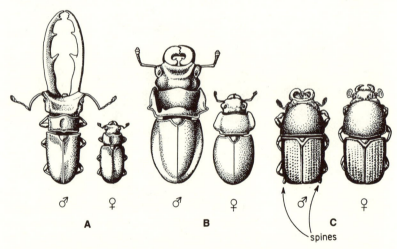

5.12   Sexual dimorphism in the fighting equipment of insects. The male is on
the left in each of three pairs (not drawn to scale). (A,B) Males and females of
two lucanid beetles in which the male has vastly larger jaws. (From Otte and
Stayman, 1979. Copyright Academic Press.) (C) Only males of the bark beetle
Scolytus quadrispinosus have small spines on the rear of the abdomen. They
use these to prop themselves against the walls of a nest burrow when facing an
intruder in a contest for control of a female. (From Goeden and Norris, 1964.)

Table 5.4   Examples of sexual dimorphism in morphological features useful in combat. In all cases males exhibit the larger or more powerful characteristics.

Dimorphic feature	Reference
Mandibles	
Many beetles, especially the Lucanidae, but also the Cerambycidae, Phengodidae, Platypodidae	Arrow, 1951; Otte and Stayman, 1979; J. E. Lloyd, 1979b
Some agaonid fig wasps	W. D. Hamilton, 1979
The eumenid wasp *Synagris cornuta* and a few halictid and andrenid bees	Roubaud, 1910; W. D. Hamilton, 1979
"Horns" and other prominent outgrowths of the head and thorax	
Many beetles, especially the Scarabeidae but also the Tenebrionidae, Ciidae, Brentidae, Platypodidae, and Scolytidae	Arrow, 1951; Otte and Stayman, 1979; Eberhard, 1979, 1980; W. D. Hamilton, 1979; L. Brown, 1980
Some platystomatid and drosophilid flies	W. D. Hamilton, 1979
Some pentatomid bugs	W. D. Hamilton, 1979
The cockroach *Gromphadorhina*	Chopard, 1950
Legs	
The coreid bug *Acanthocephala*	Mitchell, 1980
The coreid bug *Acanthocoris*	Fujisaki, 1981
The phasmid *Cryocelus australe*	Lea, 1916
Abdominal forceps	
Certain species of earwigs	Diakonov, 1925

occur in insects, males usually have the more developed wings and the larger mate detecting devices, the greater "eagerness" to copulate (even at the expense of often misdirecting their courtships), and the more powerful combat equipment to monopolize potential mates. These dimorphisms in morphology and behavior tend to assist males in the rigorous competition to fertilize eggs.

6

Competition in the Attraction of Females

CHAPTER 5 discussed exceptions to the typical pattern of searching males, notably in species in which females respond to attraction signals given by males. In this chapter we analyze how signaling males try to attract as many mates as they can. We outline some possible correlates of signal mode and environmental factors to show how the nature of the message and the way it is transmitted appear to be well suited for the physical environment of the signaling male. Our working hypothesis will be that males gain by projecting their signals long distances in order to contact a relatively large number of females. But the social environment of a "calling" male is at least as important as its physical surroundings. In a number of species, groups of signaling males may form; in extreme instances, hundreds or even thousands may produce their songs, displays, or pheromonal messages in mass aggregations. The formation of these groups has been taken by some authors as evidence that the participants are working together for a common goal or for some mutual benefit. It is a special challenge for the evolutionary biologist to explain why unrelated males appear to help one another secure mates, because individual selection should not often favor cooperation among reproductive competitors.

Acoustic Signaling in *Syrbula admirabilis*

To provide a focal point for our discussions of competition and cooperation by signaling males, we begin with a sketch of the calling behavior of an acridid grasshopper, *Syrbula admirabilis* (Otte, 1972). Males of this species are scattered throughout the grasslands of Texas and, as is typical for acridids, males produce stridulatory signals to attract receptive females. A calling male perches on a grass stem and rubs the femur of a hind leg against a fore wing. The femur has a line of pegs that is scraped across a ridge in the wing (the file), causing the wing to vibrate. Each leg stroke generates a new pulse of wing vibrations that are transmitted to the air as a pulse of sound (Fig. 6.1). A male's calling song comprises 10 to 50 such strokes, and a male may produce a cluster of up to ten songs in a short period, each stridulation separated by an interval of a few seconds.

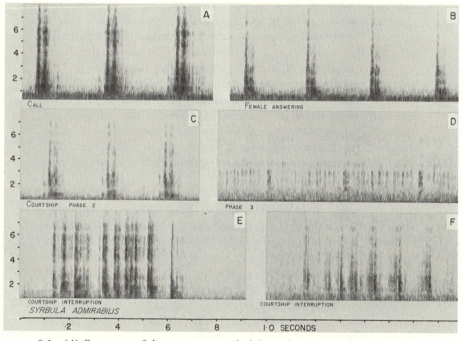

6.1   (*A*) Sonogram of the acoustic signal of the male grasshopper *Syrbula admirabilis*. (*B*) The female's answer. (*C* and *D*) Courtship stridulations by males at different stages of an encounter. (*E* and *F*) Calls given by a male that has lost contact with a female. (From Otte, 1972.)

TIME (MIN)

0	5	10	20

20	30	40

40	50	60

Syrbula admirabilis

6.2 The pattern of singing by a group of 13 males of *Syrbula admirabilis*. Males tend to give calls at times when other individuals are singing. Periods of overlapping songs are underlined. (From Otte, 1972.)

Receptive females in the field are solicited by a number of singing males. A female has the option of approaching the caller of her choice and answering him, when she gets close, with a signal of her own. This call is similar to the male's song but less intense. Males have no difficulty discriminating between male and female songs. Upon hearing a female's stridulation, the male hurries to locate the female and engage her in what is often a long and elaborate courtship.

Sometimes a male that has found a female loses contact with her, either because she has voluntarily moved off or because of interference from other would-be courters that have detected her presence. When this happens, the male gives a special "courtship interruption" signal—a much more rapid series of stridulations than the calling song. This sometimes slows or stops the female's retreat and the male may then relocate her.

A number of males of S. *admirabilis* may be within earshot of one another. These individuals exercise one of two options upon hearing a neighbor's song. The first is to remain silent and wait for the sounds of a female answering the calling male. The silent male may then dash to the female ahead of the signaler, stealing her from his rival. The second option is far less obviously competitive. Some males join in singing with an initiator of a calling bout, creating a *Syrbula* chorus. As Fig. 6.2 shows, males do not sing constantly; there are brief bursts of song activity, when several males may stridulate more or less simultaneously, followed by periods of silence.

Singing by males of S. *admirabilis* raises some questions. Why do males employ acoustic messages rather than some other communication mode in attempting to attract females from a distance? Why do males sing at times when other males are calling? Is it possible that the males are cooperating to produce a louder, more effective attractant call? Or is the overlap in calls another manifestation of reproductive competition among males? We shall consider first the costs and benefits of the various modes of long-distance signaling and then analyze the relative contributions of cooperation and competition to the patterns of group signaling by males.

The Costs and Benefits of Acoustic Signaling

In species with dispersed females the problem for the signaling male is how to transmit an attractive message a suitable distance. A successful signal must travel through space readily and provide information about the location and nature of the male that convinces females to leave their resting places and move to the male. Although it is possible to communicate in water with either subsurface or surface waves (R. L. Smith, 1979a; Wilcox, 1979), and on land with leaf shaking (P. D. Bell, 1980a,b; G. K. Morris, 1980) or by thumping the substrate (Beer, 1970), airborne vibrations are more often used to communicate over long distances. Acoustic signals are one of the three major modes of long-range communication among insects, along with visual and olfactory signaling (Table 6.1).

Table 6.1   Comparison of the relative benefits and costs of the major modes of communication. (From Alcock, 1979a.)

| Variable | Channel | | | |
	Chemical	Auditory	Visual	Tactile
Range	Long	Long	Medium	Short
Ability of signal to reach receiver				
Rate of transmission	Slow	Fast	Fast	Fast
Flow around barrier	Yes	Yes	No	No
Night use	Yes	Yes	No[a]	Yes
Information available				
Fadeout time	Slow	Fast	Fast	Fast
Locatability of sender	Difficult	Fairly easy	Easy	Easy
Cost to sender				
Broadcast expense	Low	High	Low-medium	Low
Risk of exploitation	Low	Medium	High	Low

[a] Except bioluminescent signals.

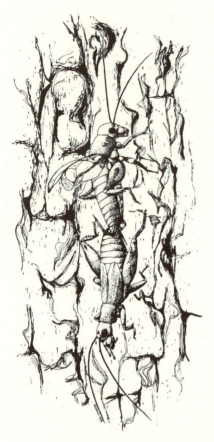

6.3   A male short-tailed cricket, *Anurogryllus arboreus* (above) copulating with a female and at the same time singing from his perch on a tree trunk. (From T. J. Walker, 1980.)

Acoustic signals are not restricted to use during daylight hours, unlike most visual displays, nor are they as severely constrained by wind conditions, as are airborne pheromones. This is not to say, however, that sound transmission is independent of environmental conditions, for vegetation and uneven terrain act as physical obstacles that block or distort acoustic messages (Wiley and Richards, 1978). Those species that live entirely within dense vegetation may find effective acoustic signaling difficult. Alexander (1962) notes that crickets in cluttered environments have less complex systems of songs than those that inhabit more open areas. An advantage of calling from an elevated perch, as *Syrbula* does, is that some obstacles and sound absorbers are avoided. Males of the cricket *Anurogryllus arboreus* sometimes leave their burrows to call from perches on vegetation or tree trunks (Fig. 6.3). By selecting a raised perch, a male is able to broadcast his message to an area 14 times as large as one covered by a male singing at the same intensity within the burrow (T. J. Walker, 1980).

In addition to sound-reflecting and sound-absorbing barriers, the acoustic signaler must also contend with such things as heated air, which can distort a song (Morton, 1975). This may be one factor favoring aerial stridulation in certain grasshoppers that fly up above hot grasslands to sing. Wind too can be an obstacle to sound transmission, as the noises generated by high winds can obscure the signal. Some cicadas may sing primarily in dawn and dusk periods to avoid competing background noises (wind speed is generally low at these times of the day) (Crawford and Dadone, 1980; Young, 1981). There are other hypotheses concerning restricted calling periods, but it is plausible that the cicada which sings in a quiet environment will be able to project his message to more females.

As anyone who has listened to a noisy cicada or katydid knows, these animals are able to project their message considerable distances despite the difficulties they face. The major long-distance acoustic insects are found in the Orthoptera and among the cicadas (but see Kay, 1969, on an acoustic moth and Jansson, 1973, on stridulatory corixids). They are usually large insects that possess bulky muscle masses to generate powerful signals in the 80 to 100 db range. The muscles controlling the prominent hind legs of grasshoppers are large to enable the animal to leap long distances; secondarily, these muscles have become involved in control of the stridulatory apparatus, of which there are many types in the acridid grasshoppers alone (Otte, 1970, 1977). In the cicadas the sound-producing device consists of an area of thin cuticle (the tymbal) on the abdomen of the male, which is underlain by a number of air sacs and connected to the large tymbal muscles. Almost the entire abdomen is devoted to sound production. When the tymbal muscles contract, they pull the tymbal in, producing a click; cicada calls comprise a series of clicks generated in extremely rapid fashion through the fast cycle of contractions and relaxations of the tymbal muscles. The resultant sounds can reach 100 db (Smith and Langley, 1978).

There are other means to amplify the sound the male produces. The French mole cricket, *Gryllotalpa vineae*, sings so loudly that humans can hear the monotonous trill 600 m from the singer. The calling male stations himself near the entrance to his shallow burrow, which is shaped like a double megaphone (Fig. 6.4). This structure acts like an acoustic horn, increasing by some 35 percent the efficiency with which the work of wing scraping is converted into sound (Bennet-Clark, 1970; Nickerson, Snyder, and Oliver, 1979). Moreover, the design of the burrow serves to concentrate the sound in a

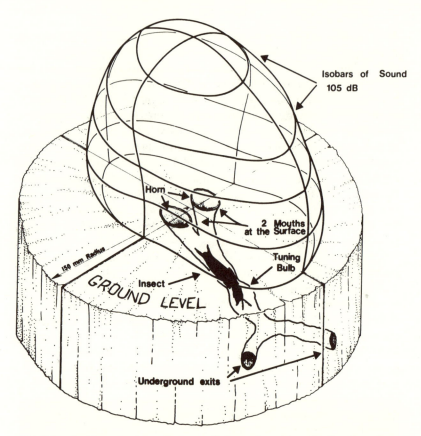

6.4   The double-megaphone burrow of the mole cricket *Gryllotalpa vineae* makes male calling more efficient and more effectively directed. (From Bennet-Clark, 1970.)

relatively narrow disc above the length of the burrow. Without the tunnels the sound would be more or less evenly distributed in a semi-hemisphere around the singing insect. By compressing the distribution of sound waves, the animal's burrow amplifies the intensity 1.4 times in the zone above the burrow; the song is 105 db well above the entrance. Female mole crickets fly to their mates. The male, projecting a powerful song upward, makes it more likely that a female flying some distance above the ground will be able to detect his signals and so will be induced to land and enter the burrow, where copulation occurs.

The double-megaphone burrow is only one of a variety of sound-amplifying devices used by various species of crickets. The shared problem of all crickets is that their vibrating wings are small relative to the wavelengths of sound produced; as a result, acoustic interference occurs between the alternating pulses of sound emanating from the sides of the vibrating wing area. A common solution is to employ acoustic shields or baffles, which separate the outputs from the two sides of the sound-generating disc and prevents them from canceling each other (Forrest, 1982). The tunnel walls of the mole cricket serve this purpose. Some crickets that call from the surface of the ground hold the broad femurs of their hind legs at the side of the down-curved wing edges, forming a shielded sound cabinet. Other crickets that live in vegetation use leaves as baffles (Fig. 6.5). The tree cricket *Oecanthus burmeisteri* offers the most dramatic example (Prozesky-Schulze et al., 1975). The male gnaws a pear-shaped hole in a leaf and then perches so that his tegmina (wing covers) are pressed against the edge of the hole. In this position the leaf surface acts as a sound baffle that amplifies the male's song about three times.

If it is to be a useful long-distance signal, an acoustic message must not only be relatively loud but must also provide information that enables the receiver to locate the source. The "ears" of female crickets, grasshoppers, and other acoustic insects have properties that enable them to track singing males (see Michelsen, 1979). What is especially

6.5 A male *Oecanthus* cricket using a leaf as a sound baffle to increase the efficiency of its signaling. (From Forrest, 1982.)

interesting is the possibility that males of some species may produce songs with special properties that foster quick contact between a signaler and an approaching respondent. For example, the bog katydid's song contains both a 15 to 20 khz component (audible to humans) and a 33 khz component in the ultrasonic mode. Morris, Kerr, and Gwynne (1975) have shown that experimentally altered taped songs in which the ultrasound was deleted are less readily localized by females than those in which both the "audio" and the ultrasonic components are retained. They suggest that because sounds in the two frequency ranges attenuate differently over distance, the female could compare the intensities of the two modes and use their relative loudness as a cue to the location of the singer.

The advantages, however, of noisy, easily located acoustic messages are, to varying degrees, offset by the disadvantages associated with attracting unintended receivers such as predators (Burk, 1982) and rival males. Predators may find and kill the signaler; rivals may steal females that the calling male had attracted, as in the case of *Syrbula* grasshoppers (Otte, 1972). The presence of signal exploiters imposes evolutionary constaints on the nature of the signal males can use and the pattern in which the songs are given. There are many male Orthoptera that have apparently secondarily lost the ability to sing (Otte, 1977), and still others that employ their calls most sparingly. Rentz (1975) has proposed that the tropical katydids rarely sing, because foliage-gleaning bats may use katydid calls to locate their prey, a suggestion confirmed by recent experiments by J. Bellwood (unpublished). Likewise, the loss of calling songs in certain crickets may have evolved because of pressure from acoustically orienting parasitic flies (T. J. Walker, 1974a, forthcoming a).

The Attraction of Females

Loudness is not the only feature of an acoustic signal to determine male reproductive success. Females may perceive signals to which they do not respond. To attract a female, the song must carry useful information. Male grasshoppers (Otte, 1972), crickets (Alexander, 1962; Pollack and Hoy, 1979), and cicadas (Leston and Pringle, 1963) all take advantage of the rapid fadeout time for sounds to produce distinctive songs whose temporal properties announce the species to which the singer belongs (Fig. 6.6). Species specificity is one aspect of a male's song that females may prefer, as it reduces the risk that they will approach a male of another species and produce defective hybrid offspring.

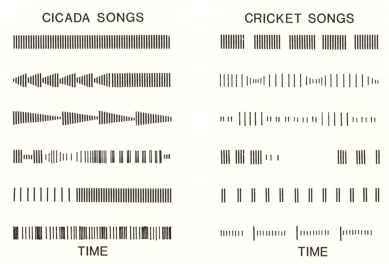

6.6 Species specificity in the acoustic signals of a group of cicadas and crickets. Each line represents diagrammatically the temporal pattern of calls of a different species, with the width of the dark bars proportional to the duration of each sound and the height related to the amplitude or intensity of the sound pulse. (From Alexander, 1962; Alexander and Moore, 1962; Leston and Pringle, 1963.)

The signals of male insects may provide much more information than merely species membership. Relative loudness and locatability may themselves be song parameters that influence the attractiveness of a song to females. A relatively easily located call may reduce the time required by a female to find a male, and loudness may be correlated with body size—which, if partly heritable, may be a desirable trait to pass on to one's male offspring. Forrest (1980) has shown that for the Florida mole cricket *Scapteriscus acletus* there is considerable variation in the intensity of calls given by males, with larger individuals capable of producing louder calls. By placing males of different sizes in bucket traps and counting the number of trapped females attracted to each male per night, Forrest found a ten-fold increase in the number of attractees to louder-singing males (+ 5 db) relative to the softer singers in his sample (although they were no more than 20 m apart). When electronic artificial crickets were used to produce artificial mole cricket calls of two intensities (one 6 db greater, or twice as loud as the other), the capture rate averaged about *six* times as many females for traps with the louder songs. In both experiments, therefore, females were attracted in greater numbers to the louder calls than would be expected if attraction were a simple function of the larger sound field created by the more intense calls. Flying female mole crickets actively select the louder males as mates.

The position of a singing male relative to his calling rivals may also influence his attractiveness to females. Males of *Anurogryllus arboreus* that secure the lowest position on a tree trunk when several males are calling together enjoy a reproductive advantage (T. J. Walker, 1980). Perhaps females are more likely to approach the lowest male because they move up the trunk from its base. Although male *Anurogryllus* do not appear to jockey for positional advantage in any overt fashion, clustered male katydids (*Neoconocephalus nebrascensis*) often change their calling perches, even during one night's chorusing (Meixner and Shaw, 1979). It seems likely that these individuals are attempting to secure the optimal position in the group from which to broadcast their messages. Finally, Doolan (1981) has shown experimentally that male bladder cicadas, whose song perches are less than 1 m apart, are not so likely to be closely approached by a receptive female as males that are singing from more distant points. She suggests, as a proximate explanation, that the songs of two close neighbors interfere with each other, reducing their attractiveness to females.

Visual Attraction Signals

Like acoustic signals, many elements of visual signals used in courtship have been shaped by sexual selection. Visual signals have many of the same advantages and disadvantages as acoustic messages. Potentially they can be broadcast reasonably long distances. The signal can be highly complex and information rich, as when a variety of colors is transmitted in a single distinctive message or when body parts are moved in a variety of positions creating a rapid sequence of signals. The receiver of a visual message can instantly locate the sender. On the debit side, visual signals are more subject to obstruction by physical barriers than acoustic or pheromonal cues and—with the exception of bioluminescent signals—are restricted to diurnal use only. As with acoustic messages, the very ease with which the signaler can be perceived and located can be dangerous because of predators capable of intercepting the message and tracking the sender. Color patterns on the wings and other body parts have the potential advantage of acting as a continuous advertisement for the male, requiring no special activity or energy expenditure to generate the signal. A male butterfly uses his wings for locomotion, but in so doing the appearance of the wings also conveys information that a receptive female may use to locate and approach a mate (Rutowski, 1980).

In addition to the passive use of body surfaces as signal generators, some insects have evolved specific movements that act as visual displays designed to attract females from a distance. Males of some Hawaiian *Drosophila* and orchid bees defend perch sites on the trunks or limbs of selected tropical trees (Spieth, 1968, 1974a,b; Kimsey, 1980). There they regularly adopt a distinctive perching position, which in the fruit flies is characterized by bobbing of the abdomen coupled with wing waving movements that show off the strikingly banded wings. The male bees elevate their strongly striped abdomen and hop on and off their perches on smooth-barked tree trunks (Fig. 6.7). These displays are given spontaneously by isolated males, suggesting that they may be used to lure females that the males cannot see but that may be in the vicinity of the displaying individuals.

The flight movements of individual males in swarms of various flies (true Diptera), caddisflies (Trichoptera), and mayflies (Ephemeroptera) may also serve to attract females from a distance. For example, males of certain mayflies have a striking flight display in which they

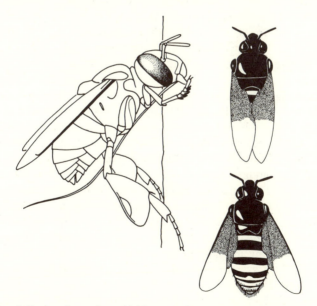

6.7   Visual displays by male orchid bees. Male *Euglossa imperialis* hold their body in a distinctive position while perched on a tree trunk from which they launch repeated display flights, hopping on and off the perch. Males of *Eulaema meriana* display by opening their half-black, half-white wings to expose their yellow-and-black-striped abdomen. (From Kimsey, 1980.)

flutter up several meters and then drop down, light reflecting from their silvery wings, to repeat the activity again and again (Spieth, 1940). This may help the male advertise his location in the swarm.

Probably the most dramatic visual displays, however, are produced by male fireflies, nocturnal beetles that violate the general rule that visual messages are employed only in the daytime. Although strictly speaking the light flashes of many fireflies are not attractant calls (they do not draw a female closer), they are designed to persuade receptive females to announce their position as they perch in foliage and so are functionally related to attractant signals.

Male fireflies flying over a meadow or marsh generate a characteristic signal pattern (Fig. 6.8) that varies from species to species with respect to color, rate of flashing, length of flash pulse, and intensity of flash. The color of the flash has been correlated with the time the species is active. The group of fireflies that locates mates primarily in twilight conditions has flashes dominated by yellow hues, while those that begin calling later, when it is darker, employ predominantly greenish flashes. Lall and coworkers (1981) argue that this significant correlation occurs because of the advantages of avoiding masking by ambient light levels. The twilight species must contend with a considerable amount of reflected light from vegetation, light in the green wavelengths. Greenish flashes thus become less conspicious and less easily detected by a perched female, and twilight-active fireflies utilize yellow flashes.

Producing a message the female can readily perceive is one problem for males. Inducing the female to flash back in response to the signal so that the male can find her and mate is another. In *Photinus macdermotti,* males produce a flash pattern made up of two blips of light about 2 seconds apart. Receptive females respond with an answering flash about 1 second after the last of the male's two signals. (The timing of the male flash pattern and the female response is temperature dependent.) The interval between the male's two flashes is critical. If a female is experimentally subjected to flashes less than 2 seconds apart, she will not respond. This is useful to her, as it prevents her from reacting to males of the wrong species, some of which also use a two-flash system but with a different interval between flashes (J. E. Lloyd, 1966).

Thus both the color and the temporal patterning of the flash system may contribute to success in attracting a response from a female, enabling the male to locate potential mates. Still another attribute of long-distance male signaling that is related to this goal is exhibited by males of some *Photuris* fireflies. Females in this genus are highly

6.8 Flash patterns of a number of firefly species, each of which generates a distinctive pattern of signals. (Drawing by D. Otte; from Lloyd, 1966.)

predatory insects that mimic the "come-hither" signals of receptive females of *Photinus* fireflies. *Photinus* males that respond to the false signals are sometmes captured and eaten (J. E. Lloyd, 1965, 1975). Nelson, Carlson, and Copeland (1975) have shown that females of one species of *Photuris* respond sexually to the typical flash pattern of their own species only when they are virgins. Once having copulated,

6.9   (*This page*) A male *Photuris* firefly (above) copulating with a perched female. (Photograph by J. E. Lloyd.) (*Opposite page*) The male produces pulsed species-specific flashes when cruising in the tops of pines but can descend and mimic the flickering and glowing flash patterns of males of *Photinus* and *Pyractomena*. (From J. E. Lloyd, 1980. Copyright 1980 by the American Association for the Advancement of Science.)

the readiness of the female to answer calling, conspecific males ceases. She attempts instead to lure flashing males of the prey species. J. E. Lloyd (1980b) has discovered that males of several species of *Photuris* sometimes mimic the flash patterns of a number of other fireflies, including the prey species on which their females feed (Fig. 6.9). He believes that they may do so to deceive a mated, hunting female *Photuris* into answering back, thereby revealing her location. The male may then move to the female and induce (or force) her to mate again, regardless of her earlier mating history. If Lloyd is right, male *Photuris* may employ one signal type to attract virgin females and another to mate with nonvirgin females (perhaps later in the season when virgins have become scarce or nonexistent).

Male fireflies may even emit mimetic female signals to help them locate a mate. In the species *Luciola lusitanica* a receptive female engages in a flash dialogue with a male she has detected. As he gives the male signal, she answers with a distinctive flash again and again as he draws nearer. Sometimes, however, the female stops answering before the male has reached her. He may then alight and begin to give the

signal that females use to respond to males. This attracts other males whose signals may stimulate the true female to begin answering again. The pseudofemale can then perhaps relocate his would-be mate and reach her before the newly arrived males (J. E. Lloyd, 1971).

Olfactory Attraction Signals

Whether pheromone-using males are as ingeniously devious as fire-flies is unknown, perhaps because the human olfactory sense is less effective in perceiving the olfactory world of other species. It is known, however, that male insects rarely release long-distance chemical attractants (Thornhill, 1979a; Greenfield, 1981). This is curious, because males often use pheromones in short-range courtship and female insects frequently employ long-distance pheromone communication with success.

Pheromones have many advantages as long-distance information transmitters (see reviews in E. O. Wilson, 1975; Shorey, 1977). Chemical scents of some female moths may be carried on the winds, creating a pheromone trail several kilometers long that can be detected and followed by males. The message can flow around obstacles in the environment, and the energetic cost of using this channel of communication appears to be relatively low. In one species, the boll weevil, in which males do use pheromonal long-distance signals, the entire lifetime production of the monoterpene sex attractant has been estimated to be only 0.2 percent of the adult male's body weight (Hedin et al., 1974). In contrast, over half the daily respiratory budget in some crickets is devoted to acoustic signaling; starved crickets lose about 5 percent of their body weight daily as a result of calling (Prestwich and Walker, 1981; see also Mac Nally and Young, 1981).

Balanced against these advantages, however, are some disadvantages such as the dependence of the signal releaser upon restrictive wind conditions, the difficulty of tracking the signal relative to acoustic and visual means of communication, and the slow fade-out time of chemical cues that makes it less easy to send complex messages in rapid sequence (but see Conner et al., 1980).

Among the male insects that apparently attract females with wind-dispersed scents are a few moths (see, for example, Leyrer and Monroe, 1973), a number of flies (Fletcher, 1968; Nation, 1972; Alcock and Pyle, 1979), some beetles, notably the boll weevil (Hardee, Mitchell, and Huddleston, 1967; W. H. Cross, 1973) and bark beetles of the genus *Ips* (Silverstein, Rodin, and Wood, 1966; Birch, 1978), as well as some mecopterans (Bornemissza, 1966a,b; Thornhill, 1979b; Byers and Thornhill, 1983), and certain Hymenoptera (Alcock, 1975; Velthuis and de Camargo, 1975a,b; Gwynne, 1978). Harvester ant males scent mark vegetation with secretions from their mandibular glands. These odors attract females to a mating site (Fig. 6.10).

Although few male pheromones of this sort have been analyzed in the detail associated with the study of sex pheromones of female insects, they probably are similar in having the properties of volatility and species distinctiveness. On the one hand, if the odor is to be disseminated a considerable distance in a reasonable time, it must be of sufficiently low molecular weight to be easily carried by wind currents

6.10   Large numbers of female *Pogonomyrmex* ants flying upwind to a scent-marked tree. (From Hölldobler, 1976.)

(Wilson and Bossert, 1963). The highly floral scent of the males of *Xylocopa varipuncta* can be detected by humans many meters downwind of the hovering male (Marshall and Alcock, 1981). Female boll weevils are able to track a male by his pheromone over a distance of 80 m (Hardee, Cross, and Mitchell, 1969; Hardee et al., 1969).

On the other hand, if several species are releasing pheromones in the same area, calling males may enjoy an advantage if their signal is species specific. This favors the use of substances of relatively high molecular weight or distinctive pheromonal bouquets composed of several components (Silverstein, 1981). The fewer the atoms in a pheromone, the fewer the variants possible and the greater the likelihood of similarity between the pheromones of different species (especially if these are simple long-chain hydrocarbons as is true for many insect pheromones). Females may find a distinctive signal more attractive because it enables them to reduce the time and energy they must spend searching for an appropriate mate and decreases the risk of hybridization.

Breed, Smith, and Gall (1980) have shown that females of the cockroach *Nauphoeta cinerea* can determine the social status of males by pheromonal cues they release, and females prefer the scent of dominant males (see Fig. 14.3 later). It has been suggested that slight individual differences in the structure of the pheromones released by males of the same species may allow females to recognize individual males and assess their performance over time before reaching a decision on a mate (J. E. Lloyd, 1981). Pheromone-releasing males may sometimes compete with one another in the manufacture of a pheromonal variant and in the strategic release of their pheromonal stores over time. Even in species in which pheromone type and production are uniform from male to male, competition for positional advantage is almost certain to occur, as it does in acoustic and visual signalers. Selection of points from which to disseminate the scent must influence the length of the pheromone trail and thus the number of females alerted to the male's presence. Moreover, males may be able to detect the pheromone plumes of rivals, station themselves within an already established plume, release their own pheromone, and intercept females that had earlier begun to travel upwind toward the original caller.

The Use of Arrestant Pheromones

Not all pheromone dispensing males remain in one location releasing a highly volatile substance as an attractant. There are a number of species whose males mark a series of perches with a durable arrestant pheromone that helps them locate females. In some small parasitic bees of the genus *Nomada* (Tengö and Bergström, 1977), various bumble bees (Darwin, quoted in Freeman, 1968; Free, 1971; Kullenberg et al., 1973; Svensson, 1980a,b), and at least one wasp (Litte, 1981), males patrol long routes from one marking site to another. When a male reaches a marking point, he alights and applies a mandibular gland product to the area and then flies on to the next point and repeats the process (Fig. 6.11). The male may mark each site only once each day, which suggests that a durable pheromone of low volatility is used (Alford, 1975). Male bumble bees have been called "veritable perfume brushes" because with their proboscis they apply the mandibular scent to their bodies and groom vigorously, smearing their hairy surfaces with the odor. The whole body then becomes a brush with which to mark those plant parts they choose as attraction points for females. In the laboratory it has been shown that virgin female bumble bees can detect the marked substrate and will alight there to await the arrival of a male (van Honk, Velthuis, and Roseler, 1978). As a patroller cruises his circuit, he checks and rechecks his marking points and is quick to copulate with females he finds waiting for him (Free, 1971).

6.11   Scent-marking male of the territorial bumble bee *Bombus nevadensis.* The male is painting the plant with a mandibular gland secretion. (Photograph by J. Alcock.)

The patrol routes of different males often overlap considerably. By inspecting another male's marking points, the "freeloader" increases the number of sites at which females may be found along his route without investing the time and energy required to produce and apply the pheromone marks. Male parasitism of this sort may be especially prevalent if males vary in the quantity or type of pheromone they can produce and if females use this variation to discriminate among potential mates. Males unable to provide the preferred scent marks may attempt to intercept females attracted to the marks of other "superior" males.

Male "Cooperation" in the Attraction of Females

We have emphasized that signaling males tend to be highly competitive in an effort to transmit their messages farther or more effectively than their rivals. Competition also manifests itself in the attempt by

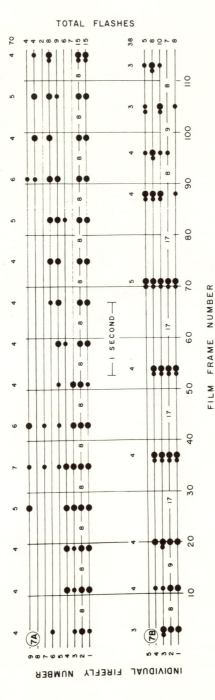

6.12 Synchronous signaling by male fireflies. A record of the flashes given by groups of neighboring males of the southeast Asian species *Pteropteryx malaccae* as they called from a tree. (From Buck and Buck, 1968. Copyright 1968 by the American Association for the Advancement of Science.)

some males to exploit the signals of others—as male bumble bees may do, and as males of *Syrbula admirabilis* certainly do (Otte, 1972). The long-distance caller that attracts other males would seem to suffer, because the newcomers may intercept females or interfere with courtships that the original signaling male had begun. At the same time, males that join a group of calling individuals appear to put themselves at a reproductive disadvantage by entering an area where the competition for attracted females will be especially severe. Neither signaler nor joiner seems likely to benefit from the formation of a group of mate-attracting males.

Despite the logic of this argument, examples of group signaling by male insects are far from uncommon and may involve gatherings of dozens, hundreds, and even tens of thousands of males that join to produce their calls, displays, or pheromonal signals. The apparent cooperation among males reaches its most extreme levels in the synchronously signaling species. In the spectacular flashing displays of certain southeast Asian fireflies (Fig. 6.12) thousands of males of *Pteropteryx malaccae* occupying a riverbank tree pulse their light flashes simultaneously in an awe-inspiring nighttime performance (Hanson et al., 1971).

As Alexander (1975), Otte and Smiley (1977), J. E. Lloyd (1979b), and others have noted, attempts to account for the formation of male aggregations in general and the occurrence of synchronous signaling in particular have often foundered on the shoals of group selectionist thinking. In the past these phenomena have been said to be adaptations designed to facilitate the union of the sexes (and thus, by implication, the propagation and preservation of the species).

The individual selectionist asks why a particular male may better advance his own genetic interests by signaling in an aggregation rather than by avoiding his competitors and attempting to attract females in places where he would be free from potential sexual interference. Because there are considerable costs to individuals that participate in group signaling, it is far from obvious how joining behavior and cooperative calling could have evolved. Yet it is clear that in a broad range of species males are attracted to a location with calling conspecifics.

Males Attracted to Other Males' Signals

Table 6.2 presents a list of species in which males are attracted by the signals produced by their potential rivals. In some instances the attraction leads to the formation of dense aggregations, as in the Asian

Table 6.2   Examples of species whose males actively seek out other males by tracking their calling signals.

Group	Reference
Males that track acoustic signals	
Cicadas	Alexander, 1975; Doolan and Mac Nally, 1981
Katydid	G. K. Morris, 1971
Crickets	Cade, 1979b, 1980
Grasshoppers	Otte, 1972
Mole crickets	Ulagaraj and Walker, 1973; Forrest, 1980
Males that track olfactory signals	
Beetles	Hardee, Cross, and Mitchell, 1969; Hardee et al., 1969; Wood, 1973; Birch, 1978
Otitid and trypetid flies	Fletcher, 1968; Alcock and Pyle, 1979; G. Dodson, 1982
Pentatomid bug	Harris and Todd, 1980a
Mecoptera	Bornemissza, 1966b; Alcock, 1979b; Byers and Thornhill, 1983; Thornhill, manuscript a
Wasps	Alcock, 1975; Litte, 1981
Bees	C. H. Dodson, 1966, 1975; Velthuis and de Camargo, 1975a,b; Kimsey, 1980
Ants	Hölldobler, 1976
Males that track visual signals	
Empidids and other swarming flies	Downes, 1969; Alcock, 1973; Sullivan, 1981
Fireflies	J. E. Lloyd, 1971, 1979b
Mecoptera	Thornhill, 1979c; Byers and Thornhill, 1983

fireflies and periodical cicadas; but in others the density of males, although increased by their attraction to one another's signals, remains comparatively low. Individual scent-marking bumble bees, for instance, often develop overlapping patrol routes but the males in flight rarely meet. The population as a whole is highly dispersed in comparison with the Asian fireflies or periodical cicadas.

It is not invariably the case that males that locate other males and move near them join the group in order to begin signaling there themselves. As we have seen in *Syrbula*, some joiners remain silent, awaiting the moment when they can attempt to interrupt a courtship and secure a receptive female. In scorpionflies some males approach other males to steal their nuptial gifts (Bornemissza, 1966b; Alcock, 1979b; Thornhill, 1979c). But in many insects a male that joins a group adds his signals to the messages being transmitted there. A good example is provided by the scolytid bark beetle *Ips pini* (Birch, 1978). Males of this species find and attack a potential host tree by burrowing under the bark to create a nuptial chamber. As they work their way through the wood, the pioneers create fecal material and debris that contains within it a bouquet of chemical substances that attract females and more males to the host. The newcomer males burrow into the tree and add their own pheromonal mixtures to those already being released at the location. Still more colonizers arrive. Eventually when very high concentrations of the signal are present, flying males are inhibited from landing on the infested tree and the population stabilizes. That males can be attracted to bolts of wood painted with the attractant chemical in the appropriate concentration indicates that this cue alone is sufficient to generate aggregations of males (and females).

The Benefits of Aggregating

In the insect literature, when members of one sex produce a signal that attracts others of the same sex, the cue is said to be an aggregation signal such as the aggregation pheromones of bark beetles (Wood and Bedard, 1975; Coulson, 1979) and the aggregation beacon of some fireflies (Buck and Buck, 1966). But the assumption that male signals have evolved to attract other males deserves close scrutiny. When the activity of one male leads another male to approach it and join in signaling, does the joiner gain or does the original signaler? Or do both individuals benefit when one produces a cue to which the other responds?

Only the joiner benefits when he does not add his signal to those of the other male(s) present but instead steals nuptial gifts, or lurks nearby and intercepts receptive females attracted by the signaler. To speak of the songs of male crickets or *Syrbula* grasshoppers as aggregation signals simply because other males are attracted to these calls is misleading at best. Silent satellites exploit a communication signal that has evolved because of its positive effects on fitness for the signaler and the *legitimate* receivers, which are conspecific females (Otte, 1972). Silent males impose a cost on the system that the callers cannot easily avoid if they are to produce a strong signal that many females can detect.

aggregation

   Pheromone-releasing males can also be parasitized by silent rivals, with aggregations forming purely as an incidental result of a communication system whose function is to attract females to the signaling males. Males of some species of philanthine wasps form loose groups of individuals that mark their perches with a sex attractant pheromone. Other males commonly come to these locations and move from one signaling male to another (Simon-Thomas and Poorter, 1972; Alcock, 1975). Perhaps these visitors are trying to find a calling perch of their own, but it seems just as likely that they are searching for attracted females in order to steal them from the perch-marking males.
   A puzzling case of probable male-male exploitation of this sort involves some species of tropical euglossine bees that assemble in small swarm flights. Some of the individuals in the group (often only a small fraction) will, prior to assembling, have spent many hours laboriously scraping fragrance compounds from orchids in the forest (see Fig. 6.13 and C. H. Dodson, 1966, 1975). These materials are stored on the animal's hind legs, and it has been suggested (but not verified) that they serve as precursors of a sex attractant pheromone (Michener, 1974; Kimsey, 1980). If pheromones are implicated in the attraction of females to a swarm, and if only the orchid fragrance collectors provide

6.13 An orchid bee, *Eulaema* sp., collecting scent materials from a plant. (Photograph by C. W. Rettenmeyer.)

these pheromones, the noncollectors may gather near pheromone dispensing individuals in order to intercept females drawn by the scent.

It is possible, however, that collector males use an orchid derivative solely to promote the aggregation of males if females are much more likely to visit a large rather than a small swarm of males. Nevertheless, if all the males have the same probability of mating, the collector would once again appear to be exploited by the noncollectors because he will have spent considerable time and energy at orchids while other males were swarming, feeding, or resting.

Another explanation for the tendency of some males to join others in signaling is that the visitor males are trying to disrupt the sexual contacts of their rivals. Because fitness is relative, an individual can theoretically gain genetically not only by raising his own reproductive output but by reducing a competitor's success. The spiteful individual, however, incurs a time, energy, and risk cost in return for a gain (the reduction in fitness of a rival) that is shared by all the other nonrelatives of the victim in the population. And these beneficiaries secure the benefit at no cost to themselves. For this reason, it is generally felt that spiteful behavior will rarely evolve (Rothstein, 1979; Tullock, 1979). In fact, what at first glance appears to be a purely spiteful action may be a tactic that sometimes translates into improved mating opportunities for the initiator. For example, males of the fly *Physiphora demandata* occur in relatively dense mating aggregations, with individual males calling from perches in shrubs and trees. A perch owner that is courting a female runs a moderately high risk of attracting an intruder male before he can secure a copulation, because courtship is often prolonged. The intruder disrupts the courtship and while the resident male is dealing with the visitor, the female decamps. Frequently neither the intruder nor the original courter can relocate the female after their conflict, and as a result the visitor's behavior has the appearance of spite. Occasionally the intruder, after fleeing from the perch owner, comes across the female resting nearby and is able to initiate a courtship of his own (Alcock and Pyle, 1979). Perhaps this is the major motivation for courtship interference.

Signaling by Joiner Males

Even when the male that joins another does not remain silent but begins to signal, there is no reason to assume that the two males are cooperating in the attraction of females. This is particularly clear in the case of certain fireflies whose joiner males employ their signals specifically to disrupt a courtship in progress. The sexual communication between males and females of *Photinus macdermotti* (and many other species) is linked, as previously mentioned, to a timing mechanism in the nervous system of both sexes (J. E. Lloyd, 1966). If the two light flashes of a male firefly occur too closely together, a female of *P. macdermotti* will not answer but will instead reset her "timer" to measure the interval between the second flash and the next signal she receives.

Males of *P. macdermotti* that observe an ongoing courtship may fly to the area and begin flashing, sometimes synchronizing their calls with those given by the original male. Indeed, they have the ability to skip their first signal and join on the second flash given by their rival. This may help them draw the female away from the first male; if so, the advantage to the interloper of synchronized calling is clear (J. E. Lloyd, 1979b). Less obviously advantageous and more difficult to explain are calls given by the intruder halfway through the interval between the two flashes produced by the original signaler (Fig. 6.14). These asynchronous calls are unlikely to attract the female. They seem spiteful in the sense that they only block the female's responsiveness to the first male. The injection flash makes the first male appear to be a member of the wrong species and takes advantage of the

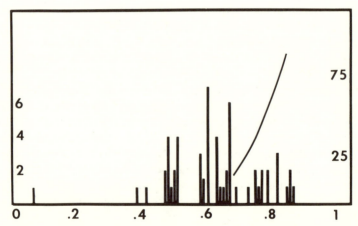

6.14 The timing of injection flashes by males of *Photinus macdermotti* that apparently attempt to disrupt communication between another male and a female. A disrupter male inserts his signal flash between the two precisely timed flashes of the rival. The number of recorded injection flashes (*left vertical axis*) peaks at the midpoint between the double flash given by a rival. The likelihood that a female will give her answering flash (*ascending curve and right vertical axis*) is strongly reduced by midpoint injection flashes. (From J. E. Lloyd, 1979b.)

fact that females are programmed not to answer males that produce signals too close together. Although the injection flashes appear only to reduce the mating chances of a rival, they may in the long run improve the courtship disrupter's chances of mating. They buy time for the intruder, and the longer he can prevent the other male from mating, the better his own chance of initiating a courtship with the female. Again, what appears to be spiteful behavior may in fact be competitive behavior that improves the initiator's chances of copulating.

Whatever the benefit of the joiner's signals to its own reproductive success, the point is that the original signaler does not gain by attracting the interloper firefly. Therefore the male's signal does not have the *evolved function* of drawing other males together. The boll weevil has been said to have an aggregation pheromone, because almost as many males as females are attracted by the male sex pheromone (W. H. Cross, 1973). But an increase in the size of a group of calling males does not improve the original signaler's mating chances. In fact, there is evidence that isolated males attract more females than a group of simultaneously signaling males (Hardee, Cross, and Mitchell, 1969; Hardee et al., 1969). Thus the joiners are probably taking advantage of the release of a sex pheromone designed to communicate between male and female. They may use the cue to locate a site where they have some chance of attracting and intercepting receptive females, even though this adversely affects the fitness of the male that was first to signal from the location.

There are other cases in which males apparently do attract more females per signaler by joining forces and calling together. As Bradbury (1981) has shown, however, an enhanced signal created by overlap in simultaneously released messages cannot reach an exponentially greater number of females than an individual signal. Even though the augmented signal released by a group of males is more powerful and will contact a greater total number of females, the physics of signal transmission and the geometry of female distribution patterns combine to mean the average number of females contacted *per male* cannot increase as the intensity of combined signals increases. In fact, for almost all kinds of signals and all reasonable distribution patterns of females, the average number of females per male within range of the augmented signal falls as group size increases.

There is a distinction to be made between females that can perceive a signal and those that respond to it. It is possible that females will be more likely to approach a group of calling males than an isolated signaler. If females find a more powerful signal more attractive, then the average number of females per signaler that are drawn to a location

could rise as group size increases. If so, a male might mate more frequently if he were in an aggregation than if he were to call by himself.

There has been at least one experimental demonstration that two
calls (given synchronously) are more than twice as effective as one in
attracting females. Morris, Kerr, and Fullard (1978) placed female
katydids (*Conocephalus nigropleurum*) in a test arena with a pair of
speakers in the walls, one on either side of the circular container. Four
times as many females exited through the opening by the loudspeaker
broadcasting two songs simultaneously than through the exit by the
single-song speaker (Fig. 6.15). Katydid females may gain by going to
areas with dense populations in order to compare males more easily.
Large males are preferred because they provide larger spermatophores, which the female eats after mating (Gwynne, 1982). Because
larger males sing more loudly than smaller ones, females may be able
to use the relative strength of acoustical signals to guide them to a
male that will offer a larger spermatophore. In fact, the reason that females of *C. nigropleurum* were more attracted to two calls projected
simultaneously may have been that they interpreted the double signal
as one. Females of this species may be programmed to move toward
the louder of two signals in order to locate the larger of two males.

The relation between male density, strength of the summed signal
generated, and attraction of a disproportionate number of females also
applies to some *Ips* bark beetles. Here too, because of the nature of interspecific competition in *Ips*, females may gain better access to valuable resources by going to a dense population of calling males of their
species rather than to scattered or isolated individuals. Experiments
have shown that when *I. pini* and *I. paraconfusus* occupy a host in
about equal numbers, the survival chances of the broods of both species are relatively reduced (Birch, 1978). For this reason, if a female
has a choice between a fallen tree that has been colonized by a moderately large number of males and females of her species as opposed to
one with only a small number of conspecific attackers, she will probably do better by settling in the more fully occupied tree. Such hosts are
unattractive to colonizers of other competing species of *Ips*. But a tree
with relatively few members of her species may be taken eventually
by many colonizers of another *Ips*, with damaging effects on her
brood.

A strong female preference for large aggregations of males means
that males can gain both by calling in conspecific males, especially
when the population density is low in a host, and by joining a group in
response to signals coming from it, especially when the population is
growing rapidly. The formation of aggregations of signaling males may

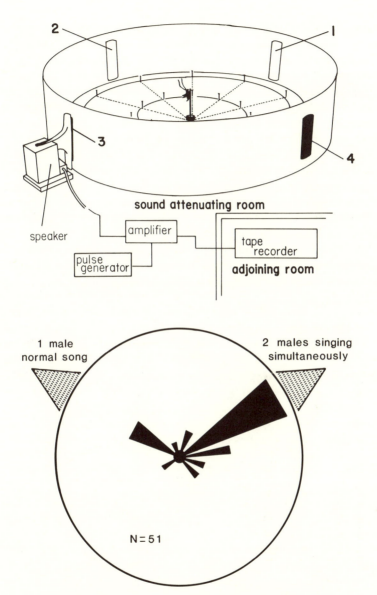

6.15   (*Above*) The experimental arena designed to test female choice in *Conocephalus* katydids. Females had the option of leaving through various exits. (*Below*) Generally, females left via the exit closest to the speaker playing the tape of two males singing. (From Morris, Kerr, and Gwynne, 1975; Morris, Kerr, and Fullard, 1978.)

provide reproductive gains for all the conspecific cooperators by making the site unattractive to male colonizers of other species (Birch and Wood, 1975) and more and more attractive to female mates. As the probability of damaging interspecific competition declines, the benefit of attracting additional conspecific males also declines. As brood density increases, *intra*specific competition among larvae can become as damaging to male reproductive success as *inter*specific interactions, because of the limited amount of food available in the cambium of the host tree (Fig. 6.16). Thus there will come a point when established residents of the tree will no longer gain by attracting additional colonists, but dispersing individuals may still exploit cues associated with the aggregation to locate the site and settle there because their other options are even less attractive (Alcock, 1982a).

Female bark beetles may be attracted to aggregated males because their chance of converting a food resource into surviving offspring is thereby improved. There are other reasons why a female might prefer to mate with a member of a group rather than an isolated male, even

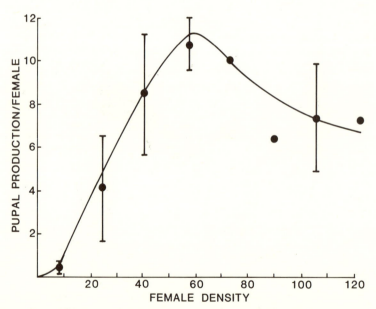

6.16   Effect of density of colonists in a host tree on the reproductive success of females of the bark beetle *Dendroctonus montanus*. Individual reproductive success first rises and then falls sharply with increasing density of settling females. (From Raffa and Berryman, in press. Copyright © 1983, the Ecological Society of America.)

if doing so did not improve her chances of gaining access to a resource. In many insects the male's only contribution to his mate may
be his genes. If genetic quality varies among males and if females can
assess this attribute in any way, then by going to a group of males a
female may be able to acquire a superior partner more efficiently. Janetos (1980) has shown that in theory the best strategy for a female to
use in picking a mate (under most conditions) is to examine a set
number of males (at least five) and to choose the best among them
(see also Downhower and Brown, 1981). Employment of the "best-of-
n-males" strategy is greatly facilitated if males are forced to gather together and demonstrate those qualities relevant to female choice. This
reduces the travel costs for the mate searching female, as she need not
go from one dispersed male to another. In addition, she need not possess an elaborate memory system to record the location and attributes
of five or more potential mates before returning to select the best one.
By examining the interactions among clustered males, a female can
make an immediate selection of a socially dominant male who may
provide her with superior gametes. If females do gain better mates by
visiting groups of males, an isolated male may fail to attract visitors
and will be forced to become a group member if he is to have any reproductive opportunities at all.

A different sort of advantage of group-joining males may be a reduction in the risk of predation. An individual in a group is able to hide
more effectively among his companions (W. D. Hamilton, 1971), reducing the risk that he will be one of the unfortunate ones consumed
during a period of signaling. The impressive synchrony in emergence
and chorusing of the periodical cicadas probably has evolved because
of its antipredator advantages (Lloyd and Dybas, 1966). If a group-
joining male can live longer, he may well increase the total number of
chances to mate during his lifetime relative to isolated, shorter-lived
males. Both T. J. Walker (1969), writing on chorusing crickets, and
Spieth (1974a), on lek-forming Hawaiian drosophilids, have suggested that predator pressure may have shaped the group-forming
tendencies of these insects. Improved avoidance of predators and increased access to females are not necessarily mutually exclusive outcomes of aggregating, but may instead be complementary results of
cooperative joining by males. Nevertheless, in the one careful test of
the attractiveness of aggregated versus solitary callers, Cade (1981a)
has shown that the average number of females and parasitic flies attracted *per male* was the same whether the males were clumped or
isolated. Table 6.3 summarizes the major hypotheses on the cooperative benefits of aggregation by males.

Table 6.3   Possible cooperative benefits of joining a group of calling males. Both the individual that joins and the males already present gain reproductively when the group grows in size.

Improved access to females; females prefer to mate with an individual in a group of males because—
  (1) The production or survival of offspring is improved in an area with many conspecifics.
  (2) Females can better assess alternative resources offered by different males in return for copulation.
  (3) Females can better assess the genetic quality of potential mates.

Improved chance of survival by the male because—
  (1) Members of a group better detect and evade approaching predators.
  (2) Members of a group saturate the feeding capacity of their predators.

The Timing of Signal Releases

A member of a group of signaling males is faced with the problem of when to release his signals in relation to his neighbor's messages. The overlap between one individual's signals and those given by others around him may be (1) absent, leading to alternation or interspersing of calls, (2) incomplete (partial synchrony), or (3) total (complete synchrony). Let us examine each of these possibilities, with the recognition that their costs and benefits are dependent upon the signal mode, the feature of the signal that is attractive to females, and the number of males in the group.

The alternation of signals is practiced only when acoustic or visual displays are employed and only when the density of calling males is relatively low. Because of the long fade-out time of pheromonal messages, a male that attempted to alternate his scent signals with those of another individual would be usurped by those that began to release the pheromone first. (Males might strategically alter the amount of pheromone released in relation to the scents they perceive about them, but such tactics would take place in the context of at least partial synchrony.) As density rises, even in species using brief signals with rapid fade-out times, the attempt to insert a signal into the "quiet" intervals between another's calls would be doomed to failure.

Males of the katydid *Pterophylla camellifolia* form alternating choruses (Shaw, 1968). As expected, the groups constitute only a few individuals, with 1.5 to 15 m between nearest neighbors. Interacting members of the group are a leader and a follower, with the leader consistently initiating a bout of singing that occurs when the wings are rubbed together (Fig. 6.17). At the proximate level, hearing a call evidently stimulates the follower to join in; but on the other hand, perception of a chirp produced by another male also delays the production of a chirp by the listener (Jones, 1964). As a result, if two males are singing, their calls may at first overlap somewhat; then they begin to increase the interval between their chirps, placing their own bursts of wing rubbing in the intervals between the rival's chirps.

Not only do some male katydids have the capacity to adjust their rate of chirping, they also can alter the length of the chirp (by adding an extra wing rub or two per chirp) and they can vary the intensity of the call as well (Jones, 1966). Some males faced with a nearby competitor may sing louder and longer in an attempt to enhance the relative strength (and attractiveness?) of their call. Close rivals may lengthen their chirps markedly and engage in disruptive solo singing, speeding up their chirp production even though this breaks the alternating pattern. Presumably their goal is to demonstrate that they possess more powerful song-producing abilities. If one katydid can "convince" the other male that he is unlikely to be able to match the more powerful singer in attracting females, then the second male gains by moving away. In katydids and other calling orthopterans, males usually move outside the range of mutual acoustic interference, freeing subordinates from direct competition for mates with dominant males (Latimer, 1981). At the same time the more powerful singers, freed from the close-in presence of rivals, may be able to reduce their investment in singing and thereby conserve energy and reduce the risk of attracting predators.

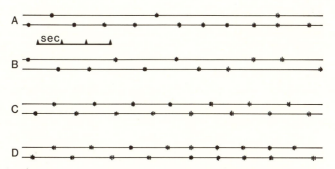

6.17   Alternation in the calls of two individuals of *Pterophylla* katydids. At the outset (*A*) one male (*lower line*) sings more often than the other, but by (*C*) they are alternating. (*D*) Note the bout of perfect synchrony in calls. (From Jones, 1966.)

Synchrony in Calling

Although it is possible to attribute cooperation in alternating callers, as this activity requires a certain amount of coordination between the duetting pair, it is still easier to do so when groups of males signal in synchrony with one another. Yet synchronous signaling can, like alternation, be the outcome of attempts by males to avoid disruption of the important components of their messages. In locations in which acoustic or visually displaying males occur in high density, the opportunities for successful alternation are reduced. For example, in another katydid, *Neoconocephalus nebrascensis,* as many as 10 or 15 males may be clustered within a few meters of one another (Meixner and Shaw, 1979). With so many potential singers nearby, the possibility of initiating a duet often fails because there are too many other individuals trying to add their calls to any paired interaction. The solution may be to call precisely when the other males in the group broadcast their signals. Males of *N. nebrascensis* call synchronously—although there does appear to be a leader, one male in the group that is likely to initiate a bout of synchronous singing by starting to chirp after a period of quiet.

One way in which males of *N. nebrascensis* could potentially avoid overlap would be to divide up the total calling period available to them, each male taking a certain amount of time. This pattern (Fig. 6.18) actually does occur in a few orthopterans, but it requires special circumstances (T. J. Walker, forthcoming b). First, there must be constraints (such as high energetic demands of signaling) that limit the total time any one male can call per day. Second, the period of the day during which females may respond to calling males must be relatively great, so that many males can fit their bouts of singing within this period. Third, females must be *uniformly* available during the calling period—for if more were receptive to signals at any particular part of the time, selection would favor males that timed their calling bouts to coincide with this interval.

For a great many species these restrictive conditions are not met. Males are forced to call in an overlapping fashion if they are to have any chance of attracting females. A variety of factors, such as falling temperatures and an increased risk of predation at certain times, can limit the period of safe and profitable calling to a relatively brief interval (Doolan and Mac Nally, 1981; T. J. Walker, forthcoming b). Female choice itself can act in the same direction if females prefer to select mates when they can hear and compare the simultaneous songs of as many males as possible. The end result may be that males can only

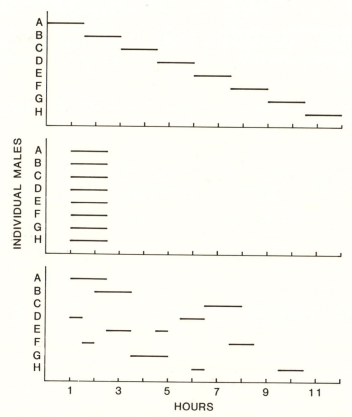

6.18   Possible temporal patterning of male signaling in a group of eight orthopterans occupying the same area. The top graph shows complete segrega-
tion of signal periods with each male calling for a separate 90-minute period. In
the middle graph there is complete overlap in signaling, with all males calling
simultaneously (the predicted pattern if females are available only during this
period). The bottom graph shows "random" distribution of male signaling
times, with partial overlaps between some individuals. (After T. J. Walker,
forthcoming b. Reprinted by permission of Westview Press, Boulder, Colorado.)

acquire a mate during a 30-minute spell each day, as is true of a
cricket (T. J. Walker, 1980) and an Australian cicada (Doolan, 1981).
Temporal packing of this sort severely reduces the likelihood of escaping competition by singing in quiet periods during the calling minutes
or hours. Thus partial synchrony in calling may occur initially as a side
effect of the attempt of males to make the best of a difficult situation,
in which a limited calling period and an abundance of competitors

make isolated calling impossible. Complete or nearly complete synchrony can evolve if, as we have just suggested, there is a premium on projecting a clear signal that advertises species membership (or some other critical attribute of the male).

There are other possible advantages of synchrony in message generating. If female choice forces males to call together, females may prefer to visit the group of males that is the most highly synchronized. At the proximate level, this could be mediated by a power preference on the part of females. Ultimately females with such a preference might gain reproductively by being able to make more precise comparisons between fully synchronous calls as opposed to those that only partially overlap.

From the male's perspective, if there is a "dialogue" between an attracted female and a male the male signaler may be better able to perceive a reply in the quiet interval between synchronous songs or the nonluminous period between synchronous visual flashes. In *Syrbula admirabilis* there may be competition between the initiator of a signal bout and those that join in at much the same time. The joiners may attempt to sing a little longer than the initiator in order to prevent him from hearing a female's answering call (Otte, 1972). This in turn selects for initiators that are able to sing as long as the joiners are likely to sing. The result, synchrony in calling, could be the evolutionary product of the conflict among males to obscure their rivals' hearing while trying to prevent blocking from happening to them (Otte and Loftus-Hills, 1979).

Likewise in the fireflies, the time interval between a male's signal and a receptive female's reply is a species-specific characteristic that enables a male to locate answering females. An unsynchronized male's visual perception may be blocked by flashes of other males around him, or his ability to discriminate between male and female flashes may be reduced by the visual signals of neighbors. Males that time their calls to overlap precisely with those around them avoid these difficulties.

These possibilities do not end the list of benefits for the synchronous signaler (Otte, 1980). Again, as we have already suggested, the male that calls in synchrony may be able to detect not only females that he has attracted but also those that are responding to other nearby males. This gives him the chance to interlope on budding courtships. Moreover, if signaling males alter their calls upon detection of a female, a synchronized male may be better able to tell when a neighbor falls out of synchrony. This may produce a cue that a female is close by and may be stolen.

Thus modern hypotheses to account for the evolution of synchronous singing or flashing fall into two categories (Otte, 1980). In noise reduction hypotheses, the synchronous male is said to be better able to transmit his own messages and better able to detect a response if his signal and vision or audition are not obscured by the signals of other members of the group. In the interloping hypotheses, the synchronized male is said to be better able to detect and interfere with early courtships and usurp females that have responded to signals other than his own. There is no reason why both kinds of benefits might not be secured by some synchronous calling males.

We conclude that for insects the social environment is fully as significant as the physical and predatory environment in shaping male signals. Competition among males for access to as many mates as possible favors individuals able to project their message a long distance. This requires the ability to cope with the special obstacles to message transmission in the insect's environment and has a trade-off in increased risk of attracting predators and parasitic males. A host of factors, including climatic restrictions, predator pressure, and female preference, can limit the time available for effective signaling and lead to varying degrees of signal overlap among males in a given location. Perfect synchronous overlap superficially appears to require a high degree of cooperation among males, but can be the product of these limiting influences coupled with selection favoring males that offer an uncluttered signal while remaining sensitive to opportunities to steal mates attracted to neighboring rivals.

7

Selection and Defense of Mating Sites

THE LOCATION that a male selects to broadcast his attractant signals or to search for mates will affect his reproductive success. Searching males frequent places where receptive females are relatively numerous. Depending on the species, these locations include sites from which adult females emerge, or places where egg laying occurs, or sites where females gather food for themselves or their offspring, or "neutral" landmark points such as distinctive topographical features chosen by females purely for the purpose of securing a mate. The first portion of this chapter documents this diversity in encounter sites and attempts to explain why male insects vary in their selection of places in which to search for females. The second section examines the related issue of why males vary in the degree to which they defend a potential encounter site from other males.

The Diversity of Favored Mating Sites

We have said that male insects almost always focus their mate location efforts at one of four possible sites: the zone of emergence, the foraging area, the oviposition site, or a distinctive landmark. There are

Table 7.1   Examples of insect species whose males search for emergence sites in order to copulate with virgin females.

Order	Source of virgin females	Reference
Hymenoptera		
Bombus spp. (some bumble bees)	Natal nests in grass	Alford, 1975; J. E. Lloyd, 1981
Nasonia vitripennis (pteromalid wasp)	Parasitized fly pupae	King, Askew, and Sanger, 1969
Xylocopa virginica (carpenter bee)	Natal nests in wood	Gerling and Herman, 1978
Andrena erythronii (andrenid bee)	Natal nests in ground	Michener and Rettenmeyer, 1956
Melittobia spp. (eulophid wasp)	Parasitized fly pupae	Buckell, 1928; Matthews, 1975
Schizaspidia tenuicornis (eucharid wasp)	Parasitized ant nests	Clausen, 1923
Trigonosis cameronii (sphecid wasp)	Brood cells on rocks	Eberhard, 1974
Epsilon sp. (eumenid wasp)	Brood cells on rocks	Smith and Alcock, 1980
Rhytidoponera metallica (ponerine ant)	Natal nests	Hölldobler and Haskins, 1977
Diptera		
Deinocerites cancer (culicid mosquito)	Tidal crab holes	Downes, 1966; Provost and Haeger, 1967
Ephemeroptera		
Many species	Aquatic habitats	Needham, Traver, and Hsu, 1935; Spieth, 1940
Lepidoptera		
Colias spp. (pierid butterfly)	Food plants of larvae	Silberglied, 1977
Pieris protodice (pierid butterfly)	Food plants of larvae	Rutowski, 1979
Heliconius charitonius (heliconiid butterfly)	Pupal cases of females	Bellinger, 1954
Coleoptera		
Hypothyce mixta and other scarab beetles	Emergence burrows in ground	A. Milne, 1960; Barfield and Gibson, 1975

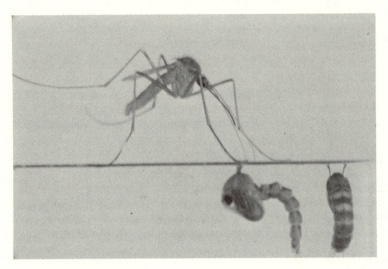

7.1 A male of the mosquito *Deinocerites cancer* holding a pupa in order to mate with the virgin female upon her emergence. (From Provost and Haeger, 1967.)

exceptions: for example, in some populations of monarch butterflies males contact females in the overwintering aggregations for which this species is famous. Sexual activity occurs for several weeks before the insects leave the wintering area for northward migration (Hull, Wenner, and Wells, 1976; Tuskes and Brower, 1978). Similarly, males of several species of vespid wasps take up residence in the fall at hibernation sites that will be used by future foundress females. They sometimes succeed in mating with arriving females (Lin, 1972; Litte, 1979). These cases support the rule that searching males tend to gather in that part of their environment where receptive females are concentrated. In species that overwinter in groups, opportunities for mating may be greatest at the hibernation or sheltered overwintering site. More typically, however, receptive females are most highly clustered in places where they emerge, feed, or oviposit—and male behavior reflects this reality. This was recognized for flies some years ago (Anderson, 1974) and is true for other insect groups as well.

We have already discussed how male digger bees and lovebugs patrol or wait near an emergence area for the chance to contact a virgin female. Table 7.1 gives a selected list of additional examples of insects whose males hunt for emerging or recently emerged females (see Fig. 7.1). The swallowtail butterfly *Parnassius phoebus* is representative of this group. Males fly circuits through areas containing the food plant of their species. They appear to be able to detect freshly emerged females, probably on the basis of odor cues, because they have been observed *in copula* with females so recently out of the chrysalis that their wings have not yet expanded (Scott, 1973a). The ten-lined june beetle is even more similar in its behavior to the digger bee, *Centris pallida*. Males fly to a spot on the ground just before an emerging female breaks through the surface. If, as is often the case, several males arrive more or less simultaneously, a vigorous fight for possession of the female ensues. The winner copulates with her, and the losers fly off (M. Birch, personal communication).

In some species males are able to find recently eclosed females without ever emerging to the surface. Meloid beetles of the genus *Hornia* parasitize ground-nesting *Anthophora* bees. The bees form large, dense colonies of females, each of which constructs a multi-celled nest of underground brood pots similar to those of *C. pallida*. The female meloid makes her way underground along nest tunnels, ovipositing in the brood cells she discovers. Her offspring consume the pollen and nectar provisions gathered by the female bees. After they mature, a female's daughters cut a hole in the brood cell but remain within the cell until contacted by a male. Emerging meloid males leave their natal cell and wander along the burrows until they find a brood pot containing a virgin female (Linsley and MacSwain, 1942).

Mating at Oviposition Sites

In many other insects, males search for females that are about to lay their eggs (Table 7.2). There are species whose females lay their eggs in excavated burrows in wood or in the ground; in some of these, including certain bark beetles, a cockroach, and various wasps, the male stations himself in or by the nest entrance in order to mate with the resident female(s). For example, males of *Dendroctonus* bark beetles search for host trees that are in the process of being attacked by burrowing females. They use the odor cues associated with nest construction to track a suitable host and then search for a burrow that contains a single, unpaired female. After a vigorous, even violent, courtship, a male may succeed in copulating with the female in her nest (Wood, 1973; Ryker, in press).

Many female insects oviposit in dung or carrion; their males often

Table 7.2 Examples of species whose males locate mates at sites in which females oviposit.

Order	Oviposition site	Reference
Hymenoptera		
Oxybelus spp. (sphecid wasp)	Burrows in sand	Peckham, Kurczewski, and Peckham, 1973
Dynatus nigripes (sphecid wasp)	Burrows in sand	Kimsey, 1978
Nomadopsis spp. (andrenid bees)	Burrows in soil	Rozen, 1958
Diptera		
Scatophaga stercoraria (scatophagid fly)	Dung pats	Parker, 1970a,b
Odontoloxozus longicornis (neriid fly)	Decaying spots on cactus	Mangan, 1979
Rhagoletis completa (tephritid fly)	Walnuts	Boyce, 1934
Drosophilia melanderi (drosophilid fly)	Mushrooms	Spieth and Heed, 1975
Somula decora and other syrphid flies	Moist pockets in trees	Maier and Waldbauer, 1979
Lepidoptera		
Aglaius urticae (nymphalid butterfly)	Nettles	R. R. Baker, 1972
Papilio indra (papilionid butterfly)	Serviceberry bushes	Eff, 1962
Coleoptera		
Ips spp. and many other scolytid bark beetles	Burrows in various host trees	Goeden and Norris, 1965; Borden, 1967; Birch, 1978
Bolitotherus cornutus (tenebrionid beetle)	Bracket fungus	Pace, 1967
Rhinostomus barbirostris (brentid beetle)	Fallen host trees	Eberhard, 1980
Anthonomus grandis (curculionid weevil)	Cotton bolls	Cross and Mitchell, 1966; Cross, 1973
Monochamus scutellatus and other cerambycids	White pine trunks	Hughes, 1981; Linsley, 1959
Orthoptera		
Gryllus spp. *Anurogryllus* spp. (crickets)	Burrows in ground	Alexander, 1961; T. J. Walker, 1980
Locusta migratoria (acridid grasshopper)	Moist soil	Parker and Smith, 1975
Cryptocercus punctulatus (wood cockroach)	Chambers in wood	Ritter, 1964

Table 7.2   (*continued*)

Order	Oviposition site	Reference
Odonata		
Calopteryx maculata and many other damselflies and dragonflies	Submerged aquatic vegetation	Waage, 1973; Corbet, 1980
Siphonaptera		
Spilopsyllus cuniculi (pulicid flea)	Baby rabbits	Rothschild, 1975
Homoptera		
Pemphigus bursarius and other aphids	Poplar tree bark	Dunn, 1959; Kennedy and Stroyan, 1959

race to freshly deposited feces or recently killed animals in order to be present when receptive egg-laden females arrive (Fig. 7.2). There are many kinds of dung flies in addition to the well-known *Scatophaga stercoraria,* whose behavior has already been sketched. There is an equally impressive collection of carrion insects, notably certain beetles, that gather at dead animals and copulate there. For example, males of the remarkable silphid beetles (*Necrophorus*) (Milne and Milne, 1976) first find a small dead mammal, then release a sex pheromone, and eventually cooperate with their mate in burying the carcass upon which their offspring will feed.

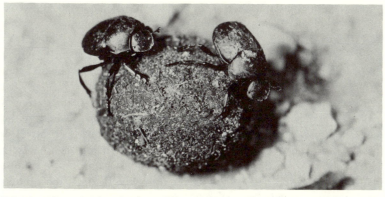

7.2   A male and a female dung beetle rolling a ball of dung to a future nest site. Mating in dung beetles often occurs within the nest burrow; an egg is then laid on the dung ball. (Photograph by J. Alcock.)

A female insect need not lay her eggs in dung or carrion, nor have a discrete burrow, in order to be contacted near the time of oviposition. Dragonflies and damselflies are classic examples of insects that mate at the egg laying area (Corbet, 1980). Although weedy ponds, streams, river edges, and the like are places in which emerging females can be found, males almost never search for teneral individuals, nor do they pursue foraging females that often hunt for prey long distances from water. Rather, they come to aquatic habitats that have some characteristic attractive to egg-laden females. Depending on the species, males may be spaced more or less evenly along an entire shoreline or they may be concentrated at places with submerged or floating vegetation or fast-moving water. As the gravid females come to the water, they are intercepted by the waiting males and copulation occurs.

Mating at Foraging Sites

Females of many insects collect food at some distance from a nest or oviposition site during a portion of their adult lives. This provides opportunities for males to contact mates at foraging areas (Table 7.3). Mating at food plants is particularly common among the bees. For example, males of some carpenter bees travel rapidly from one flowering shrub to another until they detect a pollen- and nectar-collecting female. They then attempt to capture the female, whether she is receptive or not, and sometimes succeed in securing a mate (Anzenberger, 1977; Velthuis and Gerling, 1980).

Other male carpenter bees defend a patch of flowers attractive to females, and the same is true of various megachilid bees (Jaycox, 1967; Alcock, Eickwort, and Eickwort, 1977; Severinghaus, Kurtak, and Eickwort, 1981). A female of *Callanthidium illustre* that enters a male's territory is quickly detected and grasped by the male, which darts at her as she attempts to enter a flower to extract its pollen; copulation always ensues if the male secures a firm grip on her (Fig. 7.3).

Flowering plants are not the only resource at which males gather to await the arrival of a conspecific female in search of food. Males of the tsetse fly find large game animals attractive, because receptive females in search of a blood meal attack these mammals (Jaenson, 1979). In the mosquito, *Eratmapoides chrysogaster*, males hover about an unfortunate jackrabbit's ears, from which female mosquitoes are extracting blood. The males pounce upon the females, which become receptive immediately after they have had a full meal (Gillett, 1971).

Table 7.3   Examples of insect species whose males locate mates at foraging sites frequented by females.

Order	Foraging site	Reference
Hymenoptera		
Xylocopa spp. and many other bees	Flowering plants	Anzenberger, 1977; Alcock et al., 1978; Eickwort and Ginsburg, 1980
Pseudomasaris vespoides (masarid wasp)	Flowering penstemons	Hicks, 1929
Diptera		
Glossina pallidipes (glossinid fly)	Large mammals	Jaenson, 1979
Melophagus ovinus and other hippoboscid flies	Sheep and other hosts	Prouty and Coatney, 1934; Bequaert, 1953
Somula decora and other syrphid flies	Flowers visited by females	Maier and Waldbauer, 1979
Cerotainia albipilosa and other asilid flies	Open woodlands	Lavigne and Holland, 1969; Scarbrough, 1978
Coleoptera		
Chauliognathus pennsylvanicus (cantharid beetle)	Flowering weeds	McCauley and Wade, 1978
Hippomelas planicosta (buprestid beetle)	Creosote bushes	Alcock, 1976
Pyrota postica and other meloid beetles	Creosote bushes and other flowering plants	R. B. Selander, 1964
Scarabeus spp. and many other scarab dung beetles	Mammalian dung pats	Halffter and Matthews, 1966
Thanasimus dubius (clerid beetle)	Trees infested with bark beetles	Vité and Williamson, 1970
Oryctes rhinoceros (scarab beetle)	Palm tree leaves	Zelazny, 1975
Orthoptera		
Bootettix argentatus (acridid grasshopper)	Creosote bushes	Otte and Joern, 1975
Hemiptera		
Pachybrachius bilobatus (lygaeid bug)	Seeds	Sweet, 1964
Acanthocephala femorata (coreid bug)	Sunflower stalks	Mitchell, 1980
Euschistus conspersus (pentatomid bug)	Blackberry plants	Alcock, 1972

7.3   Resource defending male, the megachilid bee *Callanthidium ilustre,* copulating with a female attracted to a flowering penstomen that the male controls. (Photograph by J. Alcock.)

Females of the clerid beetle *Thanasimus dubius* consume neither nectar nor blood, but the bodies of small bark beetles. Males as well as females are attracted to the odors produced when a tree is being attacked by a mass of beetles. Both feeding and mating by the clerids occur on the bark of the tree (Vité and Williamson, 1970).

A special category of species in which mating occurs at the feeding site is that in which males collect or produce food as a nuptial gift for their mates. For example, males of some empidid flies form small aerial swarms with virtually all individuals carrying a small insect that they have killed as a nuptial offering (Kessel, 1955). Receptive females come to the swarm, select a male, and are rewarded by the opportunity to copulate and feed simultaneously (Fig. 7.4). In some cases females never search for prey on their own but subsist solely on gifts offered by a succession of mates (Downes, 1970).

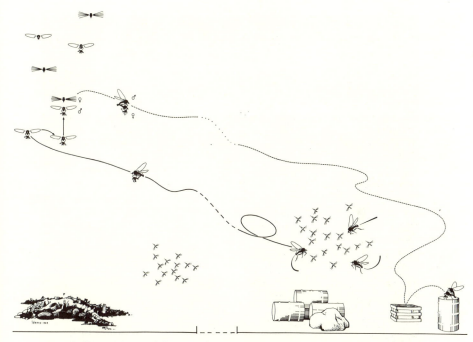

7.4   Mating swarms of the empidid fly *Rhamphomyia nigripes* form above specific swarm markers near aggregations of the prey species, mosquitoes, which in turn occur near conspicuous landmarks in their environment. In the arctic tundra, oil drums and garbage heaps are selected by the flies. (From Downes, 1970.)

Mating at Landmarks

The most puzzling of the major mate-location sites are the landmark mating areas. Some insects do not meet their partners at any of the seemingly logical locations (emergence, oviposition, or feeding sites) but instead congregate at distinctive topographical or physical features of the environment (Table 7.4). The landmarks may be as insignificant as a clearing in a forest, a bare patch of soil in a field, a stretch of white water in a stream, or a clump of bulrushes (Downes, 1955; E. O. Wilson, 1957; H. T. Nielsen, 1964; Edmunds and Edmunds, 1979; Tozer, Resh, and Solem, 1981). Alternatively, the landmark may be as monumental as the peak of a mountain, where mating aggregations of ants, beetles, flies, wasps, and butterflies have been observed (O. Shields, 1967).

Table 7.4 Examples of insect species whose males locate mates at conspicuous landmarks visited by females solely or primarily for mating, not for feeding or egg laying.

Order	Landmark	Reference
Hymenoptera		
Xylocopa varipuncta and other carpenter bees	Tops of bushes, trees, hilltops	Marshall and Alcock, 1981
Bombus confusus and other bumble bees	Tall weeds, shrubs	Alford, 1975
Pogonomyrmex spp. (harvester ants)	Hilltops	Hölldobler, 1976
Vespula spp. (vespid wasps)	Tall trees	MacDonald, Akre, and Hill, 1974
Pheidole sitarches (myrmecine ant)	Bare spots in weedy areas	E. O. Wilson, 1957
Eucerceris spp. (sphecid wasps)	Bushes or shrubs	Alcock, 1975; Steiner, 1978
Apis mellifera (honey bee)	Above hills, prominent trees	Ruttner, 1956
Diptera		
Ceratolaemus sp. (bombyliid fly)	Rock faces	Mühlenberg, 1973
Physiphora demandata (otitid fly)	Bare branches of trees/weeds	Alcock and Pyle, 1979
Drosophila planitibia (drosophilid fly)	Tree ferns	Spieth, 1968
Cephenemyia spp. (oestrid bot fly)	Trees/shrubs on hilltops	Catts, 1964
Tabanus bishoppi and other tabanid flies	Tall trees, hilltops	J. A. Chapman, 1954; Blickle, 1959; Corbet and Haddow, 1962
Culicoides nubeculosus (ceratopogonid midge)	Dark wet patches of open ground	Downes, 1955
Mansonia perturbans (culicid mosquito)	Dark spots on light soil	H. T. Nielsen, 1964
Lepidoptera		
Papilio zelicaon and many other butterflies	Hilltops	O. Shields, 1967; Scott, 1968
Pararge aegeria (nymphalid butterfly)	Sunspots	Davies, 1978
Ochlodes snowi (hesperiid skipper)	Gullies	Scott, 1973b

Table 7.4   (*continued*)

Order	Landmark	Reference
Coleoptera		
Pteropteryx malaccae (lampyrid firefly)	Trees along riverbanks	Hanson et al., 1971
Hippodamia spp. (coccinellid ladybugs)	Trees on hilltops	Hagen, 1962
Homoptera		
Quesada gigas (and other cicadas)	Sunlit, white tree trunks	Young, 1980
Trichoptera		
Hectopsyche albida (leptocerid caddisfly)	Over open water near prominent bulrushes	Tozer, Resh, and Solem, 1981

Many of the swarming flies aggregate near landmarks. Males of the mosquito *Aedes taeniorhynchus* gather in the evening over a prominent tree or bush. By marking the mosquitoes with a colored dust, Nielsen and Nielsen (1953) found that certain individuals arrived at the aerial swarm site each evening at the same time and stayed for only a few minutes. Other males then took their place, so that the swarm persisted for a time. These authors saw no copulations and doubted that the aggregation formed for sexual purposes. For many landmark-based mating systems, however, the frequency of copulation is extremely low (Catts, 1967; Alcock, 1981a). Failure over a brief period to observe mating is not proof that copulations do not sometimes occur at a landmark site.

A classic hilltopping species (that is, one which mates at mountain peaks) is the swallowtail butterfly *Papilio zelicaon* (O. Shields, 1967). During the breeding season in southern California a small number of males of the swallowtail arrive at the top of Dictionary Hill in the late morning and during the midday period. Instead of forming an aerial swarm, they alight on shrubs and other elevated perches and scan their surroundings for females to pursue. Those females that fly to the hilltop (around noon) usually prove to be recently emerged and uninseminated, whereas in the population as a whole unmated females are very rare. There are no flowers or larval food plants at the summit to attract visiting females, only males that chase and court them. After copulation the females leave the area at once—further evidence that they are not drawn to the hilltops by food or oviposition resources.

The Evolution of Favored Encounter Sites

Mate location behavior appears to be evolutionarily labile, sensitive to and shaped by the ecological pressures peculiar to a species. This is shown by the diversity of favored encounter sites that exists within families, and even genera, of insects. Thus in the carpenter bees there are species whose males wait by nests for emerging virgin females, while in others males search for mates at flowers, and in still others males wait at hilltops or other landmark points in areas devoid of attractive flowers or nest sites (Marshall and Alcock, 1981). How can we explain this behavioral diversity?

G. A. Parker (1978b) has explored this question in some detail. He notes that receptive females of an insect species are unlikely to have ever been distributed randomly in their environment. Assuming that males are free to search wherever they choose, they should move about so that their densities match those of receptive females in the various habitats frequented by the species. There is likely to be one area with the most females, in which the bulk of mating occurs. In this favored area receptive females will have the shortest waiting time for copulation because male density will be greatest.

If prompt mating confers a reproductive advantage on females, over time there will be a still greater concentration of potential mates in the major encounter site. Selection will also favor males that search in this portion of the environment, resulting in a positive feedback loop that promotes the evolution of an encounter site convention, in which almost all mating takes place in sites which at an earlier time may have been preferred by only a plurality of females.

This scenario explains why most encounter sites are focused on resources attractive to females, such as food or egg laying areas, or in emergence sites. But how do males evolve a preference for a neutral landmark outside areas containing emerging or resource-using females? Parker argues that if females are attracted to a favored area primarily because of the advantages associated with prompt mating, then males at the periphery of the area may benefit because they tend to meet incoming females first. Male aggregations on the edge of a nesting or foraging area may be centered on landmark points if these features are used as orientation guides for females moving through the area. Alternatively, males may initially have preferred elevated landmarks simply because they afforded better views of approaching females.

Parker's thesis suggests that encounter site conventions arise as a product of the ecological factors that concentrate females spatially and determine the pattern of female receptivity. Let us examine this proposition in more detail by presenting a series of "rules" that appear to govern male decisions about where to search for mates.

Single Mating by Females and Mate Location in Emergence Sites

Searching for mates in emergence areas should be adaptive if (1) females mate just once or at most a few times at long intervals in their lives, (2) females are receptive shortly after emergence, and (3) emerging females tend to be clustered spatially. Under these circumstances males can expect a large genetic gain by being first to reach a virgin female; and if such females are clumped, a male can potentially contact numerous mates in a short period if he locates an emergence site.

The conditions appear to apply to both the digger bee, *C. pallida,* and the lovebug, *P. nearctica,* whose females mate just once immediately after emerging from an area that contains large numbers of other about-to-emerge females. The same elements occur in various small parasitic wasps from many different families, in which males generally emerge before their sisters and copulate with them as they eclose (Clausen, 1923; Matthews, 1975; van den Assem, Gijswijt, and Nubel, 1980; Grissell and Goodpasture, 1981; Ryan, Mortensen, and Torgersen, 1981). The single- or seldom-mating pattern applies also to females of most of the other insects appearing in Table 7.1.

The advantage enjoyed by being the first to reach a virgin in a species with single-mating females is so great that there are numerous cases of male insects, like *C. pallida,* that have evolved the capacity to detect *preemergent* females (Table 7.5). One of the most remarkable involves the fig wasps of the genus *Blastophaga* (Hill, 1971). These wasps participate in an exceedingly complex symbiotic relationship with their host fig plants. Mated female wasps fly to and force their way into a fig that is a hollow, fleshy sphere. In some species the figs are either female (containing female flowers only) or male (containing pollen-bearing anthers and "gall flowers"). If the wasp enters a female fig, she will pollinate the flowers but will not be able to oviposit successfully (because of the length or design of the female flower). Since her wings have been destroyed in the struggle to enter the fig, she is doomed to die without reproducing. But if the wasp happens to

Table 7.5   Examples of insect species whose males are able to detect virgin females prior to their emergence as full adults, in order to be first to reach a potential mate.

Order	Location of preemergent female	Reference
Hymenoptera		
Colletes spp. (colletid bee)	Buried beneath sand	Hurd and Powell, 1958; Bergström and Tengö, 1978
Idarnes spp. (agaonid fig wasps)	Within fig gall	Hill, 1971; W. D. Hamilton, 1979
Dusmetia sangwani (encyrtid wasp)	Within scale insect host	Schuster, 1965
Pristiphora geniculata (tenthredinid sawfly)	In emergence tunnel beneath soil	Forbes and Daviault, 1965
Goniozus gallicola and other bethylid wasps	Within pupal cocoon	Gordh, 1976; Mertins, 1980
Megarhyssa spp. (ichneumonid wasp)	Within exit tunnel in wood	Nuttall, 1973a,b; Crankshaw and Matthews, 1981
Bembix rostrata (sphecid wasp)	Buried beneath sand	Schöne and Tengö, 1981
Coeloides dendroctoni (braconid wasp)	Under bark within exit tunnel	DeLeon, 1935
Diptera		
Opifex fuscus (culicid mosquito)	Within pupal case	Haeger and Provost, 1965
Arachnocampa luminosa (mycetophilid fly)	Within pupal case	A. M. Richards, 1960
Clunio marinus (chironomid midge)	Within pupal case	Olander and Palmén, 1968
Lepidoptera		
Heliconius spp. (heliconiid butterfly)	Within pupal case	Bellinger, 1954; L. E. Gilbert, 1976
Coleoptera		
Pleocoma dubitalis (scarab beetle)	In burrow beneath loose soil	Ritcher and Beer, 1956
Leptothea galbula (coccinellid beetle)	Within pupal case	A. M. Richards, 1980

choose a fig that contains gall flowers, she will be able to insert her ovipositor succesfully into the flower. There she lays an egg and adds a chemical substance that induces formation of a gall, which her larva consumes.

When the bizarre, wingless male progeny mature, they gnaw their way out of the gall in which they have developed and proceed to search for galls containing encapsulated females. Upon locating one, the male chews an opening in the gall, inserts his abdomen through the opening and copulates with the female in her chamber. Only after mating does the female exit. She then collects pollen from the male flowers and departs through a tunnel chewed by males to search for a new fig in which to replay this drama (Fig. 7.5).

Phenomena of this sort are not restricted solely to wasps. A butterfly analog is *Heliconius charitonius,* whose males perch upon the female chrysalis and attempt to penetrate the pupal case with their genitalia. Some males may copulate with the pharate female (one that has completed development but not yet emerged; see L. E. Gilbert, 1976). In the crabhole mosquito *Deinocerites cancer,* females lay their eggs in the water in the deep burrows of crabs, where they undergo transformation into larvae, pupae, and then adults. Mature males wait in the darkness on the surface of the water to detect olfactory cues given off by pupae containing individuals near eclosion (see Fig. 7.1; Provost and Haeger, 1967). Males approach and grasp these pupae by their pupal horns (the respiratory tube that breaks through the water's surface) and they play their long, filamentous antennae on the captured pupa (presumably to determine the species, sex, or age of the individual within). A male will remain with a pharate female and when the pupal skin breaks preparatory to her emergence, he attempts to copulate with her, sometimes successfully. Males of another mosquito with similar characteristics, *Opifex fuscus,* search visually for pupae in the coastal rock pools of New Zealand. The searchers go so far as to stick their heads underwater to scan for potential "premates," which they also capture and hold by a pupal horn (Haeger and Provost, 1965).

The eagerness of male fig wasps, crabhole mosquitoes, and *Heliconius* butterflies to mate with females even before they have fully emerged is exceeded by a species of thrips (a minute plant-sucking insect). Mature males of *Limothrips denticornis* occur in association with groups of prepupal females on grass stems (thrips pass through larval, prepupal, and pupal stages before metamorphosing into active adults). Lewis (1973) reports that female prepupae already have well-developed spermathecae containing viable spermatozoa, indicating that they mate while still immature in every other way.

Multiple Mating by Females and Mate Location in Oviposition Sites

We have argued that single mating and early receptivity by adult females create conditions that favor males which search emergence areas. The converse of this hypothesis is that in species whose females

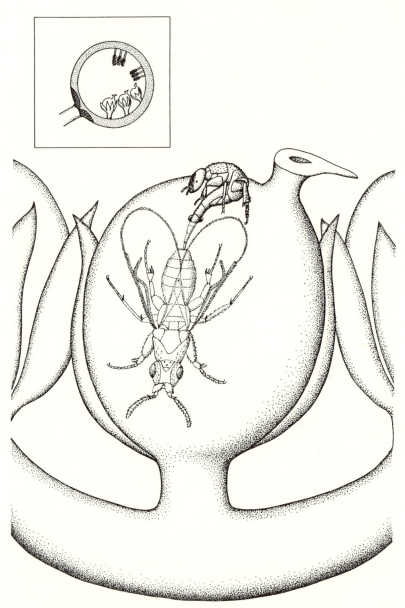

7.5 Wingless males of some fig wasps mate with virgin females before they have emerged from their fig brood chambers. (Drawing by M. H. Stewart.)

copulate multiply, we expect males to search for mates in localities where females oviposit. We have remarked that multiple-mating females tend to use the sperm of their most recent mate when they fertilize their eggs. If sperm precedence occurs, a male gains by mating with a female just before she lays her eggs. He should, therefore, search at or near preferred oviposition sites, particularly if such areas are limited in size or number and are readily identified. This ecological factor will promote clumping of receptive females, which rewards males that search in oviposition areas.

The insect order whose members most consistently mate at oviposition sites is the Odonata. This taxon is characterized by females that mate just prior to or during egg laying. Although sperm precedence has been documented for only one species (Waage, 1979a), it almost certainly is widespread in the group (Parker, 1970c; Waage, forthcoming). Because female odonates oviposit in submerged aquatic vegetation and often have reasonably restricted habitat requirements, males can identify specific locations that are likely to attract relatively large numbers of receptive females. Attempts to mate with emerging females would yield little genetic gain in species in which sperm precedence occurred, because of the likelihood that the female would mate again prior to her first oviposition bout. Likewise efforts to find and mate with foraging females would be nonproductive, not only because of the high probability of sperm displacement but also because females searching for food away from water are usually more widely dispersed and more difficult to locate than gravid females near discrete oviposition sites.

The same general argument applies to insects like the dung fly *Scatophaga stercoraria,* various orthopterans, and some weevils whose females alternate bouts of oviposition with periods of feeding. Here too, receptive, about-to-oviposit females may be clustered on limited suitable habitats (such as a fresh mound of dung or the fallen limb of an appropriate tree). Because the females mate multiply and may use the last male's sperm, mate location elsewhere is disadvantageous to males.

Males whose females mate just once may also hunt for mates at oviposition sites if their females simply refuse to copulate until they are ready to begin nesting or egg laying. This could be advantageous if male activity makes nest construction easier or egg laying safer or more productive. For example, some bark beetle females might prefer to mate with a male that had excavated a nuptial chamber in wood rather than one that met them as they emerged, because the nuptial chamber could be used as a safe starting point for construction of egg laying galleries. The same point applies to those crickets in which males dig a burrow that their mates can use as an oviposition site.

In other species "continuous receptivity" by females may create the social environment that makes it genetically advantageous for males to remain in association with a female at her nest; this in turn enables the female to secure the benefits of having a nest guard to repel parasitic intruders and other nest enemies while she is foraging for brood provisions (Colville and Colville, 1980; Hook and Matthews, 1980). In several species of *Oxybelus* (Fig. 7.6) and *Trypoxylon*, males search for nests that are not guarded by rivals (Peckham, Kurczewski, and Peckham, 1973). In the case of *T. politum*, for example, the female constructs an elongate mud tube in which a number of cells will be provisioned with spiders (Cross, Stith, and Bauman, 1975). Males search for nests that are in the process of construction. If the tube is unclaimed by another male, the finder will remain with it until the nest is completed. He guards the nest entrance while the female collects mud for tube construction and then spiders for her brood. Generally the male copulates with the female each time she returns to the nest, presumably supplanting any sperm that she may

7.6   Nest-guarding males of *Oxybelus sericeus* copulate with the nest owner each time she returns with prey to provision her burrow. (From Hook and Matthews, 1980.)

have received while away from the area. Colville and Colville (1980) have shown for a closely related species that the presence of a male reduces the incidence of parasitism on the nest and, in fact, females do not begin provisioning until a male is in residence. Similarly in *O. subulatus* the male guard repels conspecific males and certain nest parasites, while copulating with the female each time she returns to the nest with prey (Peckham, 1977).

The Male Search for Females Clustered in Foraging Areas

Given the numerous advantages of contacting females either immediately after they have emerged or just before egg laying, why do males of some species search at foraging areas for females that have already emerged but are not necessarily on the verge of laying their eggs? Selection may favor males that search for feeding females if their rate of encounter with receptive females is much higher at the foraging location than it would be at either the emergence or egg laying locales. Relatively high rates of mating with foraging females can occur if food resources are clumped or easily identified relative to emergence or oviposition sites.

There are many solitary bees whose males search for emerging females, when the nests containing these individuals are numerous and clumped, and an equally great number of cases in which males find mates at flowers, when emerging females are scattered. Members of the same genus may differ in mate location tactics in accordance with the degree of nest clumping (Table 7.6).

There are exceptions, however, to this pattern. Females of the megachilid bee *Hoplitis anthocopoides* sometimes form nesting aggregations on rock faces, yet males do not search for females at these sites, despite the fact that newly emerged virgins are receptive and will mate only once in their lives. Eickwort (1977) notes, however, that nest clusters are probably an evolutionary novelty in this species, the result of human agricultural practices that maintain stable, long-lasting populations of *Echium vulgare*, the sole food plant of the bee. *E. vulgare* is an early-successional species, which normally would appear briefly in a disturbed site and then disappear, favoring bees with a high capacity for dispersal. If we accept this argument, it is likely that in the past, cells containing virgin females were scarce and scattered. If male behavior was shaped under these ecological conditions, the failure of modern males to search for emergence sites becomes more understandable. Instead, males patrol one or a few plants of *E.*

Table 7.6   Selected examples of the relation between opportunities for multiple matings at emergence sites and the mate locating behavior of male bees.

Nests are scattered[a]; virgin females emerge in isolation from one another. Males search for females at flowers.

Colletidae
    Ptiloglossa jonesi   Cazier and Linsley, 1963
    Caupolicana yarrowi   Hurd and Linsley, 1975

Oxaeidae
    Protoxaea gloriosa   Cazier and Linsley, 1963

Megachilidae
    Anthidium spp.   Haas, 1960; Jaycox, 1967; Alcock, Eickwort, and Eickwort, 1977; Severinghaus, Kurtak, and Eickwort, 1981

    Anthidiellum spp.   Turell, 1976
    Callanthidium illustre   Alcock, 1978

Andrenidae
    Perdita texana   Barrows et al., 1976
    Nomadopsis puellae   Rozen, 1958

Anthophoridae
    Centris cockerelli   Hurd and Linsley, 1975
    Triepeolus sp.   Alcock, 1978
    Peponapis pruinosa   Matthewson, 1968

Halictidae
    Augochlorella endentata   Eickwort and Eickwort, 1973

Melittidae
    Meganomia sp.   Rozen, 1977

Nests are aggregated[b]; receptive females emerge in numbers within a relatively small area. Males search for females in the emergence area.

Colletidae
    Ptiloglossa guinnae   Roberts, 1971
    Colletes cunicularis   Bergström and Tengö, 1978

Andrenidae
    Calliopsis andreniformis Shinn, 1967
    Andrena erythronii   Michener and Rettenmeyer, 1956
    Melitturga clavicornis   Rozen, 1965
    Nomadopsis anthidius   Rozen, 1958

Table 7.6   (*continued*)

Anthophoridae	
Centris pallida	Alcock, Jones, and Buchmann, 1976
Tetralonia minuta	Rozen, 1969
Emphoropsis pallida	Bohart et al., 1972
Halictidae	
Nomia melanderi	Bohart and Cross, 1955
Melittidae	
Meganomia binghami	Rozen, 1977

[a] Fewer than 10 nests in close proximity to one another, or nests not located by observer, suggesting a secretive, isolated nesting pattern.
[b] More than 10 nests, and generally many dozens or hundreds within view of an observer at one time.

vulgare and may meet virgin females drawn to the plant on one of their early flights after emergence.

There are other examples of males that, by controlling a food resource, gain access to dispersed receptive females. Within the Mecoptera, females broadcast their eggs widely through their environment, often simply dropping them on the forest floor more or less at random (Byers and Thornhill, 1983). The task of locating emerging or ovipositing females becomes difficult under these circumstances. We have already mentioned that in some species of *Panorpa* males find and advertise a dead insect with an attractant pheromone. In these and other species in which the males practice nuptial feeding, females presumably benefit from refusing to copulate until they have found a male with a food gift. By mating at the feeding site, and not upon emergence or at oviposition, the female gains a resource "for free." Thus as soon as some males begin to offer nuptial gifts, selection acts in a positive feedback cycle to favor females that mate only when fed and males that use food to lure females to them.

Mating at Landmarks When Females Are Consistently Dispersed

Although in most species receptive females are likely to be clustered at an emergence or oviposition site or foraging resource, in some insects females are consistently dispersed. In still others, females may be clustered in certain areas but are not receptive at this phase of their life cycle. If the resources that can be exploited by egg laying and feeding females are widely dispersed, it follows that females will evolve the capacity to travel long distances in search of suitable nesting or foraging areas. In a species with highly mobile females with large home ranges, males may be forced either to travel widely themselves (an energetically expensive option) or to wait at a single location in hopes of spotting a passing female.

Landmarks provide some advantages as good waiting points. A clearing in a tropical forest offers a male orchid bee a less obstructed view of his surroundings than he would have in the depths of the forest (Kimsey, 1980). A tall weed or a perch on a hilltop elevates a male above obstacles and so gives him a broader field of vision. Moreover, wide-ranging females may tend to visit the prominent features in their environment as a means of learning landmarks to help orient their travels. If for any of these reasons, more matings take place at a landmark than at other kinds of locations, selection may favor females that travel to these points in order to reduce the interval between emergence and mating. Males would also be favored that gather in areas where receptive females are most numerous, eventually establishing an encounter site convention of the sort envisioned by Parker (1978b).

The advantage of reducing the prereproductive period of adult life would be especially strong in rare species whose members were few and scattered. Lepidopterists have long recognized that prime locales to collect relatively rare butterflies are hilltops or mountain peaks (O. Shields, 1967; Scott, 1968). The same relation between rarity (and therefore a probable high degree of dispersion of males and females) and the use of hilltops as encounter sites occurs in some bees, wasps, and flies (Alcock et al., 1978; Alcock, 1979c). Bot flies of the families Oestridae, Cuterebridae, and Gasterophilidae (Catts, 1964, 1967, 1979, 1982) provide classic examples of rare, dispersed insects that use hilltop mating sites (Fig. 7.7). These insects are scarce in collections, which indicates that their population densities are low. The adult flies emerge from hosts, such as woodrats or deer, that are themselves often widely distributed through the environment. The fly *Gasterophilus intestinalis* victimizes horses, which do occur in groups, but the adult undergoes a lengthy pupation in horse dung and could easily emerge to find no hosts in the immediate vicinity (under natural conditions wild horses are wide-ranging animals). If no hosts are evident, females may choose hilltop encounter sites in order to be inseminated promptly; if they were to search for a horse without first having mated, they might find a host with no male flies in attendance and thus could not take reproductive advantage of their discovery.

7.7   Hilltopping insects. (*Above*) A mountain ridge in southern Arizona. The palo verde tree that grows on the highest point of the ridge is the focus of intraspecific competition among territorial males of many insects, including (*below left*) the bot fly *Cephenemyia jellisoni* and (*below right*) the hairstreak butterfly *Atlides halesus*. (Photographs by J. Alcock.)

The same relation between a parasitic mode of existence and a landmark-based mating system occurs in other flies, including those that attack singing crickets (Lederhouse et al., 1976). An interesting exception is the bombyliid *Heterostylum robustum*. This bee fly parasitizes the alkali bee, which forms huge aggregations. In one nesting location judged to contain a half-million bees, Bohart, Stephen, and Eppley (1960) observed male bee flies searching for and finding emerging females in the nesting area. Although rare in number relative to the bees, the flies were sufficiently abundant to make mate location in the emergence site a profitable option for males.

Further evidence that waiting at landmarks is a default strategy (Bradbury, 1981) of males unable to find clusters of receptive females elsewhere is provided by the ladybird beetles (Fig. 7.8). Those species that aggregate to mate at mountain peaks feed on ephemeral populations of winged aphids rather than on nonmigratory aphid colonies and other sessile homopterans (Hagen, 1962). Because of the transient nature of their prey populations these ladybird beetles must be highly mobile; landmark mating may evolve because of the difficulty males have in locating areas with a concentration of females. In contrast, males of those ladybird species whose females remain in prolonged association with a long-lived colony of aphids mate on the plants where the aphids are found.

Landmark Encounter Sites in Common Species

Not all insects that employ landmarks as mating sites are unusually rare or highly dispersed. Many of the midges, mosquitoes, and other swarming flies are relatively abundant, yet rely on characteristic environmental markers as a cue for swarm formation and mating encounters. Parker's (1978b) suggestion that males waiting at the edge of a resource area perhaps gain a reproductive advantage because of a female preference for rapid mating may apply to many of these species. Consider the case of the otitid fly, *Physiphora demandata* (Alcock and Pyle, 1979). This is a widespread and abundant insect whose multiple-mating females oviposit in a variety of rotting organic materials, including animal dung. Males do not seem to contact females at the oviposition area, but instead defend territorial perch sites on exposed twigs and stems of prominent vegetation near oviposition sites. Perhaps because females are so catholic in their selection of oviposition substrates, they can avoid male suitors that attempt to monopolize egg laying sites simply by moving to one of the many alternative areas. Under these conditions males may be "forced" to wait near oviposition areas for receptive females to come to them. Male competition for superior perches may then make territorial ownership a prerequisite if a male is to gain access to a receptive female.

7.8 An aggregation of ladybird beetles clustered on a small tree trunk. Mating occurred at the site. (Photograph by J. Alcock.)

The Defense of Encounter Sites

Parker's arguments (1978b) about where in their environment males should search are based on the assumption that males can distribute themselves freely. This assumption, as Parker is aware, does not apply to all insect species. For example, a territorial male of *P. demandata* does not tolerate the presence of another male on the 10 to 15 cm segment of the twig or stem on which he perches. If another male alights, the resident quickly advances, fore legs waving. If the intruder does not decamp at once, the two flies initiate a fight that may involve release of chemical signals, proboscis touching, and finally outright wrestling in which they seek to throw or chase their rival from the perch (Alcock and Pyle, 1979).

In the insects, aggression by males is not uncommon. The fighting male may gain the exclusive right to mate with females that come under his control (the benefit), but his aggressiveness also carries with it (1) the risk of death or injury at the hands (or jaws, or legs, or wings) of a competitor, (2) the risk of death due to increased exposure to parasites, predators, or conspecific cannibals (Gwynne and O'Neill, 1980; Cade, 1980, 1981a; O'Neill and Evans, 1981; Burk, 1982), (3) the risk of injuring a genetically related individual and so reducing the chance that this relative will transmit genes shared by descent (see van den Assem, Gijswijt, and Nubel, 1980, who suggest that the intensity and frequency of aggression among male chalcid wasps rises with the likelihood that males are in sexual competition with *unrelated* males), and (4) the loss of time and energy that would otherwise be allocated to the search for mates.

Although death and injury are dramatic disadvantages of aggression, the fourth category of costs may be the most significant in terms of natural selection. Males of the digger bee *C. pallida* fighting for a digging site occasionally miss the emerging female altogether, for she can crawl out and fly off while they wrestle on the ground. T. J. Walker (1980) suggests that because the calling time of the cricket *Anurogryllus arboreus* is limited to 30 minutes each evening, the time is too valuable to be spent in disputes with other males for a territory site on a tree trunk (see also Doolan, 1981).

Just as important as time are the energy expenses of fighting. So many calories are required to fuel the patrolling and repelling flights of the territorial dragonfly *Plathemis lydia* that males at best can claim a territory for a few hours of their life (Campanella and Wolf, 1974). Male bot flies of the genus *Cuterebra* fly at extremely impressive speeds in their territorial contests (although not at the hundreds of miles per hour once attributed to them). Catts (1967) has shown that tethered males, while robust in appearance, cannot sustain more than 3 minutes of continuous flight. They hold their territories for a very short period, then are forced to relinquish the area to a rival. Because there are disadvantages as well as gains to be made from aggressive attempts to monopolize access to females, male insects should not employ this strategy except in situations where the benefits are likely to exceed the costs.

The Economics of Territorial Defense

A major principle of ethology, derived primarily from studies of birds and mammals, is that individuals aggressively defend a space only when the territory is relatively small and contains a resource in limited supply (Brown and Orians, 1970; Carpenter and MacMillen, 1976). The evolutionary logic of this rule is that in general the smaller the defended area, the lower the cost of territorial patrolling and repulsion of invaders. But a territorial male can gain relative to his rivals only if the area he owns has or attracts more than its share of a desired resource, such as receptive females. If a nonterritorial male can find as many receptive females as a territorial male, the nonaggressive male enjoys a selective advantage, for he does not pay the costs of fighting and so is likely to live longer and reproduce more than his territorial competitors.

The territories of male insects usually are small, as expected, and either contain one or more receptive females or have the capacity to attract numbers of mates. Males of C. *pallida*, for example, defend a digging site that is only a few centimeters square but is likely to contain a valuable virgin female who will fertilize all her eggs with her mate's sperm. The costs of territorial defense for this species are relatively low because the investments of time, energy, and risk are slight (the defended area is guarded for only a few minutes, it is so small that few other males can find it, and the risk of predation while digging is almost nil). An alternative strategy of defending a large portion of an emergence site against all males could yield more than one mate, if the site could be protected for several hours or days. But males make no effort to defend a sizable patch, perhaps partially because it is difficult to determine exactly which location will produce the most emerging females on any given day. Also male density is sufficiently high that the defender of a large territory would spend all his time and energy trying to drive agile and fast-flying rivals from his turf.

Alcock and his colleagues (1978) reviewed 36 records of territorial bees and sphecid wasps; there were no cases of individual males defending an area of more than 10 m². For the large majority, the territory was less than 2 m². This was true for both landmark and resource-based territories. Defended perch sites at landmarks do not themselves contain nesting, emerging, or foraging females, but their value may consist in either providing a superior vantage point for the scanning male, as suggested earlier, or helping the male appear more conspicuous to females, as Kimsey (1980) argues for the smooth-barked trees preferred by certain male orchid bees. Given the rarity of observed matings for many of the landmark species, it is often difficult to demonstrate that males holding peaktop or other prominent encounter points do achieve greater reproductive success than males excluded from these areas.

The picture is clearer for some other kinds of territoriality by male insects. Most defended locations contain one or more receptive females that are nesting or emerging, or else the area contains a clumped food resource that attracts numbers of receptive females. Thus, for example, male wasps belonging to the genera *Sphecius, Trigonopsis, Epsilon,* and *Euodynerus* (Fig. 7.9) all defend territories ranging from 0.5 to 10 m² that contain clusters of emerging receptive virgins (Lin, 1963; Eberhard, 1974; Smith and Alcock, 1980; D. P. Cowan, 1981). In each case there are visible cues that enable a male to identify limited areas from which relatively large numbers of potential mates will emerge. Thus male aggression can be economically centered on defensible territories of high productivity.

Territoriality that is limited to a very small area is widespread among various parasitic wasps and fig wasps whose males guard a patch of a few square centimenters. A male wasp that controls the small area about a stink bug egg mass or mantis egg case will have about 15 females in his territory (Eberhard, 1975; Grissell and Good-pasture, 1981). In the pteromalid wasp *Nasonia vitripennis* a male that has just mated will mark the spot with a pheromone, perhaps to keep track of potentially productive places in his environment and to focus his defense on these points (van den Assem, Jachmann, and Simbolotti, 1980).

Male insects that defend foraging or egg laying resources which attract females also generally protect relatively small areas against intruders. Single male tephritid flies of some species of *Rhagoletis* guard individual walnuts (Boyce, 1934; Bush, 1966). Females lay their eggs exclusively in walnut husks (the larvae are specialist feeders); damaged fruits are the preferred territories of males, because they are the preferred oviposition sites of females. After egg laying the females be-

7.9 Males of cicada killer wasps defend territories several meters square at sites with an above-average number of emergence holes. Because males emerge from the soil before females, locations with many emergence openings eventually will yield a relatively large number of additional wasps, most of which will be receptive females. (Drawing by M. H. Stewart.)

come receptive, conferring a reproductive advantage to the territory owner of a walnut that attracts gravid females.

Males of the cactus fly *Odontoloxozus longicornis* hold territories of a few square centimeters about an oozing crack in a saguaro cactus. These rare and scattered cracks attract females about to oviposit. Mating occurs after egg laying, and the possessor of a superior location acquires a small harem of females. Females that oviposit in a territory never reject the male owner. In fact, females usually refuse to mate with nonterritorial males that court them away from the oviposition points at feeding sites on the cactus (Mangan, 1979).

Insect territories are not always tiny. For example, Campanella (1975) reports a mean territory size of 250 m² for the dragonfly *Libellula luctuosa*. He notes, however, that one male's territory may be occupied also by a number of subordinates, which are not chased away by the dominant resident male. Moreover, territory owners generally restrict their defense to a smaller core area, only winning a large majority of their chases when the intruder comes within 5 m of a central point. Even so, the defended area is still substantial in this and other odonates (see, for examples, Waage, 1973; Ubukata, 1975; Pazella, 1979).

The high degree of mobility of dragonflies may permit these insects to patrol large areas relatively economically. Nevertheless, male odonates do not defend territories at all unless they are likely to increase encounters with females. Many damselflies cannot easily identify an unusually productive oviposition site because the ovipositional resource—algal mats, for instance—are widely and evenly distributed in an aquatic habitat. If males cannot predict where females are likely to lay their eggs, the costs of territorial defense will outweigh the reproductive gains associated with territorial possession. Males of *Enallagma civile* exhibit little or no aggression at a pond with a uniform distribution of egg laying substrates (Bick and Bick, 1963). By contrast, males of *Calopteryx maculata* are intensely aggressive in their defense of several square meters of stream that happen to contain concentrations of plant rootlets in shallow water. This oviposition material is patchily distributed, and superior sites can be identified visually by both males and females. This permits males to find and defend locations where they will have above-average chances to mate (Waage, 1973).

The Timing of Territoriality

The time of day that male insects exhibit territoriality is a function of the temporal availability of mates. Males of *Calopteryx maculata* guard their streambank territories only in the middle of the day, when females come to the stream to oviposit. It is common in the odonates for there to be a temporal clustering of receptive females during midday, and male territorial aggression reaches its peak during this time (see Fig. 4.10).

Among those Hymenoptera in which females mate upon emergence, the only territorial behavior males exhibit is often during a few hours in the morning when emergence is occurring (see Fig. 4.9). The desert bee *Protoxaea gloriosa* provides a dramatic example of a species whose males switch from territorial to nonterritorial behavior in the course of the day. In the midmorning, males vigorously drive intruders from their territories centered about a patch of flowers or a flowering shrub. Female emergence occurs during this time and receptive females sometimes visit flowers and mate there. After the morning emergence time has passed, males leave their territories and by midafternoon actually aggregate by the dozens to the hundreds, forming a compact "sleeping cluster" in the desert shrubbery (Fig. 7.10). They spend the night in completely nonaggressive association with their competitors, departing early the next morning to establish their territories anew (Hurd and Linsley, 1976).

In contrast to *P. gloriosa,* males of the bee *Anthidium maculosum* defend a patch of flowering food plant throughout the entire day. Foraging females are continuously receptive in this species and the foraging period extends virtually from dawn to dusk. As a result, there is always the chance of encountering receptive females at flowers and therefore always the possibility of a reproductive payoff for a territory holder during the daylight hours (Alcock, Eickwort, and Eickwort, 1977).

The Influence of Male Density on Territorial Behavior

Male insects not only act in ways that tend to increase the gains to be expected from territorial behavior but also attempt to lower the costs of aggression by reducing the time and energy spent on territory maintenance. For example, as male density increases the intrusion rate will rise, and with it the costs of repelling intruders from a territory. One way to cope with these increased costs is to reduce the size of the defended area—or, if the costs continue to rise, to abandon territorial defense of the area altogether.

Male insects, under the appropriate circumstances, appear to do both. In the Japanese dragonfly *Cordulia aenea,* Ubukata (1975) found that when one or two males were present at a pond they patrolled 20 to 50 m of shoreline. As the male population increased to eight males, the size of the territory fell to 15 m and then to 10 m when more than twelve males were present. By reducing the area they defended, the males lowered the time and energetic expenses of repelling intruders (see also Pazella, 1979). The same relation between male density and territory size has been established for a megachilid bee, *Hoplitis anthocopoides* (Eickwort, 1977). At the onset of the

7.10 The bee *Protoxaea gloriosa* alternates between territorial and nonterritorial behavior. (*Above*) A male hovering above the site from which he will repel all male intruders. (*Below*) A group of males in a peaceful sleeping cluster, which forms in the afternoon after the daily mating and territorial period has ended. (Photographs by J. Alcock.)

7.11   Seasonal variation in territory size in the megachilid bee *Hoplitis antho-copoides*. (*A* and *B*) Territories are large at the beginning of the flight season, when few competitors are present and few food plants are in flower. (*C* and *D*) Territories shrink at the height of the flight season when male density is high, and then (*E* and *F*) increase again at the end of the flight season when density falls. (From Eickwort, 1977.)

flight season when males were few, individuals patrolled areas of 10 to 12 m^2 that contained the food plant visited by females. But as the number of males emerging grew, and with it the number of intruders at territories, the defended areas shrank in size until they enclosed only a single flower plant (0.2 m^2). As the flight season drew to a close and the male population decreased, territories began to expand again, eventually reaching the dimensions of the plots defended early in the mating period (Fig. 7.11).

The complete breakdown of territorial behavior under conditions of very high density has been observed in the wasp *Philanthus zebratus* (Evans and O'Neill, 1978). Males in one population formed an aerial swarm above the nesting site, where they contacted females flying to and from their nests, occasionally capturing a receptive individual. In more diffuse "colonies" males defended small perching areas on the ground in the nesting/emergence site.

Grasshoppers provide additional examples of species whose males have the capacity to adopt or abandon territorial defense in relation to the costs of dealing with intruders. Otte and Joern (1975) observed a

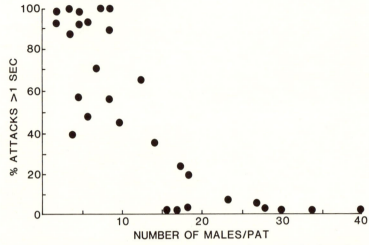

7.12   The relation between territoriality and male density in the dung fly *Scatophaga stercoraria*. The percentage of prolonged attacks (by territorial males on intruders) drops as male density increases. When there are about 20 males per dung pat, the flies no longer attempt to hold territories. (From Borgia, 1980.)

dense population of *Bootettix argentatus* in which males made no effort to defend individual creosote bushes. An average of four or five males occupied each bush without aggression. By contrast, in a low-density population in which there was an average of only one individual per four bushes, males were territorial and attacked intruders that trespassed on their bush (Schowalter and Whitford, 1979).

Finally, Borgia (1980) has also demonstrated that dung flies are territorial at low but not at high densities. When there are few competitors, males attempt to exclude others from a portion of a dung pat and enjoy exclusive access to females drawn to the dung. But when the intruder rate rises, males largely ignore one another and instead scramble to be first to grasp a female arriving at the pat (Fig. 7.12).

Avoidance of Damaging Combat

Fighting males obviously run the risk of being injured or killed. Inasmuch as death greatly reduces a male's opportunities to copulate, selection has favored males that can assess accurately their chance of defeating a rival. This is often done through the use of threat displays that enable individuals to resolve conflicts without strenuous fighting or sometimes without even making physical contact. Males of a South African dragonfly have laterally expanded red and white tibiae, which they employ in a dramatic flash display that settles many territorial disputes (Robertson, 1982). A male of the digger bee *C. pallida* sometimes need only turn and face a rival, with head raised and legs spread wide to send the other male on his way (Fig. 7.13; Alcock, Jones, and

7.13 (*Above*) Visual threat display by a male of the bee *Centris pallida*. By spreading his legs wide and elevating his body, a large male may deter a small one from attempting to gain control of a digging site containing a virgin female. (Photograph by J. Alcock.) (*Below*) Acoustical threat by a male field cricket. The male on the right is stridulating at his opponent. (From Alexander, 1961.)

Buchmann, 1976). Even the tiny pteromalid wasp *Nasonia vitripennis* has a visual threat display in which a male confronting an opponent spreads his mandibles wide and elevates his wings (King, Askew, and Sanger, 1969). Acoustic signals, as well as visual ones, may serve as effective threats. In most singing grasshoppers, crickets, and katydids, the sound of a male's aggressive call is generally sufficient to keep another male from approaching too closely. Pairs of rival males of an oedopine grasshopper studied by Loher and Chandrashekaran (1970) also engage in alternating bouts of stamping on the ground, threatening one another via substrate vibrations, but not actually touching at this stage of a contest. It is possible that olfactory signals constitute threats in some species such as the Hawaiian *Drosophila* (Spieth, 1968) and the otitid fly *Physiphora demandata* (Alcock and Pyle, 1979); territory owners may confront intruders by raising their abdomen and producing a bubble of anal fluid whose scent may contribute to their defense of the site.

The threat displays of insects generally convey information about a male's size and vigor. Males of the digger bee *C. pallida* vary greatly in weight and size; by rearing up and spreading his forelegs, a male demonstrates his size to his opponent. If one bee judges the other to be significantly larger, it will be to his advantage to depart without fighting. When fights do occur, invariably the larger male throws his opponent on his back and generally thwarts his movements (Alcock, 1979d). Because the probability that a smaller male can force a larger one to leave his digging site is so low, small males can save time and energy (and reduce the risk of injury) by conceding defeat without contact if they determine that their opponent is larger than they are.

The large-male advantage is common among aggressive insects (see Table 9.1 further on) and it is in these species that threatening postures are often used to resolve conflicts. Among some highly territorial Hawaiian *Drosophila,* competing males approach each other and stand head to head (see Fig. 3.12). Spieth (1981) has suggested that the enlarged heads of males of certain species have evolved as communicators of body size. He has observed smaller individuals avoiding physical combat after having engaged in the preliminary face-off. The culmination of the evolutionary thrust in favor of large heads is represented in the bizarre eyestalked flies; members of eight families of Diptera have evolved this feature (D. K. McAlpine, 1979). Some males of the eyestalked platystomatid *Achias australis* are one-fifth the size of others. In conflicts at courtship locations two opponents will approach, then strike each other with the forelegs while holding their heads close together. In this position the eyestalks are parallel to

7.14 A fly with eyestalks. These peculiar structures may have evolved in the context of threat displays as size indicators for an opponent. (Photograph by E. S. Ross.)

one another, enabling each male to assess the head width of the other. Because head width is strongly correlated with thoracic length (and thus body size), a male can quickly determine the size of his opponent (Fig. 7.14). McAlpine has found that most fights last no longer than 6 seconds.

Just as important as size in male combat is the question of who had possession of the disputed resource at the onset of the encounter. The original owner usually enjoys an advantage. For example, in the cricket *Teleogryllus oceanicus* territorial males under laboratory conditions won 84 percent of 1,205 fights with intruders (Burk, forthcoming). Why should a territory holder be challenged so feebly in the vast majority of cases? One explanation may have reasonable broad application in those insects with relatively long-lived males. Presumably it takes time to discover a vacant territory, or one so weakly defended that it can be easily taken from its original owner. Therefore a resident male, having spent the days required to find and acquire a territory, will usually be older than an intruder searching for a territory. If the older resident concedes to the newcomer, he will have to invest considerable time once again to find another territory vulnerable to takeover, at which point he will be older still. Perhaps past his prime, the territory seeker may be unable to fight (or search) effectively and so may never again acquire a territory. If the resident is ousted, then his reproductive chances decline dramatically.

On the other hand, if the young intruder does not press his challenge but leaves to search for a vacant or vulnerable territory, he still has a fair probability of encountering a suitable location that he can acquire economically. If he stays to fight, he will have to cope with a territory holder who can afford to sacrifice a great deal in his effort to retain his site, because the resident will lose so much should the intruder win (Parker, 1974a).

An alternative explanation for the "prior-possession advantage" has been advanced by Maynard Smith and others (Maynard Smith and Price, 1973; Maynard Smith, 1976a). They argue, on the basis of game theory, that it is mathematically possible for *any* difference in the attributes of two opponents to be used *arbitrarily* as the means for settling disputes (Maynard Smith and Parker, 1976). Such a system of conflict resolution theoretically is evolutionarily stable and could be adaptive in situations where fighting is costly and it is difficult for males to assess the combat ability of rivals. An asymmetry exists between two males if one holds a territory and the other does not. It is possible that this attribute alone, which need have no correlation with resource holding power or fighting ability, could be used as the cue that resolves disputes over ownership of a resource.

A species in which the resident male *always* wins is the speckled wood butterfly *Pararge aegeria* (Davies, 1978, 1979). Males of this species claim sunspots on the forest floor as mating territories, shifting their perches as the spots move slowly across the ground with the movement of the sun overhead. Males without territories search for sunspots but never strongly challenge a male that occupies one. Maynard Smith points to this case as a possible example of the use of an "uncorrelated asymmetry" to resolve territorial claims. Because the resident is never deposed, it looks as if the butterflies are simply programmed to concede defeat if confronted by a territory owner, regardless of which individual is the superior fighter.

Critics of the uncorrelated asymmetry hypothesis argue that speckled wood butterflies that have found a sunspot are more likely to be superior resource holders than later-arriving searchers, and that therefore the difference in territorial status is correlated with fighting ability (Austad, Jones, and Waser, 1979). The ability of the residents to get to a sunspot before their rivals indicates that they are either more skillful searchers or that they possessed sufficient energy reserves to enable them to begin searching earlier than their competitors. Either trait is plausibly linked to aerial agility or flight endurance. Alternatively, the male perched in the sunspot may be better rested or closer to the thermal optimum for the species and therefore capable of more agile or more sustained flight than the intruder. Flight ability is likely to resolve any long-term dispute among butterflies, because the male that can exhaust his opponent can claim a perch territory without fear of interference from the rival male. If an intruder is likely to be bested in the long run, it pays him to save his time and energy and to invest it in the search for an unoccupied sunspot.

The point is that quick acceptance of the superiority of a resident male need not be an arbitrary decision by the rival male, but it may be one that reflects an accurate assessment of the probable fighting ability of the two individuals. Moreover, the rapid resolution of male-male encounters by speckled wood butterflies occurs in part because of the relatively low value of any given territory. Sunspots are numerous on sunny days, and therefore the searching male is likely to benefit by avoiding a costly prolonged dispute with a territory owner for something that he should find elsewhere with relatively little effort.

There are other factors besides size and prior possession of the contested resource that can influence who is likely to win a dispute. These include such things as the past experience of a competing male (has he won recently or lost?) and the frequency with which he has copulated (as shown for crickets by Alexander, 1961). Breed and colleagues (1981) have also found that in the case of cockroaches the male that initiates a conflict is more likely to be successful than the responder. These and doubtless many other attributes create differences between contestants that play a role in settling fights among male insects promptly and with a minimum of physical contact with its attendant risk of exhaustion, injury, and death.

Escalated Conflicts

Not all disputes among male insects are resolved quickly and painlessly. Threat signaling may give way to an escalated series of actions by an intruder and a territory holder, leading to an ever more violent confrontation. Although, for example, the vast majority of interactions between two male field crickets end quickly, males can do more than merely threaten each other acoustically. They may move together to jostle, kick, or butt their enemy. Either individual may withdraw at any point, or escalate the conflict still further until reaching the most serious level. At this stage the two males will grasp each other's mandibles to form a "jaw lock." Each then struggles to throw the other neatly on his back. When this happens, the outwrestled male withdraws (Alexander, 1961).

Heightened and prolonged aggression generally occurs when two opponents are relatively evenly matched in size, experience, and the other components that determine fighting ability and motivation (Alexander, 1961; Breed and Rasmussen, 1980; Sigurjónsdóttir and Parker, 1981). For example, Davies (1978) was able to induce his speckled wood butterflies to fight with one another by removing the territory owner, permitting another male to enter and occupy the territory, then after a while releasing the original resident male. The new resident, possibly having acquired the psychological advantage derived from the experience of territory *possession,* fought vigorously with the original owner, who was motivated by his past experience of owning the territory. The two males engaged in an upward-spiraling flight, rising several meters in the air, descending, and flying up again. Similiar contests have been observed in a number of other territorial butterflies (see Fig. 7.15 and R. R. Baker, 1972; Alcock, unpublished). The rapid vertical ascents required by these maneuvers presumably are physically demanding for the participants. By observing the reaction of an opponent, a male can assess his rival's capacity for swift, sustained flight. Closely matched males may require many tests before it becomes clear who is in better condition, at which point one male withdraws. Fighting butterflies generally make little contact with each other (but see Eff, 1962), perhaps because wrestling, bumping, and hitting do not provide information relevant to their disputes. In a spiral ascending flight a male may be able to determine whether he would outrace his competitor to a passing female if both were perched nearby. The slower male gains nothing by remaining and therefore should abandon the dispute and the area.

Contests of endurance and speed rather than physical strength have evolved in insects other than butterflies. On the same mountaintops in central Arizona where various butterflies engage in ascending spiral flights for possession of scanning perch territories, there are territorial tarantula hawk wasps and bot flies. These insects also employ ritualized pursuit flights to determine ownership of desirable perching areas. Most interactions are resolved in favor of the resident after one or two chases of less than a minute's duration, but determined intruders can provoke a prolonged series of flights that can last many minutes. In all these species, as in the swallowtail *P. zelicaon,* females are captured as they fly at high speed over the perching area.

In other insects information about an opponent's pure physical strength is highly pertinent to a combatant, in that it helps the male determine whether he will be able to physically remove the other indi-

7.15   Males of the great purple hairstreak, *Atlides halesus*, sometimes engage in complex aerial chases in which they fly great distances upward, spiraling about each other as they go. These contests are for control of a perch territory in a hilltop tree. (Drawing by M. H. Stewart.)

vidual from a desired resource, (a nest site, a rich food supply, or a female herself). In *C. pallida* wrestling matches occur between males of roughly equal size as they seek control of a site from which a female will emerge. Likewise, large scarab and brentid beetles often grapple with one another in the presence of a potential mate, employing elaborate maneuvers to pick an opponent up and drop him from a plant, or flip a rival from his perch, or turn him on his back (Eberhard, 1979,

1980). Otte and Stayman (1979) make the point that in these beetles visual assessment of a rival may be difficult because of perceptual limitations of the beetles (or because of their nocturnal activities). If so, a direct physical trial may be a superior method of determining the resource holding capability of an opponent (Fig. 7.16).

Resource Value and Investment in Fighting

Prolonged tests of endurance and strength are most likely to occur not only when the combatants are evenly matched but also when the resource at stake is unusually valuable (Sigurjónsdóttir and Parker, 1981). As we suggested earlier, males of the speckled wood butterfly avoid lengthy fights because the territory, a sunspot, is not valuable enough to repay the costs of intense aggression. We can predict that a male's willingness to invest in costly aggression should be proportional to the genetic gain associated with winning a dispute *minus* the genetic gains to be derived from pursuit of an alternative option (West-Eberhard, 1979). If the winner of a fight is guaranteed access to many receptive females, males should fight more vigorously than if the winner has only a slightly improved chance of mating. Fierce fights should also occur if the alternative (a search for an unoccupied territory or undiscovered female) is unlikely to yield a return. The butterfly owner of a sunspot has a low probability of mating per unit time spent on the territory (Davies, 1978), and an intruder has a high probability of finding an unguarded sunspot fairly quickly. Aggression among males of the speckled wood butterfly is muted. In contrast, male sugarcane beetles, *Podischnus agenor,* fight furiously with one another for possession of a nest burrow in a cane stalk (Fig. 7.17). The male that constructed the burrow is in a good position to attract a mate; the alternatives for the loser, to build a burrow of its own or to find a weakly defended burrow, require large time investments (Eberhard, 1979).

That the intensity of aggression is a function of the value of the territory is also illustrated by the behavior of males of the coreid bug, *Acanthocephala femorata.* Mitchell (1980) presented a tethered intruder male to a resident bug on his sunflower stalk territory. If the plant lacked females (which fed upon its flowers), the territory owner fought the intruder by wrestling with him (using the enlarged hind legs while facing away from the rival) on only 35 percent of the presentations. But if the stalk was occupied by a group of females, the resident fought on 85 percent of the tests.

Similarly, Johnson (1982) has found that male brentid beetles fight

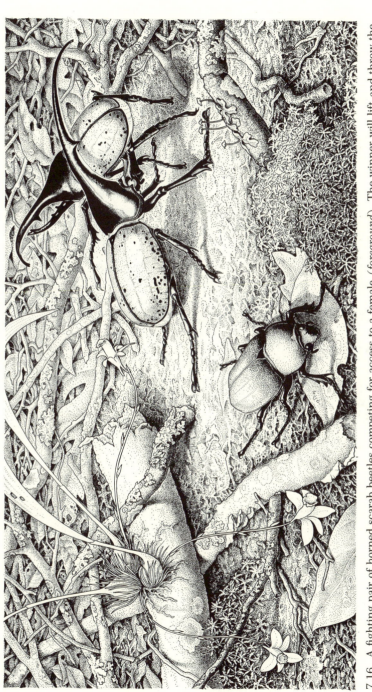

7.16 A fighting pair of horned scarab beetles competing for access to a female (*foreground*). The winner will lift and throw the opponent on his back one or more times. (From E. O. Wilson, 1975.)

7.17   A male of the beetle *Podischnus agenor* holding a rival with his horns prior to dropping him to the ground. The males were battling for possession of a nest burrow in a sugarcane stalk. (Photograph by W. G. Eberhard.)

more vigorously over large than small females. Because large females are more fecund and produce larger progeny, they repay the male for a greater investment in aggression. Likewise, male dung flies engage in escalated conflicts more often when the contested female is large (that is, fecund) and early in a bout of oviposition (still has many unlaid eggs to fertilize) (Sigurjónsdóttir and Parker, 1981).

The most extreme instances of insect aggression are those in which the opponents refuse to concede defeat but instead fight to the death. In most cases of this sort a male has no viable option other than to try to win access to a female by dispatching all rivals. In these species if a male were to withdraw and wait for the demise of his opponents, his wait would prove futile, as all the females he might contact would have been mated by this time. If the male were to leave for other areas, he would be extremely unlikely to locate new females.

Among the parasitic wasps a single male may monopolize a large number of females, provided he is able to kill his competitors. The alternative to fighting for control of the host would be to find another parasitized victim from which only females were beginning to emerge. This is unlikely to be a productive option, given the generally dispersed distribution of hosts at the proper developmental stage and the certainty that one or more males will emerge from each one (W. D. Hamilton, 1967). The older dispersing male, having expended energy in his search, would be unlikely to defeat the freshly emerged male(s) that would seek to block his access to the emerging females. Under these circumstances it is not surprising that males of *Melittobia* and other ecologically similar parasitic species regularly fight to the death (Buckell, 1928; Matthews, 1975; van den Assem, Gijswijt, and Nubel, 1980). A loser, whether he dies immediately or not, has no chance to reproduce, thus favoring all-out attempts to win at any cost.

The absence of viable alternatives to fighting is characteristic of the species whose males exhibit extreme aggression over females. In fig wasps of several genera, wingless males emerge within a fig "fruit" that also contains numbers of metamorphosing females. A wingless male that attempted to leave his birthplace rather than fight to control the area would almost surely never succeed in finding an uncontested source of receptive females. As a result, males stay and battle to be one of the survivors whose sperm will be accepted by the new generation of females produced within the fig. W. D. Hamilton (1979) studied fighting males by cutting open a fig and then adding a gall from which a female was about to emerge. He commented: "A male's fighting movements could be summarized thus: touch, freeze, approach slowly, strike and recoil. Their fighting looks at once vicious and cautious—cowardly would be the word except that, on reflection, this seems unfair in a situation that can only be likened in human terms to a darkened room full of jostling people among whom, or else lurking in cupboards and recesses which open on all sides, are a dozen or so maniacal homicides armed with knives" (1979:173).

A special case of a similar sort is supplied by the phengodid beetle, *Zarhipis integripennis* (Tiemann, 1967). Like the fig wasps, males have evolved powerful cutting jaws which they use to good effect in slicing and dismembering opponents that they encounter near receptive females. The phengodid male is highly mobile and perfectly capable of leaving a conflict to search for another unguarded female. That he does not do so may stem from the fact that he is unlikely to find another female because of the probable male bias of the operational sex ratio. The larviform female is far larger than the male (Fig. 7.18) and therefore presumably requires much more time and energy to achieve adulthood than the male. The result should be that receptive females are far scarcer than searching males. If this speculation is correct, it may explain why males refuse to leave a receptive female once they have been fortunate enough to come across one. If the probability of finding another mate is extremely low, then males may gain by risking death in order to have the opportunity to mate with a female they have found.

To summarize, we have argued that the spatial distribution of *receptive* female insects has had a profound impact on the evolution of

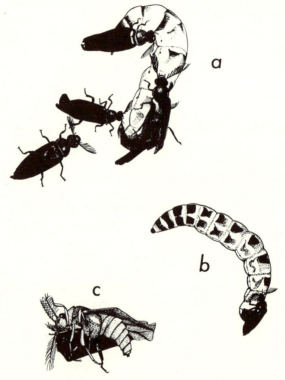

7.18   Males of the phengodid beetle *Zarhipis intergripennis* (*a*) gather in numbers near a wingless, larviform female; (*b*) a male mates with a female, (*c*) often after having fought to the death with an opponent for access to her. (From J. E. Lloyd, 1979b.)

male mate locating and aggressive behavior. The reproductive success of a searching male is a function of the frequency of contacts with receptive females and the likelihood that sperm transferred will actually be used to fertilize his partners' eggs. When females are clustered at emergence sites and mate only once, the profitability of searching for emerging virgins is high. But when emerging females are widely scattered or if females mate multiply with sperm precedence, males are more likely to gain by searching for oviposition sites or foraging areas if potential mates are clustered in these places. If there are no areas with concentrations of receptive females, males may wait for mates at landmarks (for lack of any other option).

The benefit-to-cost ratio of defending an encounter site also depends in part on the degree of clumping of receptive females. To the extent that numbers of mates can be monopolized by possession of a defendable (small) site, territorial behavior is favored. Almost all cases of territoriality by male insects involve defense of small areas that contain or attract numbers of mates. Territorial males behave aggressively in ways that help reduce the costs of defense. Territories may contract when the density of rivals increases. Owners and rivals often employ threats or displays of endurance and agility, rather than physical force, in the initial stages of a conflict over space or females. Fights to the death are rare but do occur when the value of the contested resource is high and the male has no other mating opportunities.

8

Male Mating Systems

WE HAVE NOW DOCUMENTED the gamut of possible effects of intrasexual selection on the mate locating behavior of male insects, culminating with an examination of competition for mate encounter sites. Where males choose to search for females and whether they defend these locations are key factors that help define the mating system of a given species. Chapter 3 outlined the basic mating systems of animals, discussed the differences between a populational and an individual classification system, and showed why sexual competition among males usually results in polygyny. We noted in that chapter that it was possible to make finer discriminations among mating systems, subdividing the two male options—monogamy and polygyny—into a number of categories. An ecological classification of male mating systems is presented here (Emlen and Oring, 1977; Bradbury and Vehrencamp, 1977). We illustrate each category with sketches of the reproductive tactics of various bees and wasps to show how ecological differences among species are correlated with behavioral differences. The primary cases are supplemented with examples of other species that are subject to similar ecological pressures and have independently evolved similar mating systems.

An Ecological Classification of Male Mating Systems

If we consider the mating systems of male insects from the perspective of the individual, the overwhelming majority of males are potentially polygynous in that each can inseminate many females and devotes the bulk of his energies to this end. But there are some monogamous insects. Emlen and Oring (1977) and Bradbury and Vehrencamp (1977) independently argued that the key ecological variables that determine the probability that a male will monopolize and copulate with more than one female are (1) the operational sex ratio (the ratio of sexually active males to receptive females at any one time) and (2) the spatial and temporal distribution of receptive females. If the operational sex ratio is strongly skewed toward males throughout the expected lifetime of a male, then monogamy may be his best option. If efforts to find a series or a group of receptive females are destined to fail because potential mates are not clustered in time or space, a male that has found a willing female may commit himself to her permanently. In insects there are at least two potentially different kinds of male monogamy, which we can label *female guarding monogamy* and *mate assistance monogamy*. In the first, a male may sacrifice any chance to mate with more than one female in order to prevent his current (and sole) mate from copulating with another male. In the second, a male may make the same sacrifice but with the functional goal of helping his current and only female produce more offspring than she could without his help. It is entirely possible for the two kinds of monogamy to be combined in some species (with the helpful male also preventing his mate from receiving sperm from other males), but as we shall see it is also possible both in theory and in practice for the two functions to be separate.

Usually the operational sex ratio is not so strongly and consistently male biased, nor are females so widely and evenly distributed, that the possibility of polygyny by some males is precluded. In many species, as we have seen, receptive females occur in clusters either living together in a social unit or emerging one after another within a short period in one location. This creates the potential for mate monopolization by a few dominant males that are able to defend with relatively little expense the areas that contain clustered females. The result is *female defense polygyny.*

Even if the direct defense of a group or series of mates is not feasible, a few males may still control many females. If resources, such as food or nesting sites, occur in discrete elements, the male that can defend a resource patch gains access to the numerous females attracted to the area. *Resource defense polygyny* by vertebrates is common in situations in which it seems plausible that the benefits of territoriality outweigh the costs—that is, relatively large numbers of females come to a relatively small area that one male can control economically.

In still other species neither females nor resources used by females occur in clumps that single males can easily monopolize. It is in these species that males avoid aggressive interactions with their fellows and search for receptive females one by one, racing their competitors to be first to reach a mate, *or* they defend symbolic territories at locations that contain neither groups of females nor resources of value to potential mates. The nonterritorial race to mates has been labeled an *explosive breeding assemblage* (Emlen and Oring, 1977), for it is characteristic of frogs and toads that breed in frenzied groups during an extremely restricted reproductive period. Here the advantage goes to the superior searcher rather than to the male that invests in territorial defense of an area that may be swamped with competitors (as in the explosive breeding frogs). Nonaggressive searching occurs in many insects in which there is an abundance of mates during a limited period or widely dispersed females that cannot be easily monopolized. *Scramble competition polygyny* may be a more appropriate label for these kinds of systems.

An evolutionary alternative to scramble competition polygyny is *lek polygyny*, perhaps the most poorly understood mating system (Bradbury, 1981). In lek species males defend what appears to be a purely symbolic territory, often a perch site near other males at a traditional arena or by a prominent landmark. Typically in vertebrates, and in some insects as well, this form of territorial behavior occurs when receptive females are not easy to monopolize because of their scattered distribution, high mobility, and large home ranges (Bradbury, 1981). These factors, however, are also correlated with some forms of scramble competition polygyny. One difference between the species in each category may be the longevity of females and the length of the breeding season. For a long-lived female with many days available for the selection of a mate, there may be less urgency in acquiring sperm immediately upon reaching adulthood. If so, females can perhaps afford to be more selective in picking a mate, favoring those that had demonstrated dominance in their interactions with other males because dominant males offer superior genes.

Although female longevity may more easily permit the evolution of lek polygyny, it is not an absolute requirement. Short-lived females drawn to an area with diffusely distributed (not easily monopolized) resources may be under strong selection to mate quickly. They may tend to select males that station themselves on the periphery of the resource area rather than males that search within the foraging or oviposition zone itself (Parker, 1978b). If these males compete for superior vantage points, females could select dominant mates simply by flying to the best perching areas near the resource site. Prominent elevated landmarks offer good scanning perches; perhaps this is why lek polygyny in insects is so often found in species that use landmarks as mating aggregation sites.

Table 8.1 summarizes this classification scheme for male mating systems. Finer divisions are possible, and some altogether new categories could be created. For example, lovebugs seem to incorporate elements of both female defense and scramble competition in their mating system. What is important here is not so much the labels as the underlying rationale of the approach, which is founded on the question, How might individual males leave more offspring than their rivals? Usually, but not always, this question can be reformulated to read, How might a male succeed in fertilizing more females than his competitors? There is no one answer to the question because variation in ecological pressures affecting different species leads to variation in the spatial distribution and temporal availability of females. The different solutions that have evolved can be correlated in reasonable ways with those ecological variables that affect the mate monopolization potential of males. We shall expand on this point by looking at some examples of different mating strategies and the environments in which they occur.

The Monogamous Male Honey Bee

Prior to the breeding season of honey bees the queen begins to lay large numbers of haploid eggs destined to become males (drones). At approximately the same time her workers *may* also begin to feed a few female larvae the specialized diet that leads to the development of a virgin queen bee. Although most colonies produce hundreds of drones, only colonies with a relatively large worker force create two or three queens, and only one of these survives the battles among virgin queens after they emerge. These contests determine which individual will be left with about half the work force when the old queen leaves the hive with her swarm to locate a new living site (Michener, 1974).

Because honey bees do not inbreed, males must find virgin queens produced by colonies other than their own. One option would be search for another colony and wait by the entrance of the hive for the

Table 8.1   Male mating systems in insects.

I. Monogamy: male mates with only one female per breeding season (mate monopolization potential of males is very low)

   A. Mate guarding monogamy: male remains with mate in order to prevent her from copulating with other males

   B. Mate assistance monogamy: male remains with mate in order to elevate her reproductive output

II. Polygynous mating systems: some males mate with more than one female per breeding season

   A. Mate monopolization potential of males is high due to the clumped distribution of females or resources attractive to females

      (1) Female defense polygyny: some males prevent others from gaining access to mates by defending groups of females or a series of individual females

      (2) Resource defense polygyny: some males prevent others from gaining access to mates by defending resources that attract receptive females

         (a) Defense of resources as they occur in situ

         (b) Defense of resources after they have been collected (and in some cases prepared for a female) by a male

   B. Male mate monopolization potential is fairly low, often because the emerging or resource using females are widely dispersed

      (1) Pure dominance or lek polygyny: some males gain access to mates by excluding others from certain "symbolic" mating territories preferred by selective females

         (a) Defense of a perch on a landmark site

         (b) Defense of a waiting site on the periphery of a dispersed resource area

      (2) Scramble competition polygyny: males make no effort to defend an exclusive mating territory but instead attempt to outrace their competitors to receptive females

         (a) Explosive mating assemblage: receptive females are abundant during a very brief mating period

         (b) Prolonged searching polygyny: receptive females cannot be economically monopolized because of their even distribution or the high rate of competition from intruder males

exit of the virgin queen on a nuptial flight. (A male could not enter a colony in search of a mate; he would be expelled as an alien intruder by the guard workers at the entrance.) Since hives are dispersed under natural conditions, it is possible that the waiting male could invest all his time at a hive from which no virgin would emerge (because the colony had not produced one). We suspect that under natural conditions the percentage of colonies large enough for fission is small and therefore the probability of failure of the "wait-at-nest" option is fairly high. Male honey bees do not employ this tactic but instead gather in aerial aggregations (Fig. 8.1), often over distinctive topographical features in their environment, such as a small hill (Ruttner, 1956). It is difficult to study from the ground a swarm of males flying many meters overhead, so we do not know whether males are competing in subtle ways for positional advantage in the swarm, but they do not appear to have strictly defended hovering sites within the group. Attempts to defend a territory would presumably be futile, given the high density of males at the mating site.

The single surviving virgin queen produced by a large colony makes a number of mating flights to male swarms, during which time she may permit one or more of her many pursuers to capture her and mate. Males that succeed in copulating never do so again. Successful drones achieve monogamy in a most dramatic fashion by detaching their genitalia and leaving them inserted in the female's genitalic opening as the grand finale to copulation (Fig. 8.2). The automutilated drone dies promptly.

The suicidal donation of the genitalic plug must have some benefit that outweighs its obvious cost. However, the benefit need not be very great because the cost is actually slight. The enormous imbalance toward males in the operational sex ratio (OSR) means that the likelihood of the drone's capturing two virgin queens is virtually nil. Therefore the loss of opportunities to mate again occasioned by the copulating drone's final action is small. It may be more than outweighed by an improvement in the chances that the male's sperm will be fully used by his female. To this end the genitalic plug may prevent other males from copulating with the queen. It is true that the queen usually accepts several partners on her nuptial flights. She apparently can remove the plug if it is in her interest to do so, but she can also leave it in place when she has acquired sufficient sperm, thus preventing further matings and blocking the acquisition of additional sperm when it is to her advantage. Most queens return to the hive with a "mating sign" in place (Taber, 1954).

Let us now make some plausible assumptions and see where they lead with respect to understanding the genital sacrifice of male honey bees.

(1) We assume that currently honey bee queens make 4 nuptial flights and mate a total of 16 times, 4 per flight. The estimated total that appears in the literature is 17, and it is known that queens do go on several mating flights (Woyke, 1955; Adams et al., 1977).

(2) As a second condition, we assume that if a queen did not receive a male's genitalia after mating for the fourth time, one other

8.1 A swarm of drone honey bees at a landmark mating site approaching a bottle containing a receptive queen bee. (Photograph by N. Gary.)

8.2 Mating sign. The detached genitalia of a drone inserted in the genital opening of a mated queen bee. (Photograph courtesy of N. Koeniger and F. Ruttner.)

male on average would capture her and add his sperm to the four ejaculates she had already received. Thus in an ancestral population in which drones did not provide plugs, queens would mate 20 times, 5 per flight. If a mutant plugger male appeared in the species and copulated, he would have one chance in four of being the queen's fourth partner on a nuptial flight. In this case the mutant would eliminate one competitor's ejaculate. (Again we assume that this is to the female's advantage as well, by helping her gain more control over the number of matings per flight and the quantity of sperm she receives.)

(3) Sperm from each male is evenly mixed in the spermatheca (Page and Metcalf, 1982). We assume that every sperm has the same chance of being used to fertilize an egg that will become a surviving queen (or egg laying worker). Genes within sperm that fertilize eggs destined to become sterile workers are obviously not transmitted directly to subsequent generations, so that a drone's fitness is primarily a function of the probability that his sperm will be used to produce a new queen. By reducing the number of ejaculates present in his mate's spermatheca, the mutant male would gain a small improvement in the chance that his sperm would win the fertilization lottery. In the ancestral population a mutant's gain would be about 1.25 per-

cent (assuming that all males donated equal quantities of sperm); a mutant would have one chance in four of reducing the total number of competitors from 20 to 19. The fitness gain for the mutant is therefore slight—but, to repeat, the benefit of suicidal mate guarding need not be great if the odds against remating are extremely high.

Our point is that when the OSR is consistently and strongly biased toward males, a male that leaves his mate has a low probability of finding additional receptive females. At the same time, if there is a high probability that his mate will copulate with other males, a male that remains with his mate to guard her from the competition can prevent or reduce the dilution or displacement of his sperm within his mate. In its most extreme form, as in the honey bee, mate guarding can result in monogamy, with the male (or a portion of him, at least) posthumously remaining in association with the female. Note that "monogamy" in this case refers to the mating system of those few males that succeed in mating. From a populational perspective, variance in drone reproductive success is probably greater than that of queens.

Male Monogamy in Other Insects

Male monogamy as a consequence of mate guarding occurs in insects other than the honey bee, although it is not common. In termites the typical arrangement appears to be for a male to stay with his mate for the duration of his life (E. O. Wilson, 1971). The origins of lifetime monogamy may lie in mate guarding, as it is customary for the early stages of pair formation to take place during a nuptial flight in which vast numbers of individuals are released simultaneously from neighboring colonies. When a male encounters a female after both have alighted on the ground, he runs after her "in tandem" until she selects a site in which to construct a burrow (Fig. 8.3). By closely following his potential mate, a male is in position to repel other suitors. Because the swarming nuptial flight is of such short duration and because competition for mates is so great, it is doubtful that a male that tried to copulate with several females in succession would succeed (all the more so because females generally refuse to mate until they are well established in a safe burrow). Synchrony in the availability of receptive females forces monogamy on male termites.

Female behavior is also instrumental in making monogamy the option of "choice" for some members of the ceratopogonid midges, whose males practice a form of mate guarding similar to that of honey bees. In these species the male remains in copula with his mate while she proceeds to consume him (Downes, 1978). The male genitalia are left

8.3 Tandem running in an African *Macrotermes*. The male termite closely pursues the female after they have made contact on their nuptial flight. (Photograph by E. S. Ross.)

in place as a mating plug, presumably because the female benefits by controlling the number of males that copulate with her (Fig. 8.4). Like the honey bee and termites, mate location takes place in a swarm. The operational sex ratio is probably skewed toward males, and therefore males that succeed in finding one mate have little chance of securing another. If so, they gain if their females will use a part of them to prevent additional copulations.

Both the termite and the male ceratopogonid may derive another benefit from their monogamous behavior, a benefit that blurs the distinction in some cases between mate guarding monogamy and the second major category of monogamy in insects, male assistance monogamy. If a male is unable to acquire more than one mate, he will gain if he can improve the relative reproductive success of his single partner. Imagine a situation in which a male, by remaining with a female, is able to assist her in ways that substantially raise her production of surviving offspring. For the sake of argument, we say that for a particular population females that receive help lay four times as many eggs (all of which are fertilized by a helpful monogamous mate) as unassisted females inseminated by would-be polygynist males. In this example the polygynists would have to be able to find four receptive females, on average, just to do as well as the monogamists.

The termite male may not only initially guard a mate but later may also help her in various ways, such as in construction of the original burrow. Moreover, some observers have noted that the male's abdomen may become somewhat shrunken over time (Harvey, 1934). Perhaps a male transfers nutrient and allied materials to the female during copulation(s), which then assist the female in egg production.

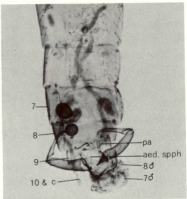

8.4   (*Left*) A female ceratopogonid midge feeding on her mate, on the left. Her mouthparts pierce the head of the male. (*Right*) The detached terminalia of another male ceratopogonid, with the hook-like claspers gripping the tip of the female's abdomen. The numbers refer to the abdominal segments of female and male; *aed* = aedeagus, *c* = cercus, *pa* = parameres, *spph* = spermatophore. (From Downes, 1978.)

Shriveling of the abdomen as a result of mating has been observed in other insects such as the lovebug, *Plecia nearctica*, which mates for many hours (Thornhill, 1980b), and meloid beetles of the genus *Hornia* (Linsley and MacSwain, 1942). It remains to be determined whether or not males are transferring life-sustaining materials from their bodies to their mates and thereby increasing the likelihood of their own prompt death while improving their partner's survival and fecundity.

The ceratopogonid midge's sacrifice certainly decreases his opportunities for polygyny but whether his body, once incorporated in the female, raises her fitness has not yet been proven. Nor has this been determined for those male mantids that definitively practice monogamy in much the same way when they are cannibalized by their mates. Here the suicidal sacrifice of the male, performed *in copula*, leaves no mating plug in place. The male's actions may have less to do with preventing his mate from copulating again than with enabling her to produce substantially more eggs than otherwise. It must be rare for a female mantis to have the opportunity to consume a meal as large as that provided by the male himself, and this could raise her fecundity (and thus his fitness).

Males need not die in order to help their mates. There are various forms of parental care that male insects can provide, although paternal behavior in insects is rather rare (E. O. Wilson, 1971; R. L. Smith, 1980). Moreover, parental behavior by males is not totally incompatible with polygyny: an individual may have a succession of mates that he assists or he may, as in the case of certain belostomatid water bugs and reduviid bugs, care simultaneously for the combined clutches of several females. But there are insects whose pair bonding and assistance behavior are analogous to the typical monogamous bird species. In a number of scarab beetles, for example, males help their mates build a burrow (Fig. 8.5) and gather dung or other materials upon which the female feeds. In dung beetles of the genus *Phaenaeus* females feed on dung for 3 to 4 months in their burrow before laying eggs. Pairs are found in burrows in May, although breeding does not occur until September (Halffter and Matthews, 1966). While there have been no studies of marked individuals, these observations are at least suggestive of male monogamy, with some individuals helping to promote development of breeding condition in their mates and thereby enjoying the genetic benefits of her reproductive gains. With the intense competition for dung as a food resource (Heinrich and Bartholomew, 1979) a male's assistance in gathering it could have a significant impact on his female's reproductive performance.

A similar form of pair bonding occurs in other beetles whose females are dependent upon a food resource for which there is severe competition. Males of *Necrophorus* spp. help their mates bury small dead rodents and birds, thus hiding them from other carrion consumers (Milne and Milne, 1976). A pair of carrion beetles constructs a brood chamber underground with a mass of rotting carrion, in the center of which they rear a brood of larvae, feeding their young a liquid regurgitate derived from consumption of the decayed meat (Fig. 8.6). Again, it is at least plausible that the male's assistance could have a positive impact on the female's reproductive success, thereby improving his own fitness. Two beetles can bury a food item at least twice as fast as one, reducing the risk that the carrion would be taken in its entirety by a mammalian scavenger and reducing the time of exposure of the carcass to blow flies and the like, whose progeny would decrease the food available to the beetles. Although a male's commitment to one female prevents him from searching for and inseminating others (a difficult task in a species whose females probably are not receptive except in the presence of a relatively rare resource, small dead vertebrates), monogamy may well have compensatory advantages for the male if the fecundity of his mate is greatly improved.

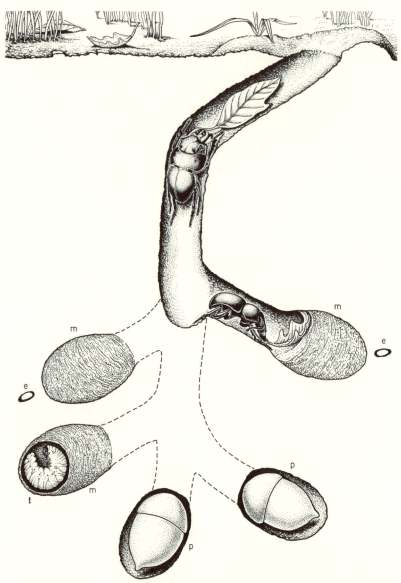

8.5 Assistance by a male scarab beetle, *Lethrus apterus*. He helps the female construct the burrow and gathers leaves to provision the brood cells. The egg (*e*) hatches into a larva (*l*), which consumes the leaf masses (*m*) and develops into a pupa (*p*). (From E. O. Wilson, 1971.)

8.6   Brood care by a carrion beetle, *Necrophorus.* The male may assist the female in regurgitating liquefied carrion to feed their offspring, which reside in a nest of carrion. (From E. O. Wilson, 1971.)

Polygynous Mating Systems

Female Defense by a Eumenid Wasp

Despite the possibilities for adaptive male monogamy, the fact remains that it is extremely rare among insects. Far more common are the various forms of polygyny, which we shall illustrate one by one in the pages that follow. The digger bee, *C. pallida,* might be used to represent female defense polygyny, as males vigorously fight for possession of minute patches of desert soil from which a virgin female may be about to emerge (see Fig. 4.1). Even more clear-cut examples exist among the bees and wasps, one of them being a small Australian eumenid wasp of the genus *Epsilon* (Smith and Alcock, 1980). In this species several females may build a series of brood cells in the same place, creating a tightly packed cluster of as many as 50 to 75 brood pots on the face of a rock overhang. (On the other hand, some females nest by themselves with the result that the site will have a mere handful of brood cells.) The males of the new generation tend to emerge prior to the females, which enables them to search for as yet unopened brood cells from which females may emerge. Large aggregations of cells are claimed by territorial males (Fig. 8.7). Small brood clusters may be visited by males, but are not taken as territories.

The owner of a territory waits on the surface of the cells. Nonterritorial visitors may be common, but the resident rarely has to do more than turn to face an intruder in order to send him on his way. Never-

8.7   A territorial male of *Epsilon* sp. copulating with a freshly emerged female at the cluster of brood cells that the male defends. (Photograph by J. Alcock.)

theless, severe fights for possession of a cluster do sometimes occur and the loser can be permanently disabled by bites from the winner. The intensity of aggression reflects the fact that a great deal is at stake in a contest over a territory; the winner has access to the females that emerge in the area under his control. As a female gnaws her way out of a brood cell, she is usually detected by the territorial male, who then stations himself by the cell. As soon as the female exits, he slips onto her back and copulates. Because females apparently mate just once (nesting females consistently reject the mating attempts of males that are strongly motivated to copulate) and because a territory contains as many as two dozen about-to-emerge females, the reproductive gains of a resident male can be great, provided he can hold the cell cluster for sufficient time to contact the virgins as they emerge.

Males of this species satisfy all the criteria for female defense polygyny. Because some brood cell aggregations contain many potential mates clumped in an area small enough to be economically defended by a single male, the mate monopolization potential of some males is great. After copulation a female's eggs can no longer be fertilized by another male, which confers a strong advantage on the resident defender of an emergence site, as the territory owner is likely to be the first to contact virgins. Males defend only the more productive of the emergence sites, where the potential return is greatest for their costly investment in territorial defense.

Among insects generally, the ecological correlates of female defense polygyny are (1) females that mate just once (because they are relatively short-lived and of low fecundity and therefore can receive all the sperm they need from a single mating), (2) females that mate shortly after becoming adults (short-lived individuals benefit from prompt mating so as to begin nesting at once), and (3) clustered distribution of emerging females that results from "communal" nesting by their parents. In some species limited nesting or oviposition substrate may generate a clumped pattern of emerging females. In *Epsilon*, however, this is unlikely to be the case. The wasp nests in rocky terrain of coastal eastern Australia, where cliff faces and overhangs are numerous; moreover, it builds its nests from the resins of eucalyptus trees, which are by far the dominant species in its forest habitat. It is possible that females may nest together in order to decrease construction time required per cell. By building where cells are already present, a female can construct her new cell adjacent to an old one, taking advantage of existing surfaces to reduce the material she must collect to complete a cell. Whatever the reason, the clumped distribution of emerging, receptive females makes female defense polygyny possible for some males.

Female Defense Polygyny in Other Insects

The direct monopolization of a cluster of females is not particularly common in insects. Some reports are merely suggestive of its occurrence. Beeson and Bhatia (1939) found fighting males at mating assemblages of *Hoplocerambyx spinicornis,* a cerambycid beetle of India, on the bark of a "particular tree." That males may be attempting to control females directly rather than defend a portion of the tree is indicated by the comment, "In captivity large males have been observed to monopolize several females driving off smaller males in much the same way as a stag or boar does" (91). If in nature males are able to find clusters of females drawn to an oviposition site, some may be able to acquire and defend a harem against rivals, thus qualifying for inclusion in the female defense category.

When defense of a harem occurs, it is usually because a group of females will emerge from a cluster of cells, parasitized eggs, or other nesting site in the manner of *Epsilon* (Fig. 8.8). The defense of a multifemale emergence site occurs in some other eumenid wasps (D. P. Cowan, 1981), some bumble bees (Alford, 1975; J. E. Lloyd, 1981), carpenter bees (Gerling and Hermann, 1978), sphecid wasps (Lin, 1963; Eberhard, 1974), and a host of small chalcidoid and other parasitic wasps in which males compete for defense of a puparium or group of parasitized eggs from which many females will emerge (see for instance Buckell, 1928; Satterthwait, 1931; W. D. Hamilton, 1967;

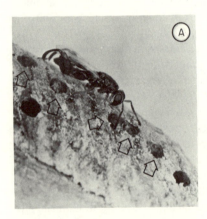

8.8 (*A*) A male torymid wasp perched on a mantis egg case, which has been parasitized by a female of his species. The arrows point to emerging receptive virgin females. (*B*) The male palpates a female's head as she emerges, and (*C*) mounts her prior to copulation. (From Grissell and Goodpasture, 1981.)

Eberhard, 1975; Matthews, 1975; van den Assem, Gijswijt, and Nubel, 1980).

A variation on the female defense theme is provided by the bumble-bee-wolf *Philanthus bicinctus* (Gwynne, 1980). The males of this sphecid do not fight for possession of areas containing a number of emerging females, for at this stage they are evidently not receptive. Females become willing to mate only after having begun to construct their long and elaborate burrows in the soil. Males search for locations containing a number of recently constructed nests and attempt to control these sites against intruders. They can identify prime sites by the presence of fresh mounds of earth, the by-products of nest build-ing.

In many other species of female defense polygynists, males are un-able to claim a site with a fixed harem of females. Instead they must defend emerging or ovipositing females one by one in a series of differ-ent locations, as in *C. pallida,* various mosquitoes, butterflies, beetles, and flies (Chapter 7). Again, moderate temporal a well as spatial clus-tering of receptive females may permit some males to practice this form of female defense polygyny.

8.9   A male of the tenebrionid beetle *Onymacris rugatipennis* that had been trailing after a female mounted her when confronted by a challenge from the marked male on the right. (Photograph by W. J. Hamilton III.)

An interesting example of the one-by-one theme is exhibited by the tenebrionid beetle, *Onymacris rugatipennis*. Females of this species burrow into the sand in the evening and mate with males that are able to locate them at this time. Thus males follow after females, often in the late afternoon (Fig. 8.9). The male, as he walks behind a female, must at times repel intruders and is also faced with competition for control of the space above a female after she has burrowed into the sand for the evening. Fights take the form of wrestling matches involving "head-butting, shoving, throwing, biting and kicking" (Hamilton, Buskirk, and Buskirk, 1976:305). Losers, those that are thrown more often or bitten, retreat and the evening victor will return to the buried female, whom he mounts and mates. The fact that females apparently become receptive for a limited period each day gives a succesful female locater and defender the opportunity for a mating every day.

Resource Defense Polygyny

Although there are insects in which direct defense of a series of females or cluster of females occurs, there are many others in which males make no effort to locate and monopolize females directly. Megachilid bees of the genus *Anthidium* provide examples of species that practice an indirect means of monopolizing mates (Alcock, Eickwort, and Eickwort, 1977; Severinghaus, Kurtak, and Eickwort, 1981). Males of *Anthidium* do not try to find and defend emerging virgin females. In fact, in at least one species they emerge after the females (Jaycox, 1967), and in all the species studied to date the focus of male-male aggression is on the food plant of their species. In the mountains of southeastern Arizona the intensely territorial males of *A. maculosum* patrol patches of a flowering mint that often grows in discrete clumps 1 to 3 m². This is apparently the sole (or at least the highly preferred) source of food, which females use to construct balls of provisions for their young. Males can identify rich patches likely to be visited by provisioning females. They attempt to control superior patches by spending much of their time in patrol flight monitoring the borders against intruders of many species, but especially conspecific males—all of which are assaulted and chased from the territory. The same treatment is given to females unless they permit the resident to mate with them. Males attempt to capture and copulate with foraging females that have entered the corolla of the mint flower; if successful, the male will tolerate the forager as she goes about collecting pollen in his territory, but if she does not mate, she will be pursued and assaulted. The attack power of a territorial male is considerable. Honey bees struck by males of *A. manicatum* regularly have a wing severed in the encounter, and even bumble bees three or four times as large as an *Anthidium* will leave an area if subjected to a male's relentless onslaught.

Why are males of this bee such determined defenders of a food resource rather than defenders of nests? In the first place, although several virgin females may emerge from a single burrow, these bees do not nest in groups, and the scattered nests are hidden in cavities where they are difficult to find (Severinghaus, Kurtak, and Eickwort, 1981). A female defense polygynist would be limited to those virgins contained within one isolated nest, assuming the male could find the nest. Even more important, females of *Anthidium* mate repeatedly and if sperm precedence occurs, mating with a virgin will yield little or no genetic gain.

Suppose we assume a "last-male-to-mate advantage," whereby the male copulating closest to oviposition has the best chance of fertilizing the egg. Why does a male *Anthidium* not fly with a foraging female to her nest and remain there to copulate with her each time she returns with provisions? This strategy would ensure that if the female were to lay a fertilized egg, the nest guarding male's sperm would fertilize it. (Behavior of this sort has evolved many times in the bees and wasps—see Rozen, 1958; Peckham, Kurczewski, and Peckham, 1973.) The genetic return for nest guarding would, however, be limited by the rate of deposition of fertilized eggs by one female. The time required to provision a cell by *Anthidium* is not known, but judging from other solitary bees a figure of 1 to 2 days is not unreasonable. If half her offspring are male (from haploid, unfertilized eggs) and half female (from diploid, fertilized eggs), each time a female laid an egg the guarding male would have only one chance in two that the egg would be fertilized. And only if an egg were fertilized would he gain. Thus a male would have to guard a nest an average of 2 to 4 days per fertilization.

Assume that he would have to invest an average of 24 hours of nest guarding (3 days at 8 hours a day) to fertilize an egg. Using the same assumption of last-male advantage, a male resident at a patch of food plant will fertilize an egg only if (1) a female visitor to his territory with whom he mates is on her last provisioning trip prior to completion of a brood ball, followed by deposition of a fertilized egg, and (2) she collects sufficient provisions in the male's territory so as to fly directly back to her nest with no stops in other territories (where she would be captured and mated by other males). The combined probability of these events is low. But if there is even one chance in 20 of a productive mating per hour of resource defense, then the male guarding a flower patch will enjoy a higher rate of genetic gain than the males that wait by a nest ($1/20 \times 24$ hours $= 1.2$ effective matings on average per 3-day period of territorial behavior at flowers > 1 effective mating at a nest).

The odds of productive matings should be improved if a male's territory is large and resource rich, thus encouraging prolonged and repeated visits by the same foraging females. Individual females of *A. maculosum* have preferences for particular territories (Alcock, Eickwort, and Eickwort, 1977). Severinghaus, Kurtak, and Eickwort (1981) have documented for *A. manicatum* that the rate of female vis-

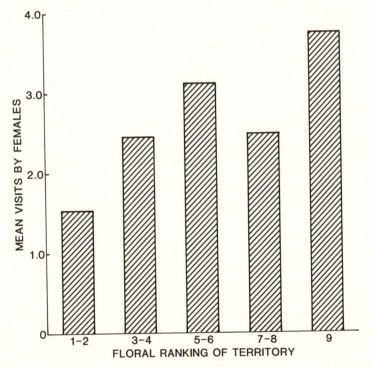

8.10   The number of flowers in a territory is related to male mating success in *Anthidium manicatum*. The more profuse the flowers, the more frequent the copulations. Territories are ranked from 1 (fewest flowers) to 9 (most flowers). Each average is based on 9 to 61 observations of 5 minutes each. (Data courtesy of L. Severinghaus.)

itation to a territory and the number of copulations secured by the owner are significantly correlated with the number of flowers in the territory (Fig. 8.10). Because some small, defendable locations contain a high concentration of attractive food plants, some males are presumably able to copulate frequently enough to more than compensate for the low probability that any one mating will yield a genetic reward.

Resource Defense Polygyny in Other Insects

Resource defense polygyny is often associated with species, like *A. maculosum*, whose females mate multiply and whose foraging resources, or nesting areas, or oviposition sites are clumped spatially. An unrelated insect whose mating system closely resembles the *Anthidium* pattern is the coreid bug *Acanthocephala femorata* (Mitchell, 1980). Males of the bug defend individual sunflower stalks, which are the feeding sites for their sap-sucking females. The male that is able to use his stout and spiny hind legs to kick or wrestle opponents from the stalk may have access to as many as six females. Territory owners are associated with three times as many females as nonterritorial males. Thus there is something about some sunflower plants that concentrates females spatially and enables some males to practice resource defense polygyny. Even though Mitchell (1980) was not able to identify the attractive features of defended stalks, she did demonstrate that defense was focused on certain plants rather than on the females per se, because solitary males will take up residence on a plant and defend it against intruders despite the (temporary?) absence of females. Whether females of this coreid mate multiply is not known with certainty, but based on the nature of the mating system and observations of polyandrous females in other coreid bugs (Hamlin, 1924), we predict that they will do so.

Although feeding sites and food itself are sometimes the focus of resource defense polygynists (see Sweet, 1964; Downes, 1970, for examples), superior oviposition areas are also defended, particularly by males of certain damselflies and dragonflies (Jacobs, 1955; Corbet, 1957, 1962; Campanella and Wolf, 1974). A representative of this group is the widespread libellulid dragonfly, *Plathemis lydia*. Females of this species are attracted to mounds of matted aquatic vegetation and mud projecting from the surface of the water. Sites with glistening surfaces are particularly likely to be approached by females, and in certain places prodigious numbers of eggs are laid (Jacobs, 1955). These locations are defended by males, although the size of a defended area and the number of subordinate males tolerated within a territory varies depending on the ecology of the area and the density of males present (Campanella and Wolf, 1974). A female that flies out over the water is likely to be intercepted by the resident territory owner and mated (copulations last only 3 seconds and are completed in the air). A male may take the female in tandem to a prime oviposition area in his territory and release her there. She will then lay her eggs while the male hovers above her.

Other insects whose males are able to monopolize discrete patches of oviposition material include certain dung beetles (Halffter and Matthews, 1966: Heinrich and Bartholomew, 1979), walnut defending *Rhagoletis* tephritid flies (Boyce, 1934), and the fungi protecting fly *Drosophila melanderi* (Spieth and Heed, 1975). The drosophilid utilizes *Amanita* mushrooms exclusively as an oviposition site, and the pugnacious males battle for sole possession of a mushroom cap.

Burrow-defending males also occur in a variety of unrelated taxa including a cockroach (Ritter, 1964) and many bark beetles (Vité and Francke, 1976; Birch, 1978). For example, a male *Ips* bark beetle first finds a host tree and then excavates a nuptial chamber in the bark. Three to five females (depending on the species) may join him. After mating they build their long and elaborate egg laying galleries out from the nuptial chamber. This pattern contrasts sharply with *Dendroctonus* behavior. Females of these bark beetles are the primary colonists, and males fly to a female's burrow to mate with her and defend the nest against intruders. Perhaps for these insects the resource base is more uniform and less concentrated than for *Ips*, so that one male cannot control an area large enough to attract and support more than one female simultaneously.

Lek Polygyny

The Tarantula Hawk Wasp

Territorial behavior of male insects is not always associated with the defense of females or the resources that females require for reproductive success. Male tarantula hawk wasps, *Hemipepsis ustulata*, defend palo verde trees, pinyon pines, jojoba shrubs, and creosote bushes in southern Arizona although females do not emerge, nest, or feed within male territories (Alcock, 1979c, f, 1981a). The plant must be on the backbone of a mountain ridge if it is to attract a territorial male, one to a tree or shrub (Fig. 8.11; see also Fig. 7.7). Females usually are nowhere to be seen on these often barren desert landscapes, and

8.11 A hilltopping wasp, *Hemipepsis ustulata*, on its territorial perch in a palo verde tree, on the alert for intruders and passing females. (Photograph by J. Alcock.)

an observer must spend hours walking along the ridge to find even an occasional hunting female, walking over the ground, inspecting crevices and crannies for her spider prey. Because the wasps probably use the scattered burrows of their victims as nests (Cazier and Mortenson, 1964), nesting females are dispersed spatially and consequently so are their progeny when they emerge the following year. In addition, although females take some flower nectar, they do so at a variety of common desert plants including creosote bush, perhaps the commonest and most broadly distributed shrub of the Sonoran Desert. Thus at all stages of their life cycle, females are not clustered spatially. This makes it difficult for human observers to find females, and by extrapolation makes it unprofitable for males to attempt female defense and resource defense strategies.

A territory owner does, however, defend his perch site vigorously and will fly out to meet every intruder. He will chase a new arrival around the tree several times—after which, in most cases, the visitor wasp moves on. Sometimes the circling pursuits are succeeded by an apparently more serious contest in which the males embark on a steep ascending flight that occasionally takes them so high that they become lost to view even when the observer attempts to follow them with binoculars. Eventually the combatants come zooming back to the palo verde; typically they circle the tree in pursuit of each other and then set off on another ascending flight, spiraling upward into the sky. Several dozen consecutive flights are not uncommon, and sometimes a resident will engage his opponent in over a hundred spiral flights in a period of a half hour or so. When a territorial takeover occurs, it is typically preceded by long bouts of spiral flights and the winner is often larger than his rival.

To judge from male behavior, the palo verdes or pines on the higher and more prominent portions of a ridge are more desirable than the lower territories. It is these sites that are occupied for the greatest portion of the flight season, are fought for most intensely, and are held by the largest males in the population (Fig. 8.12). As the breeding season advances and the number of emerged adult males increases, trees lower and lower down the slope are taken. The lower palo verdes are consistently used as perch sites by the relatively small males.

Why do males fight most for peaktop territories? Although one would predict that these territories are also preferred by females, evidence on this point is lacking. In over 300 hours of observation only three matings have been seen (Alcock, 1979c, 1981a), a puzzling result given the fact that males are not uncommon along the ridge (there should be a roughly equivalent number of females somewhere). If we simply assume that male competition must be related to female mate preference, then why might females prefer to copulate with males in peaktop territories? Perhaps because they are relatively long-lived (some males live at least 52 days), females may have the time to select males on the basis of their dominance status. By flying to a peaktop, a female can increase the likelihood of mating with a large individual with superior endurance and aerial agility. Flight competence, if it could be transmitted to one's offspring, would be a useful character.

Even though flight ability may be somewhat heritable, variation in body size is almost surely environmentally induced. A male's size is a

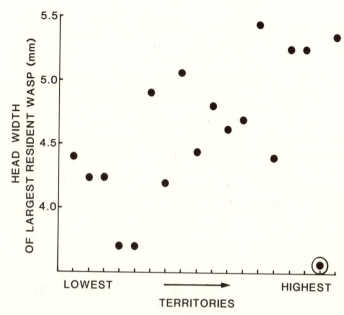

8.12 Males of *Hemipepsis ustulata* prefer territorial sites near the higher points along an ascending ridge. Trees that are held by large males tend to be located nearer the peaktop. The most striking exception (*circled point*) was a very small tree too tiny to be attractive to most males but taken for a time by a very small individual unable to claim other sites. (From Alcock, 1981a.)

function of the size of the spider his mother paralyzed and provided for him to consume. By picking a large male, a female would not in all likelihood be receiving genes to help her offspring grow large. However, by choosing a large male, a female may be selecting an individual whose mother was particularly competent at finding large prey, so competent that she was able to produce at least one unusually large son. If hunting ability has a genetic component, and there is no reason why it should not, then a preference for a large male may have a positive effect on the fitness of daughters resulting from the mating. Somewhat similar arguments on the relation between the genetic quality of the male and the fitness of his daughters have been developed by Trivers (1976) and Thornhill (1979a).

Admittedly these hypotheses are highly speculative, but there is little question that the opportunity for female choice exists in the tarantula hawk. Moreover, (1) males provide no parental care for their offspring, (2) only a limited portion of the habitat serves as a mate

encounter site, and (3) males defend perch territories in these locations, which are devoid of resources required by females. Thus, by
Bradbury's (1977, 1981) conservative definition, the tarantula hawk
qualifies as a "classical lek" species whose mating system is analogous
to that of certain birds and mammals (although male tarantula hawks
do not display to incoming females, like many vertebrate males, but
merely fly from their scanning post to capture them.)

Lek Polygyny in Other Insects

There are other lekking insects whose males defend symbolic perch
territories that are not centered on females or resources, including the
hilltopping bot flies (Catts, 1964, 1967, 1979) and several species of
landmark defending butterflies (O. Shields, 1967). In these species
food for the adults or larvae is not concentrated on hilltops, nor are
oviposition sites abundant there. The females are powerful fliers with
large home ranges, a factor that makes it difficult to defend a number
of females simultaneously or in sequence (Bradbury, 1981).

The arena at which males aggregate need not be a mountaintop or
ridge. Male orchid bees (*Eulaema meriana* and *Euglossa imperialis*)
defend perch sites in open areas created by treefalls in the tropical
rain forest (Kimsey, 1980). The bees defend smooth-barked trees that
contain neither flowers nor nesting sites for their females but serve as
platforms for the displays of the territory owners. In large clearings
numbers of males of both species can be found, although the two species differ with respect to the preferred perch sites. Males of *E. imperialis* perch at heights of around 3 m on small shaded sapling trunks
only about 6 cm in diameter, whereas males of *E. meriana* defend
trees in the 25 cm diameter range, on which they perch in full sun
only a meter or two above the ground. As in *H. ustulata,* males seem
to be competing as much for scanning sites as for display perches,
given the location of the selected trees and the tendency for males to
fly out to inspect passing bees. Females of the two euglossine bees do
not form nesting aggregations and they are long-lived, powerful fliers
with home ranges in excess of 2 km². Thus they, like the females of *H.
ustulata,* are difficult to monopolize and may have the time to inspect a number of competing males in order to select a truly dominant
individual.

Other lek-forming insects include a variety of Diptera (D. K. McAlpine 1975, 1979; Burk, 1981b; Sullivan, 1981; G. Dodson, 1982), of
which certain Hawaiian (and Australian) *Drosophila* may be the most
famous (see Fig. 3.12). Males of the lekking Hawaiian species occur

on tree ferns that do not serve as food or provide oviposition sites for females (Spieth, 1968). In some Australian *Drosophila* numbers of males may be found beneath certain woody bracket fungi that are not believed to serve as food for adults or larvae but whose smooth, white, sheltered undersurface offers a protected platform for the elaborate courtship displays of the species (Fig. 8.13; Parsons, 1977). Little or nothing is known about the mobility and home range size of these flies or about the catholicity of their feeding and oviposition preferences in relation to other species with different mating systems. But Parsons (1977) notes that the Australian lek formers resemble the Hawaiian ones in that they are relatively large species, which is suggestive of high mobility and extensive home ranges. These factors reduce the payoff to males that attempt direct or indirect defense of females. In the otitid *Physiphora demandata* and tephritid *Anastrepha suspensa,* females are known to oviposit in an unusually broad range of materials, which would make a resource defense strategy difficult to enforce. Thus males may be required to defend perch sites and to engage in complex courtship to demonstrate physiological competence to selective mates (Alcock and Pyle, 1979; Burk, 1981b; G. Dodson, 1982).

The contrast between the behavioral ecology of *Centris pallida,* the digger bee, and *C. adani,* a lek polygynist, provides a suitable comparative conclusion to this section. Territorial males of *C. adani* are

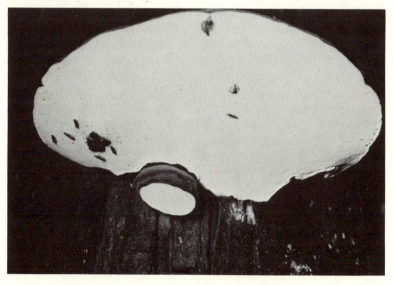

8.13   A lek of Australian *Drosophila* whose males gather on the sheltered underside of certain bracket fungi. (Photograph by J. Alcock.)

8.14   (*Above*) Landmark territorial site in a grassy field in Costa Rica. (*Below*) A solitary bee *Centris adani* defending the site. (From Frankie, Vinson, and Colville, 1980.)

regularly found on hillocks in grassy fields within second-growth Costa Rican forest during the period from January through April (Frankie, Vinson, and Colville, 1980). Each male defends a space of 3 to 4 m² centered over small clearings in the field (Fig. 8.14). The bees actively scent-mark dried grasses on the periphery of the territory, applying a mandibular secretion to the stem while crawling on it, before returning to hover on the alert. The defended area contains no nesting, emerging, or foraging females; the males seem to have symbolic territories in an open area where they wait for virgin females to arrive and make themselves available for copulation. Unlike *C. pallida*, females of *C. adani* nest as isolated individuals. The possibility of directly defending emerging virgins is therefore more limited for males of *C. adani* than for *C. pallida*. The option of monopolizing the food source is also weak, because females can gather their food at more than seven species of flowering trees that produce a mass of blooms during the breeding season of the species. The male attempting to defend a patch of the food plant would have to defend a very large area in order to achieve priority of access to many females. Some males do defend small territories in the crowns of flowering trees. The function of these territories, however, is probably not resource control but demonstration of competitive ability to selective females. The inability of males of *C. adani* to sequester emerging females or the resources they use, and the potential for selective mate choice by females given their high mobility and prolonged breeding season, combine to make lek territoriality the default strategy of males.

Scramble Competition Polygyny

The Eumenid Wasp *Abispa ephippium*

Territorial behavior by males, whether focused on females or resources that attract mates or symbolic dominance-demonstration sites, is dramatic behavior and has therefore attracted considerable attention. But polygyny resulting from territoriality is probably the exception rather than the rule for insects. In most species males avoid aggressive interactions altogether, make no effort to defend a territory, and instead invest their energies in flying or walking about in an attempt to be the first to discover and copulate with the receptive females in the area. We turn once again to an Australian eumenid wasp, *Abispa ephippium*, for an example of scramble competition polygyny (Smith and Alcock, 1980).

Males of *A. ephippium* show no aggression when they encounter one another while patrolling a section of stream visited by nest-building females. Females drink water in order to regurgitate it upon the soil, creating a clay paste that they then collect and employ in the construction of a mud nest (Fig. 8.15). As a female is drinking, she may

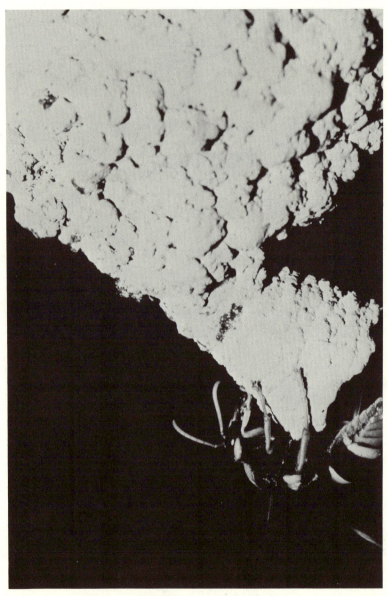

8.15 A female of the Australian wasp *Abispa ephippium* perched at the entrance to her nest. (Photograph by A. P. Smith.)

be spotted by a cruising male, pounced upon, and carried up to foliage over the stream. Females never resist copulation if captured by a male. After mating, females are released and are free to go about the business of nest construction. If they happen to be in the final phase of this process, they may complete a brood cell and lay an egg in the empty chamber. If sperm precedence occurs, the last male to mate before oviposition will enjoy a reproductive advantage. Therefore males are confronted with the same two options described for the flower-defending megachilids: (1) they may find a nest or nests and mate repeatedly with the female or females there, with a high probability that some copulations will occur just prior to egg laying, or (2) they may mate with many females at some other location, with a consistently low probability that any one copulation will be productive for the male.

Why in at least some populations do males choose to patrol water courses rather than defend nest sites? In the first place, females of this species nest in isolation from one another, so that nests are highly scattered. Although each nest is multicelled so that a few females may emerge from it, the genetic gains from copulating with virgins are reduced because females mate with any male that grabs them while they are drinking at a stream. Defending an active nest with an ovipositing female solves the potential problem of sperm competition but, as in *Anthidium,* the distribution of nesting females ensures that males will be restricted to one female as long as they remain at one nest site. If copulations can be achieved at high frequency away from a nest, the patrolling option can be more productive. The areas of Australia inhabited by these wasps are arid, and sources of water often limited. This ecological factor concentrates females spatially and makes it possible for some males to secure four or more copulations per afternoon (Smith and Alcock, 1980). Presumably the relatively high rate of male-female encounters compensates for the fact that any one mating has a low probability for a male of resulting in egg fertilization.

Why not defend a segment of a stream, thereby monopolizing a resource that attracts females? Water is a resource of uniform value to females, unlike flowers that are exploited by females of *Anthidium* and controlled by males. A water-collecting *Abispa* can satisfy her needs equally well at the tiniest puddle or the largest pool. Smith and Alcock (1980) found that almost no female returned predictably to any one spot to drink (except for a single female that had found a sheltered site where she could drink without attracting the sexual attention of males). In contrast, a pollen-collecting *Anthidium* has requirements that cannot be met equally well at any flowering plant.

Some plants are much more productive than others, and thus more attractive to females and more valuable to males. Males that monopolize rich flower patches consistently mate more often than rivals in resource-poor locations (Severinghaus, Kurtak, and Eickwort, 1981). Males of *Abispa*, faced with an even resource distribution along a stream, presumably cannot determine where females are likely to drink and therefore the costs of territoriality will not, on average, be offset by heightened mating success. Males avoid fighting entirely and cruise the stream, scanning constantly for a drinking female who may appear at any place and at any time during the day.

Our next question is why scramble competition polygyny has evolved in this species instead of lek polygyny, which is also correlated with environments in which males cannot economically defend females directly or indirectly. One explanation has been proposed by Emlen and Oring (1977) for the short-lived, explosive mating assemblage. If receptive females are available for only a short period, they may mate promptly and be less discriminating than if they had a long period in which to select a mate. As a result, a male may gain more by scrambling to contact as many (relatively) unselective mates as possible in the short time that they are about, than by investing valuable time in the defense of an area. Thus lek polygyny should be correlated with prolonged breeding seasons (and long-lived females), whereas scramble competition polygyny should be associated with shorter breeding seasons (and short-lived females).

But *A. ephippium* females do not appear to be especially short-lived, nor do they have an unusually restricted breeding season. Their dependence on water, however, enables males to encounter many potential mates at the stream. Any clumping of receptive females quickly selects for males that focus their mate locating on the concentration points (Parker, 1978a,b). Whether or not territoriality is advantageous then becomes a function of the way in which receptive females are distributed, which in turn depends on the nature of the resource or other factors that drew females to the area in the first place. Territoriality in *A. ephippium* is a costly option for the reasons already discussed; as a result, males patrol nonaggressively in a race with their competitors to detect and capture drinking females, which may appear with equal probability at any location along the stream.

Scramble Competition Polygyny in Other Insects

There are explosive mating assemblages in insects, the classic example being those mayflies that emerge en masse on one evening, mate, lay their eggs, and die all within 24 hours (Needham, Traver, and

Hsu, 1935; Edmunds and Edmunds, 1979; Sweeney and Vannote, 1982). This phenomenon is not limited to mayflies, as it has also been reported for a cerambycid beetle (Gwynne and Hostetler, 1978). Moreover, mating patterns of some ants are reminiscent of explosive mating assemblages, with clouds of reproductives released synchronously from the colonies in an area with mating taking place promptly at landmarks (Michener, 1948; Hölldobler, 1976), or in aerial swarms (Eberhard, 1978), or with males tracking perched females that release a sex pheromone (Kannowski and Johnson, 1969). In all these species there is a premium associated with rapid mating, and the opportunities for polygyny, if they exist at all, go to males that are able to locate and capture a number of females in a very short period of time.

A greater number of cases of scramble competition polygyny are available that match the *Abispa* pattern, with females occurring in relatively dense concentrations at emergence, feeding, or oviposition sites over a period of some time, but with males unable to exercise the female defense or resource defense options. One feature of these species that makes male territoriality unprofitable is the large number of males within the searching area. In some ground nesting bees and wasps, nest density reaches very high levels (Fig. 8.16). Within a small area hundreds, even thousands, of males may cruise above the

8.16 The nest entrances of a solitary bee, *Diadasia*, whose females often nest in dense aggregations composed of many hundreds of individuals. (Photograph by J. Alcock.)

emergence site at the time when receptive females may be detected (Evans, 1966; Evans and West-Eberhard, 1970). The typical behavior of these males is to patrol a portion of an emergence site, weaving low over the ground in a rapid, sinuous "sun dance"—all the while ignoring their fellow males completely and remaining alert for freshly emerged, receptive, virgin females. In some cases the distribution of these females is likely to be reasonably uniform over broad areas, making the yield for a territorial male less than it would be if mates were more concentrated within defendable areas. In other cases there appear to be few cues that enable males to discriminate between superior and inferior portions of the emergence site, even if they existed. For example, in species that nest in loose sand, emerging individuals leave no permanent mark in the soil that males can use to locate areas of high nest density from which many females may still emerge.

Although males sometimes are able to find (and defend) small spots from which a female is about to emerge, in many species this does not occur (for reasons that are not always clear). But in certain thynnine wasps, females do not become receptive until they have left the ground and crawled up onto a elevated perch from which they release a sex attractant (Fig. 8.17). Only then do the males, which patrol long circuits through appropriate habitat, detect the females and rapidly approach them (Given, 1954; Ridsdill Smith, 1970; Alcock, 1981b,c). The bottom line is that territorial males expend time and energy to repel other males from even small areas; if there are many males that will invade a territory, the owner pays a high price to control the area.

This principle may explain why males sometimes fail to defend locations against other males, even though they are able to detect preemergent females. For example, males of the ichneumonid wasp *Megarhyssa* (Crankshaw and Matthews, 1981) search tree trunks for exit tunnels that emerging females are constructing in the wood. Sometimes the male even inseminates the female before she emerges by slipping his abdomen into the tunnel before the female has departed from it (see also Nuttall, 1973a,b). Yet, peculiarly, a male that has found such a spot does not aggressively repel other males from it (Fig. 8.18) even though opportunities for female defense polygyny seem excellent. Three closely related species of *Megarhyssa* may occur on and in the same host tree, and males apparently are not adept at identifying the species of a preemerging female. This has two consequences. First a territorial male will often invest in the defense of a female of the wrong species, reducing the benefit of territoriality. Second, the fact that males of three species are present, contending for

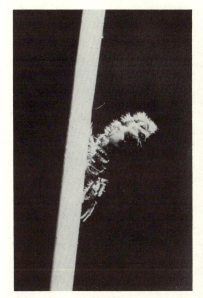 

8.17   (*Left*) Female thynnine wasp in the calling position, from which she releases her sex pheromone to attract searching males. (*Right*) An attracted male about to carry the female away to copulate with her. (From Alcock, 1981c.)

emergence holes, means that the population of potential intruders is high, and indeed up to 28 males of three species have been seen crowded around an emerging female. The high density of males raises the cost of attempts to engage in female defense and, perhaps as a result, males simply try to be first to make genital contact with a female.

Emergence sites are not the only focus of scramble competition polygyny. In some insects masses of males and females gather at a feeding site, and mating may take place there in the absence of much aggression among males. Pentatomid stink bugs and other hemipterans often form dense aggregations on their host plants (Southwood and Leston, 1959), perhaps because of mutual feeding advantages gained through joint inoculation of the victimized plant with salivary toxins or enzymes (Nuorteva, 1954; Adams and McAllan, 1958). A male that attempted to monopolize a harem of females or a rich feeding site would be confronted with a host of opponents (Fig. 8.19).

Likewise the large concentrations of male odonates of some species at the oviposition sites favored by their females may make resource defense excessively expensive. Males of the damselfly *Enallagma ci-*

8.18 A cluster of male ichneumonid wasps (*Megarhyssa* spp.) gathered about a spot from which a female is emerging. (Drawing by Kip Carter; from Crankshaw and Matthews, 1981.)

vile are representative of this large group (Bick and Bick, 1963; B. Lyon, unpublished). They gather at ponds and perch at the edge of the water with as little as a few centimeters between individuals. There are almost no aggressive interactions among males. Just as important as the high density of males in reducing the advantages of territoriality is the even distribution of the oviposition resource. Unlike females of *Plathemis lydia* that favor limited patches of moist debris raised above the water (Jacobs, 1955), females of *E. civile* can and will oviposit on a huge range of submerged aquatic vegetation, including algal mats that may spread across large sections of ponds and lakes. The location of the greatest number of ovipositing females varies unpredictably from day to day (B. Lyon, unpublished). Because there are no points to which females are especially attracted, they approach the pond from all directions. There is no special area which if controlled by individual males would promote mate monopolization. Therefore males do not invest in territoriality, but instead perch on the lookout for the gravid females flying out over the water (having come from inland foraging zones). Females copulate with whatever male is the first to capture them. The keen discrimination of the males and the rapidity of their response to females flying overhead are testimony to the strength of selection for skill in scramble competition. Fincke (1982) has demonstrated strong selection for scrambling in *Enallagma hageni,* a species whose mating system is similar to *E. civile.* She found that over 40 percent of males in a Michigan population did not obtain a single mate in their entire lifetime, but a few males were extremely successful. On the other hand, almost all females mated.

We have tried to demonstrate that an ecological classification of male mating systems can be applied to insects as readily as to birds and mammals. Male insects, like their vertebrate counterparts, appear

8.19 A large mating and feeding aggregation of the pentatomid bug *Chlorochroa sayi* gathered on a yucca pod. (Photograph by J. Alcock.)

to try to monopolize mates when it is possible to secure a disproportionate number of copulations through territorial defense of a small
area containing females or resources attractive to females. When this
is not possible because of the spatial or temporal distribution of mates,
males resort to scramble competition polygyny or lek polygyny—
although considerably more needs to be learned about the ecological
correlates of this last mating system.

9

Intraspecific Variation in Male Mating Systems

IN OUR LAST CHAPTER we attempted to assign each of the varieties of male mate-acquiring tactics to one of a handful of mating systems. In so doing we ran the risk of lumping under a single label rather different kinds of reproductive behavior. Moreover, by calling the males of a given species "resource defense polygynists" or "mate guarding monogamists," we are apt to ignore differences in the mate locating behavior of males *within* the population. The temptation to think typologically applies particularly with regard to insects, which are commonly considered to be simpleminded, rigidly programmed animals with a single battery of behavioral responses. This view is incorrect. There is a growing body of evidence on the behavioral flexibility of individual insects and on the widespread occurrence of intraspecific variation in the reproductive options of males. This chapter reviews such evidence, first presenting some examples of alternative mating strategies in insects and then analyzing the ecological correlates of variation in mating systems within a species. Although one would expect that whatever tactic yielded the greater reproductive success would over time completely replace the inferior alternative, there are genetic mechanisms that could lead to long-term maintenance of two or more mating systems within a species.

Behavioral Polymorphism

Centris pallida

Our friend the digger bee is an insect in which alternative methods of mate acquisition occur within the species (Alcock, Jones, and Buchmann, 1977). We labeled the mating system of the bee "female defense polygyny," based on the ability of some males to defend a series of small territories over locations from which receptive, virgin females would emerge. But not all males search and fight for digging sites. Some individuals invest their time and energy in the maintenance of a hovering site, sometimes by a flowering shrub or tree, typically a palo verde or ironwood tree (Fig. 9.1). Other individuals hover by mesquites or weed clumps on the edge of an emergence site over which dozens or hundreds of their fellows swirl in weaving patrol flight. The hovering males chase any flying object that comes within a meter or

9.1　Alternative mate locating behavior of the bee *Centris pallida*. Some males hover by waiting stations to pursue recently emerged females flying overhead, while others search for digging sites—see Fig. 4.1. (Photograph by J. Alcock.)

so; very rarely, they appear to grapple in flight with a male intruder. Probably as a result of repeated chases, males stationed around a tree will be spaced some distance (a meter or two) from their nearest neighbor. The hoverers therefore appear to be territorial, but even though marked males remain at one site over the course of a morning, they only infrequently return to the same post on succeeding days. Hovering males are probably scanning the sky for females passing overhead, and one occasionally sees aerial pursuits that lead to the capture of a female, with subsequent copulation.

The behavior of the hoverers does not fit neatly into any of the categories of mating systems we have discussed. These males do not appear to be defending a flower-rich territory or a dominance-demonstration territory, because they fly out of the defended area when a female passes overhead. Rather than invent a new label, perhaps we need only note that hoverers generally seem to defend a space that offers a decent vantage point from which they may dart after airborne virgin females. Digger-patrollers, on the other hand, zoom about to find positions that they defend in order to monopolize about-to-emerge virgin females. The mating tactic of the hoverers appears to be predominantly scramble competition (despite their weak territoriality), while that of the diggers can be characterized by female defense (despite the scramble to find digging sites). The essential point is that the two methods of mate location are different, although the males belong to the same species.

Panorpa Scorpionflies

Males of some species of mecopterans belonging to the genus *Panorpa* exhibit not two but three distinct mating tactics (Thornhill, 1979a, 1980e,f, 1981, forthcoming). First, as described in previous chapters, some males defend dead insects that they may encounter or collect as a nuptial gift for their females (resource defense polygyny). Males regularly find suitable gifts in such short supply that they must resort to stealing prey items from the webs of spiders (a dangerous business). After securing food, a male guards it while releasing a sex pheromone that attracts females. Visiting females feed upon the carrion and copulate with its defender.

A second alternative for a male is to feed upon dead insects (again sometimes robbing spiderwebs) to produce salivary materials. When a male has a sufficient supply of saliva within its remarkably large salivary glands, it secretes a mound of the substance onto a surface and guards it while calling to receptive females who may arrive, eat the

9.2 A male *Panorpa* scorpionfly holding a salivary secretion and releasing a sex pheromone from an everted gland in its genital "bulb" to attract a female, to whom it will present the nuptial gift. (Photograph by T. Moore.)

nuptial present, and mate (Fig. 9.2). Males of this sort, like the defenders of carrion, are engaged in resource defense polygyny but focused on a secretory product.

The third option for a male is not to defend food at all, but to search actively for females. When a male of this sort encounters a prospective mate, he darts toward her and grabs her by the wing or leg with his large genital forceps. The male that succeeds in getting a grip on the struggling female then seeks to reposition her using his notal organ (a clamping device on the dorsum of his abdomen) to hold the female while he maneuvers into position to make genitalic contact. Although females appear to resist and often struggle free, occasionally a male is able to grasp the female's genitalia with his own and inseminate her (Fig. 9.3). There is no food transferred from the male to female in these cases and the exercise has the appearance of forced copulation.

Ecological Correlates of Alternative Mating Strategies

What is there about the ecology of a species that enables males to exhibit more than one method of acquiring mates? The digger bee illustrates one of the common (but not universal) correlates of alternative mating strategies—the occurrence of receptive females in more than one location in the environment. In *C. pallida* the two classes of receptive females are (1) emerging females and (2) those that have not been mated immediately upon emergence and have left or are leaving the emergence site. The existence of two sources of females permits males to exhibit the two different tactics required to capture buried and flying mates.

In several other species males may attempt to locate mates in two or more different areas within their environment. A male of the cricket *Anurogryllus arboreus* may (1) search for a female in her burrow, (2) call to females, attempting to attract one to his burrow, or (3) leave his home and move to a tree trunk to call from an elevated perch for a mate (T. J. Walker, forthcoming a). The existence of sedentary and wandering populations of females enables males to employ more than one mating strategy. In flies it is not uncommon for males that form swarms, typically at a landmark of some sort, to be found also search-

9.3 A *Panorpa banksi* male that has succeeded in securing a forced copulation with a female to whom he has offered neither a carrion meal nor a nuptial present of saliva. The female's wing is secured in a clamp-like structure on the male's abdomen. (Photograph by T. Moore.)

ing for mates in other locations (Provost, 1958; McAlpine and Munroe, 1968). The ceratopogonid midge *Culicoides nubeculosus* is a classic swarming species, but males can also be found perched on the body of a cow ready to grasp blood-sucking females (Downes, 1955). Similarly, males of the butterfly *Poladryas minuta* engage in hilltop-ping in the morning but patrol flowers in pursuit of females in the af-ternoon (Scott, 1974). Another variation on this theme is *Gastero-philus intestinalis*, the horse bot fly, whose males may either defend a territory about a horse (an oviposition site for females) or defend a perching territory on a landmark hilltop (Catts, 1979).

A reasonably frequent combination of encounter sites for insect species includes the feeding and oviposition areas. In the palm scarab beetle *Oryctes rhinoceros*, virgin female beetles feed in the crowns of palm trees and may be mated there; nonvirgin females move to the trunks of the trees to oviposit and can sometimes be induced to mate as well at these locations (Zelazny, 1975). The presence of receptive females in both feeding and oviposition or nesting areas is not uncom-mon in the bees and wasps (Alcock, 1979d) and has also been re-ported for several species of syrphid flies (Maier and Waldbauer, 1979).

The creation of two pools of receptive females is fostered if both vir-gin and nonvirgin females can be mated. Earlier we noted that males of *Photuris* fireflies may employ totally different signal patterns, de-pending on whether they are attempting to locate virgin or already mated females (J. E. Lloyd, 1980b). Males of the Mediterranean fruit fly also appear to have two tactics, one for virgins and one for mated females. Prokopy and Hendrichs (1979) enclosed a coffee tree grow-ing on a Guatemalan plantation and introduced fruit flies into the cage. Subsequently, they monitored where males (and females, if any) were found in the tree and what they were doing. Grouping the two experiments in which males only and males plus *virgin* females were released in the cage, 76 percent of the males stationed them-selves on the undersides of leaves. Many of these individuals daubed their leaf perches with sex pheromone and often three to six males were found on adjacent leaves, one male to a leaf, a pattern reminis-cent of a lek. (See G. Dodson, 1982, and Burk, in press, on another lekking tephritid.) In contrast, less than 10 percent of the males perched on coffee berries, the oviposition site of the fly. These males tended to make relatively short visits to berry clusters, during which they rarely marked the fruits with sex pheromone. Perhaps virgin fe-males tend to mate with males on leaves because the males there in-tercept them on their way to the oviposition sites, or perhaps females prefer to mate with a dominant member of a mating aggregation—if the clusters of males truly constitute leks.

Male behavior was considerably different in an experiment in which only females that had already been mated were introduced into the cage; in this case fewer than 60 percent of the males perched on the undersides of leaves, whereas 23 percent inspected coffee fruits. Prokopy and Hendrichs (1979) suggest that males are drawn to fruits by the oviposition deterring pheromone that females place on the berries in which they have laid eggs. This cue could provide information on the relative abundance in the tree of mated, egg laying females. If ovipositing females are numerous, some males may move to fruits to search for copulating partners (perhaps mated females occasionally replenish their sperm supplies immediately after oviposition). Depending on the ratio of virgin to nonvirgin females, males may be able to adjust their mate locating tactics to match the availability of receptive females at fruits or leaves. In the last analysis, the two mate locating options exist because there are two populations of females that can potentially be mated in two distinct portions of the host tree.

The fact that mated females are often less receptive than virgins could contribute to the evolution of forced matings. In the tephritid *Rhagoletis pomonella*, ovipositing (= mated) females are much less likely to copulate than virgins. It may not be coincidental that males approaching ovipositing females dispense with courtship and slip onto the females from the rear (Smith and Prokopy, 1980). By pouncing upon her, the male may make it difficult for an unwilling partner to reject him. Likewise, in the meloid *Tegrodera aloga* some males attempt to copulate with mated females that are usually not receptive. As in *Panorpa* the male possesses an enlarged genital apparatus, which is used to hold the reluctant female firmly, perhaps increasing the probability that she will accept the mating to free herself from her captor (Pinto, 1975).

We suspect that in nature, "rapist" panorpas are often found in association with other males that are calling females while defending a nuptial present. Here the alternative strategy may consist in part in "stealing" females attracted by other males. If this speculation is correct, silent panorpas would join the better-known examples of the silent-male strategy. Some male crickets and grasshoppers do not try to attract females by singing (Otte, 1977; Cade, 1979b, 1980). They gather surreptitiously near a calling male and attempt to capture females drawn to the area by the singer. Whether the intercepted females resist copulation in these instances does not seem to be known.

Thus alternative mating tactics appear in species (1) whose recep-

Table 9.1 Insect species in which large males enjoy an advantage in aggressive disputes with smaller rivals.

Hymenoptera

Anthidium maculosum and *A. manicatum* (megachilid bees)

Euodynerus foraminatus (eumenid wasp)

Philanthus basilaris (sphecid wasp)

In all four species territorial males are significantly larger than nonterritorial individuals (Alcock, Eickwort, and Eickwort, 1977; Severinghaus, Kurtak, and Eickwort, 1981; D. P. Cowan, 1978, 1981; O'Neill, 1981).

Philanthus bicinctus (sphecid wasp)

The largest males control territories with the greatest number of nesting (receptive) females (Gwynne, 1980).

Diptera

Scatophaga stercoraria (scatophagid fly)

Larger males are more likely to own a territory and displace smaller rivals from receptive females; under all conditions of resource availability and fly density large males are more likely to secure mates than small males (Borgia, 1980, 1982).

Achias australis (platystomatid fly)

Only one large male became territorial under caged conditions; the smallest male in the group avoided combat (D. K. McAlpine, 1979).

Drosophila planitibia (drosophilid fly)

The smaller of two contestants for a lek territory almost always flees from its larger rival after a brief display (Spieth, 1981a).

Anastrepha suspensa (tephritid fly)

Larger males win more fights than smaller rivals (Burk, in press).

Coleoptera

Monochamus scutellatus

Males in the largest size class win more than twice as many aggressive encounters as they lose and spend a far greater proportion of their time paired with females than their smaller rivals (Hughes and Hughes, manuscript).

Podischnus agenor

Small males lack the horns needed for success in physical combat with their larger rivals for control of oviposition burrows in sugarcane (Eberhard, 1979, 1980).

Brenthus anchorago

In 97 fights for control of a female in which there was a size difference between the combatants, the larger male won 92 times (Johnson, 1982).

Orthoptera

Gromphadorhina portentosa

The largest male in a group of captive individuals was the most frequent victor in the dominance disputes of this aggressive species (Barth, 1968).

Table 9.1 *(continued)*

Some counterexamples: large males in combative species that are not necessarily more likely to defeat smaller rivals.

Hymenoptera

Mischocyttarus flavitarsis (vespid wasp)

Males on territories were no larger on average than a random sample of emerging males (Litte, 1979).

Diptera

Physiphora demandata (otitid fly)

Mating males in this territorial species were not larger on average than nonmating males (Alcock and Pyle, 1979).

Orthoptera

Acheta spp. (gryllid crickets)

Although large males may enjoy an advantage, other factors (such as recent copulatory success) can enable a small male to defeat a large rival (Alexander, 1961).

tive females appear in two distinct locations in the environment, such as spatially separated feeding and oviposition sites, or (2) in which virgin and nonvirgin females require different courtship tactics, or (3) when opportunities exist to intercept females approaching other males that are producing attractant signals.

Variation in the Attributes of Males and Alternative Mating Tactics

Intrinsic differences among males, as well as variation in the sources of fertilizable females, can promote diversity in male behavior. In particular, males that are relatively poor at resource collecting or resource defending may salvage some chances to reproduce by exploiting an alternative source of females. For example, some male panorpas may fail to find carrion or remove prey from spiderwebs; and even if they encounter an item, other males may steal it from them. The "weaker" males are doomed to reproductive failure unless they succeed in forced matings, in which case they will perhaps pass on some of their genes.

Similarly some male empidid flies that gather in swarms may not carry a prey item to offer to females (Downes, 1970; Alcock, 1973; Chvála, 1978, 1980). This may be so because (1) it is possible to intercept females drawn to the area by the prey-offering males and (2) some males are less able to capture prey.

Perhaps the single most important factor that affects the resource-holding power of a male is size (Table 9.1). Relatively small males are often less able to capture and defend a resource that promotes reproductive success than large rivals, as has been demonstrated for *Panorpa* males (Thornhill, 1981). Rather than compete directly with their more powerful fellows, smaller individuals may seek out other females. In the digger bee *C. pallida*, for example, some males weigh three times as much as others as a direct result of variation in the provisioning behavior of females (Fig. 9.4). Early in the cycle of constructing a nest, the female determines the size of the bee she will produce there. There are great differences in the volume of the brood pots shaped by different females. The capacity of the brood pot determines the quantity of provisions that can be stored in it, and this in turn limits the growth of the larva that emerges from an egg laid in the brood cell. Offspring that receive large amounts of food grow to be large and powerful adults, whereas those that have received less food develop into smaller bees (Alcock, 1979c,d).

Size in *C. pallida* strongly influences the fighting ability of males and thus their relative potential for defense of emerging females (Alcock, Jones, and Buchmann, 1977; Alcock, 1979c,d). In locations with dense populations of males and about-to-emerge virgin females, large males enjoy much higher reproductive success than their smaller

9.4   Size variation among males of *Centris pallida*. Large males are never found hovering; very small males rarely attempt to dig down to preemergent females. (Photograph by J. Alcock.)

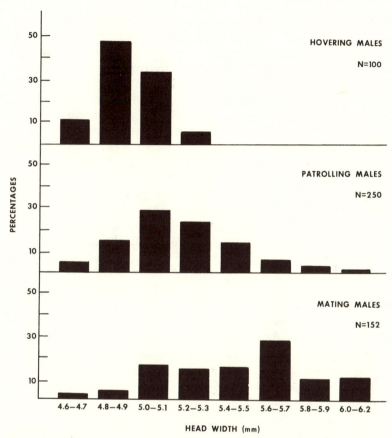

9.5 Size and mating success in *Centris pallida*. Large males enjoy disproportionate copulatory success in dense emergence sites of the bee. (From Alcock,
1979c.)

rivals. They make up a substantial proportion of the bees found mating, although they are only a small minority of the total male population (Fig. 9.5).

The success of larger males is directly related to their ability to defend their digging sites with threats and wrestling fights and to displace smaller males from their digging territories by force. In areas
with many emerging females, diggers are regularly challenged by
other males at the rate of once every minute or two at some spots.
Large males win this competition at the expense of smaller ones.

The inability of small digger bees to excavate females without interruption from larger rivals could not in itself result in the evolution
of an alternative mating tactic. What is also required is a source of fe-

males that can be acquired without wrestling over a digging site. If there were no other source, the less effective wrestler would have only two options: to wait, if there were a chance that at some future date its larger opponents would be dead or less effective fighters, or to fight to the death if there were no chance that conditions would change in the future. But in *C. pallida* there is an alternative source of mates not monopolized by the larger males: the females that emerge from scattered nests not discovered by digger males and the females that escape from diggers. The search for airborne females is largely incompatible with locating buried females. Hoverers specialize in the former, diggers focus on the latter. Large size provides no advantage and possibly even handicaps a male in an aerial chase after a swiftly flying female. Because there is a premium on speed and aerial agility, small males may do as well as or better than large ones in this specialized alternate mode of mate capture. Hoverers are consistently the smallest males in the population.

The same relation between size difference and difference in aggressive ability has been observed in many other species. Large male dung flies are able to exclude smaller ones from territories on dung pats. If one removes an owner, the site is usually taken by a smaller male. When this individual is netted, a still smaller male is likely to claim the spot (Fig. 9.6; Borgia, 1980).

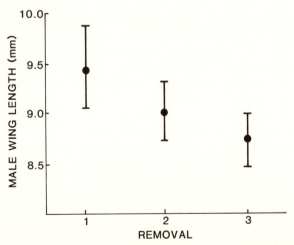

9.6 Large males of *Scatophaga stercoraria* monopolize territorial sites on dung. The original territorial males (*1*) are larger than their replacements (*2*). When replacement flies are removed, still smaller males (*3*) become residents. (From Borgia, 1980.)

9.7 Size variation among males of *Brenthus anchorago*. Two males—one small, one medium-sized—are mounted on females. Another much larger male approaches the twosomes. (Photograph by A. Forsyth.)

Small males of *Nasonia vitripennis*, a parasitic wasp, that emerge to find larger rivals on the host puparium have been observed to duck back into the puparium, presumably to mate with the females within, leaving their larger fellows outside to fight among themselves for control of the area (King, Askew, and Sanger, 1969). In the same general vein, small males of the eumenid wasp *Euodynerus foraminatus* patrol flowers and foliage in search of the occasional receptive female, while their larger rivals guard nests from which virgins will emerge (D. P. Cowan, 1978, 1981). Small males of the scarab beetle *Podischnus agenor* avoid larger opponents in part by emerging sooner and dispersing farther, thereby reducing the chances that the nest burrow they construct in a sugar cane will be stolen from them by a larger individual (Eberhard, 1979, 1980).

It is possible to avoid futile aggression and still coexist with larger males. Small males of the brentid beetle *Brenthus anchorago* regularly are found in dense mating aggregations with conspecifics much larger than they are (Fig. 9.7; Johnson, 1982). A big male uses his size and power to good effect, slamming his snout into a rival perched on the back of a female and knocking him to the ground. Even more insidiously, males may use their snouts as clubs to pound at the genitalia of copulating opponents, an action that encourages a brief mating. After persuading a rival to retract his genitalia, a large male may then pry the other male from the female. The snout power of large males gives them an advantage in the contest for mates and consequently they mate more often with females of their choice just before they oviposit. (Males guard females that are boring holes in wood preparatory to laying an egg in the hole; they often mate at intervals during this process, the last mating coming a few minutes prior to oviposition.)

Because some males weigh only one twenty-fifth as much as others, they would seem to be at a hopeless disadvantage in this species. (In human terms this would be equivalent to a 150-pound boxer matched against one weighing close to 2 tons!). But because the oviposition resource is rare and dispersed, small males do not have the option of leaving an area populated by monstrous competitors; females require fallen limbs or trunks of the tropical tree *Bursera simaruba* that are in the early stages of decay. So small males remain in mating aggrega-

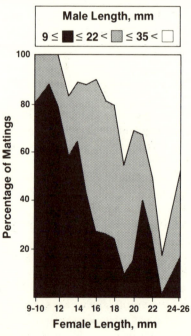

9.8 Assortative mating in *Brenthus anchorago*. Note that small males are paired most often with small females, whereas large males monopolize large females. Large females seem to prefer large males as mates; small females are left with subordinate small males as partners. (From Johnson, 1982.)

tions and exploit the two avenues of reproduction open to them. The first consists in accepting the females larger males ignore. Large males prefer large females. They have only so much time to devote to the acquisition and guarding of mates and therefore prefer to invest their effort in mating with the largest females, who lay eggs five times the size of those produced by the smallest females. The larvae that emerge from large eggs have a head start in the competition for food within the decaying log. They are more likely to secure considerable quantities of wood and so grow into large and effective fighters if male, or large and effective egg layers if female. Because competition for smaller females is less intense, males of average or small size can mate with these females without being clubbed by a larger beetle (Fig. 9.8).

There is a far more daring way in which a very small male may secure a highly desirable large mate. Because some males are tiny, they can occasionally slip undetected onto the back of a female, under the very snout of the hefty male that is guarding her, and copulate at the most propitious time for fertilization of her eggs (just after the large male has mated and just before the female oviposits). Thus smaller males of a variety of insect species have evolved tactics that, at the very least, enable them to salvage some reproductive opportunities in a social environment dominated by their larger opponents.

Age and the Resource Holding Power of Males

Size is not the only attribute that can influence a male's opportunities to monopolize females, although it is often an important character. Age, experience, and general physical condition can be just as significant in determining the outcome of a dispute between males. For example, the age of male dragonflies and damselflies is believed to be more influential than size in affecting success in aggressive interactions (Waage, 1973; Campanella and Wolf, 1974; Fincke, 1982). Relatively young and relatively old males often find themselves at a disadvantage; these individuals are therefore forced to wait or to adopt an alternative mating option. Adult male odonates often must take considerable time to acquire the food they need to come to peak fighting condition, and only males in their prime are able to challenge residents successfully or hold a territory in the face of repeated intrusions.

In the damselfly, *Calopteryx maculata*, young males that have yet to hold a territory and older ones that once did but have been supplanted by a more vigorous individual employ an alternative mating system (Waage, 1973). Making no effort to defend a superior oviposition site, the focus of aggression among some males, they instead patrol the borders of streams, always evading territory owners and assiduously avoiding combat. Their goal appears to be to find receptive females before these individuals fly out over a territorial male's preserve. If a nonterritorial male finds a female perched by the streamside, he immediately attempts to achieve the tandem position without preliminary courtship (unlike territorial males). His mating attempts rarely succeed, but sometimes a female accepts him, often under unusual circumstances. For example, the female may have been courted by a territorial male that dashed off to repel an intruder, leaving the field to a nonterritorial male waiting inconspicuously nearby; females often accept the male in these cases, perhaps having been primed to copulate by the resident male. Females mated by subordinate males must still go to the water to oviposit. They are generally detected by a territory owner, courted, and mated again with most or all of the sperm of the previous male unceremoniously removed (Waage, 1979a). Occasionally a female will lay her eggs in an undefended stretch of stream or will join a group of females ovipositing in a male's territory, with the owner either unable to detect the late arrival or unable to fertilize her because his sperm have been depleted by the previous matings of the day. If this happens the nonterritorial male may gain a measure of reproductive success, to be added to whatever productive matings he has already achieved as a past territory owner—or will achieve, should he ultimately secure a territory.

Alternative Mating Strategies and the High Costs of Success

A third major correlate of alternative mating tactics (in addition to variation in the location of receptive females and in the resource holding power of males) is the association of high costs with the strategy that appears to yield greater reproductive success. Typically the alternatives include one aggressive, territorial option and one nonaggressive, nonterritorial tactic. Selection for success in aggressive competition may lead to an "arms race" among males, with ever larger or more aggressive individuals favored (Gadgil, 1972; West-Eberhard, 1979). But the successful territorial males pay an increasing price, because of the demands and risks of their behavioral tactics and the costs of producing the morphological structures needed for their role. Large male *Centris* or hyperaggressive male crickets may be able to defeat their rivals in competition for females, but they experience time-consuming and life-threatening costs in the form of (1) time spent in aggression and risk of injury stemming from fights with conspecific males, (2) danger of parasitic or predatory attack while engaged in sexual advertisement, aggression, or resource defense, and (3) time and energy required to become an unusually large individual.

This last cost is an especially significant one. In order to become large, a developing male often must spend more time as an immature individual, with the attendant risk that it will be killed before ever having begun to reproduce. One of the benefits of being a small sugarcane beetle rather than a large one is the shortened developmental schedule that enables the smaller individual to achieve adulthood more rapidly and to enjoy some females freed from the aggressive tyranny of his more ponderous rivals.

Large body size also requires that more energy be consumed by the immature animal if it is to reach large dimensions, and still more after achieving adulthood if it is to maintain itself. As selection favors larger and larger adult males for their competitive success, the costs of securing the necessary resources may rise ever more rapidly for the immatures. We must also consider those insects in which the female parent provides materials that help determine the size of her progeny. In cases in which egg size determines larval competitive ability or the amount of brood provisions limits body size, the costs of building a large male are paid primarily by the mother. Her reproductive success is not set by the reproductive ability of any one offspring but is a function of the sum of the reproductive outputs of all the individuals that she produces. If a female can create two small males for the same price as one large one, the smaller males need not do so well as the large individual. If their reproductive success averages 55 percent of the success of a male twice their size, then their mother gains 10 percent by investing in two small males rather than one large offspring (Alcock, Jones, and Buchmann, 1977). D. Cowan (1981) has shown that provisioning females of the eumenid wasp *Euodynerus foraminatus* make smaller sons when nests are dispersed and the opportunities for female defense polygyny are reduced.

Thus, theoretically, the benefit-to-cost ratio of two alternatives might be closer than one would imagine from focusing solely on the benefits of the apparently successful strategy. The high-gain option will often also be a high-cost option, whereas the low-gain alternative may sometimes be a low-cost method. A practitioner of a low-gain tactic might live longer (or cost less to produce) because of its less demanding activities and could eventually secure some mates. But it is

one thing to show that males that practice the low-gain alternative secure *some* matings and quite another thing to prove that the competing tactics confer equal fitness on those individuals that use them. In
fact, this is very unlikely, given the extremely low probability that two
mating behaviors will yield precisely the same average genetic gain
year in, year out. So why does not one mating system completely replace the reproductively inferior one? To answer this question we
must examine the possible genetic-developmental bases of alternative
patterns of behavior in relation to evolution.

Genes, Strategies, and the Maintenance of Behavioral Variation

The major virtue of the burst of enthusiasm during the past decade for
the application of game theory to animal behavior, developed by
Maynard Smith (1972, 1974, 1976a) and championed by Dawkins
(1976, 1980), has been to demolish the expectation that all individuals in a population will evolve the same set of solutions to their ecological problems. Advocates of the game theory approach have shown
that the optimal solution is very often a function of what other individuals in the population are doing. If this is so, individual selection need
not inexorably weed out all but the single "best" behavioral option
available to individuals of the species. Instead, given certain (not uncommon) conditions, the evolutionarily stable result can be two or
more functionally equivalent, alternative behavior patterns that can
be maintained indefinitely in the population.

There are three major ways in which alternatives can coexist (setting aside a transient polymorphism, in which one trait is in the process of replacing the reproductively inferior alternative); these have
been reviewed by Cade (1980) and Dawkins (1980). In the first case,
the two alternatives are truly separate *strategies*, in the strict sense of
the word; that is, each option is linked causally to different competing
alleles and each produces the same average reproductive (genetic)
success. If one strategy were to confer higher fitness on its practitioners, over evolutionary time the reproductively inferior strategy would
be replaced. But if individuals of both types reproduce equally well,
the result can be a stable behavioral polymorphism. The only way in
which it is easy to envisage the evolutionary maintenance of two
genetically distinct options of equal *average* fitness is through
frequency-dependent selection. If as a strategy decreases in frequency
in the population, it confers greater and greater reproductive success

on its practitioners, the result will be an equilibrium of the combination of strategies (Gadgil, 1972).

A second possibility for the joint occurrence of alternative tactics in a population involves a single evolutionary stable strategy, not two. In a *mixed reproductive strategy* all individuals share the same genes for reproductive behavior, but first one and then, at a fixed time, another alternative behavior are practiced. Individuals can, in theory, possess the programmed ability to spend a set proportion of time in two distinct behaviors with the switch point set in advance, *provided* that the alternatives confer the same reproductive success. If individuals automatically switched to tactic B when continuing with tactic A would yield greater gains, selection would favor mutants that happened to persist with tactic A. Here again, frequency dependence should generally be required to promote the evolution of a mixed strategy, because of the very low probability of the reproductive equivalence of two (or more) alternatives.

But if individuals use variable conditions such as the relative age of their opponents, their relative size, the density of the population, or the behavior of their competitors to determine when to switch options, the result is labeled a *conditional strategy*. This provides a third avenue for the occurrence of a mix of alternatives within a population, even though all individuals may share the same genetic foundation for their reproductive behavior. Unlike a mixed reproductive strategy, an individual is not obligated to exhibit both alternatives, nor do the different options have to yield equal reproductive gains. Thus, for example, a male might "choose" to be territorial only if a suitable area were available in his environment. Otherwise the male might engage in nonterritorial scramble competition for females as a way of having at least some chance of mating, rather than none at all.

Let us now examine these three strategic possibilities in the context of the behavior of the bee *Centris pallida*. Because not every male switches from hovering to patrolling or vice versa (Alcock, 1979c) the alternatives probably do not represent a mixed reproductive strategy. In fact, there are no cases of male reproductive behavior in insects in which this strategy appears likely (Cade, 1980). Moreover, it is difficult to envision the ecological circumstances that could favor a preprogrammed switch in alternatives made without reference to environmental cues.

It is also unlikely that the two mating specializations in *C. pallida* have separate genetic foundations and so represent two pure strategies. There is continuous variation in size distribution, rather than two separate populations of large and small males. Furthermore, the larger individuals, which practice the digger option, often mate several times in the course of a morning, whereas small hoverers are rarely seen chasing and capturing females. The average reproductive gains of diggers versus hoverers are almost certainly different. Nevertheless, it may be that constant shifts in the direction of selection from year to year and place to place act to maintain the hovering morph. Digging/patrolling may only be adaptive in areas with dense populations of emerging females; in diffusely scattered populations, males that hover may do better. Even if this is true, a single conditional strategy could account for the variation in male behavior (Dawkins, 1980). Males that had the conditional capacity to hover, *if* they emerged in an area with few emerging females, or *if* they emerged in a place with many larger patrolling males, or *if* they were fed a small amount of food as a larva, could enjoy a reproductive advantage over males programmed to hover whatever their physical or social environment.

Likewise the three mating tactics of male panorpas represent three options within *one* conditional strategy (Thornhill, 1981, forthcoming). First, there are such great differences in the reproductive success of males defending dead crickets, versus those secreting and defending salivary mounds, versus those engaging in forced copulation attempts, that it is all but impossible that the three alternatives yield equal fitness gains. In laboratory experiments rapist males succeed in holding only a small proportion of the females they capture, and even when they copulate there is only a 50 percent chance that their mate will permit insemination to occur. Males that defend salivary mounds or carrion are not only more likely to encounter females attracted to their gifts but the females that do approach do not struggle free and will accept sperm. The attractiveness of males with salivary mounds, however, is only about half that of males with a large carrion gift.

Second, not only are the apparent reproductive gains of these options dramatically different, but there are also no compensatory cost-avoidance benefits for males engaged in the inferior alternatives. If anything, males in the rapist category are at special risk. Size in these scorpionflies influences not only resource holding power but also the ability of males to forage safely in spiderwebs (Fig. 9.9; Thornhill, 1975, 1978a). Large males can often avoid the need to collect food from spiderwebs because they have priority of access to whatever limited carrion is available. Moreover, if they do venture onto a web to feed, they are more likely to escape capture than small males. In laboratory experiments in which males of different sizes were placed in enclosures with spiders, each of which had two crickets deposited in its web, small males were significantly more often found feeding on

9.9 *Panorpa mirabilis* scorpionfly dead in spiderweb. Some males are forced to take special risks if they are to secure a nuptial prey for their mates. (Photograph by N. Thornhill.)

the spider's crickets and were more often killed than males in the large size category.

Therefore, in terms of both attractiveness to females and survivorship, large carrion-defending males are much more successful than smaller salivary-mound defenders or rapists. That males can and will switch from an inferior to a superior alternative, if conditions permit, has also been demonstrated experimentally. If 12 males are placed in a cage with 6 dead crickets, there soon will be 6 territorial males (the larger individuals) and 6 males without a cricket, most of whom will deposit a salivary mass. If the carrion owners are removed, the other males promptly abandon their salivary gifts and switch to defense of the crickets. Likewise, if males are confined to large containers in which no dead prey are available, most will produce a salivary mass that they guard. Some males do not do this (probably because of low salivary reservoirs) and instead prepare to attempt forced copulation. If one removes a number of the salivary defenders, the nonterritorial males quickly become defenders of the "abandoned" mounds (Thornhill, 1981).

Male *Panorpa* therefore have the ability to adopt any of the three options, depending on the availability of carrion and the nature of the competition. Small males that have been unable to acquire much food for conversion to saliva apparently adopt the rapist role only when competition for carrion is so intense that they cannot find or maintain a resource territory. Their only chance for reproductive success under these circumstances is matings achieved by forced copulation.

The large majority of potentially territorial insect species resemble the digger bee and *Panorpa* scorpionflies in that they usually encounter unpredictable and variable conditions, especially with regard to the amount of food available to them as larvae and also with respect to the nature of the social competition they will experience as adults (see, for example, Eickwort, 1977). Because of this, flexibility in behavior is often adaptive. Most authors are of the opinion that the bulk of examples of alternative mating "strategies" represent cases of a single conditional strategy (Cade, 1980; Dawkins, 1980). West-Eberhard (1979) has pointed out that the superiority of a facultative switch between two options as opposed to a pure fixed strategy or mixed reproductive strategy is most clear when one option (such as digging up virgins or defending carrion or holding a flower-rich territory) yields much greater reproductive rewards than the alternative. Males that are able to spend as much time as possible using the superior option productively will outreproduce those that are preprogrammed to employ only the inferior alternative or those that must spend a set fraction of their time first using one, then the other, tactic.

At the same time, the ability to switch to the secondary specialty is adaptive if this can be done when continued pursuit of the primary option will be unproductive, as when the individual has been unable to achieve large body size or when it is faced with unusually numerous rivals. The secondary option in most conditional strategies is a salvaging operation that enables the male to make the best of a bad situation (Dawkins, 1980). Persistent attempts by a relatively small, or weak, or old male to be a defender of females or of resources are likely to fail utterly. By switching to a secondary source of females, dropping out of the competitive race temporarily, or stealing females attracted to dominant males, a competitively inferior individual may have some chance, rather than none, to pass on his genes.

Although conditional strategies probably are the dominant source of variation in mating systems within a species, there are a few possible cases of mixed reproductive strategies in the insects (Brockmann, Grafen, and Dawkins, 1979) and also some examples of the coexistence of two genetically distinct options. An instructive possibility is provided by the fig wasps, which have winged and wingless male morphs (W. D. Hamilton, 1979). The wingless forms are the fighters, armed with large and powerful slicing jaws that they use to dispose of rivals for females emerging within their natal fig. The smaller, winged males use their wings to disperse from the fig in which they were born to reach other locations with emerging females and no fighting males. The two specialities are so distinctive and require such different structural adaptations that by the time a male reaches adulthood, he is fully committed to one behavioral option (Fig. 9.10). (Wings develop during the immature stage.) Thus the two behavior patterns may be two pure strategies with distinct genetic foundations rather than two facultative options of one conditional strategy. It is possible, however, that social interactions among larval fig wasps, and differences in their nutritional condition, could act as facultative switches controlling the development of wings and thus expression of the dispersal tactic.

We might expect to find a distinct genetic foundation for two options when the two alternatives are unusually different. But in the only case in which genetic differences underlying two tactics have actually been demonstrated, the options do not require dramatically different structural attributes. Cade (1981b) has conducted artificial selection experiments, which have shown that the amount of calling

9.10   An Australian halictid bee, *Lasioglossum* spp., like certain fig wasps, has two types of males, a large-jawed, flightless (fighting?) male and a small-jawed form capable of dispersing in search of females outside his natal nest. (From Houston, 1970.)

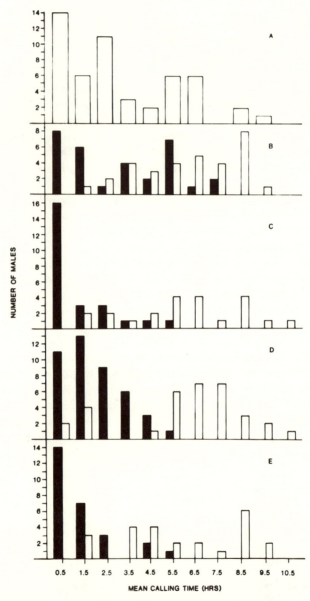

9.11 Artificial selection for singing and silent behavior in males of *Gryllus integer*. It is possible to increase the frequency of either trait through selective breeding over a period of a few generations. The black bars are the progeny of males selected because of their reduced calling time; the white bars represent the offspring of males that call for prolonged periods. (From Cade, 1981b. Copyright 1981 by the American Association for the Advancement of Science.)

time by males of *Gryllus integer* is under genetic influence (Fig. 9.11). Satellite males that wait by calling conspecifics are genetically different from callers, and the alternatives (calling versus noncalling) can be said to be distinct strategies. This example is not analogous to the fighting and dispersing strategies of fig wasps, because silent males are entirely capable of singing and occasionally do produce songs, whereas wingless males obviously cannot disperse. Although environmental conditions and male density do play a role in determining the amount of calling by a male, genetic differences contribute too. Within the population of crickets studied by Cade there are two genetically distinct conditional strategies. Genetic variation may persist because of fluctuating selection pressures created by spatial variation in the densities of parasitic flies that kill calling males. The probability of parasitic attack cannot be assessed by a male before it happens and therefore cannot be used as the basis for a facultative switch from calling to satellite behavior or vice versa. Maintenance of the competing genes may occur because of the continuously changing direction of selection; in populations of *G. integer* the males that are predisposed to call are favored when the parasitic flies are scarce, and those with a proclivity for satellite behavior are relatively successful when the flies are abundant.

Cade's work illustrates superbly how complicated the control of alternative options may be in population of insects. Few areas of insect behavior require more investigation than this topic because of the complexity of the proximate and ultimate factors involved. Thanks to the legacy of typological thinking as applied to insects, researchers until recently have not been sufficiently alert to the existence of intraspecific diversity in behavior. We are confident that many more intriguing and revealing examples of this phenomenon remain to be discovered in the insects and that female insects, as well as males, exhibit conditional mating strategies.

10

Protection of Females during Courtship and Copulation

FOR A MALE, the process of reproduction can be divided into a number of stages. The male first searches for a female or calls one to him. If a potentially receptive female is located, a precopulatory association may follow, during which the male attempts to induce the female to permit copulation to occur. Copulation itself may require many minutes or even hours before the sperm are completely transferred. After completion of insemination and termination of copulation, the male may spend additional time with his mate in postcopulatory association. As we noted in chapter 3, at any point in this process a male may experience competition from other males that will reduce his chances of successfully fertilizing the female's eggs.

Our attention thus far has been devoted to the mate locating phase of male reproduction. The presumption has been that success in reaching or monopolizing receptive females is correlated with success in passing on one's genes. But males that encounter a potential mate are by no means guaranteed the right to fertilize her eggs. A male brentid·beetle that has clambered onto the back of a receptive female may well find himself flying through the air the next moment before he has had a chance to consummate the encounter, courtesy of a swat from a rival's snout (Johnson, 1982). Even males that are in the midst

of copulating may find themselves forcibly separated from their partner, as happens in some cerambycids (Michelsen, 1963), bees (Alcock, Eickwort, and Eickwort, 1977; Rutowski and Alcock, 1980),
mecopterans (Thornhill, 1976c, 1977), and flies (Parker, 1968). The
separation may occur before the copulating male has completed transfer of his gametes. Even after inseminating a female, a male may still
fail to fertilize her eggs, if in the postcopulatory phase she mates with
another individual and unites the last male's sperm with her gametes.

The risk of postcopulatory sperm displacement and the behavioral
responses to it are the subject of Chapter 11. Here we deal with sexual
interference in the precopulatory interactions between male and female and during copulation itself. These forms of male competition
are widespread among insects and create selection pressures favoring
males that are able to prevent or reduce the chances that they will be
physically separated from a potential mate. We shall argue that most
male adaptations that serve this function can be placed in two categories: (1) attributes that make detection of an encountered female
by the finder's rivals less likely and (2) traits that make it physically
difficult for a male to take a female from her original discoverer. These
characteristics help males transfer a full complement of sperm to the
receptive females they meet.

The Concealment of Potential Mates from Competitors

Males that have found a female and are courting her or copulating
with her are likely to benefit if they are able to reduce the probability
that additional males will arrive on the scene. Such individuals may
interrupt courtship, prevent the completion of copulation, or take the
female directly from the male before he has had a chance to provide
her with his gametes (Fig. 10.1). Three categories of male responses
appear to make detection of a potential mate by other males less likely:
(1) physical removal of the female from an area with a high density of
competitors, (2) production of signals that counteract the attractant
cues associated with the receptive female, and (3) use of relatively in-
conspicuous courtship messages designed to be perceived by the female only, not by the male's opponents.

The most direct method of concealing a female is to transport her
from an area in which the risk of interference or displacement is high
to one in which it is low. Males of some Australian thynnine wasps
that have tracked a pheromone releasing female to her perch immedi-

10.1   Sexual competition in *Centris pallida*. A group of males surround a virgin female (dark-eyed individual) and struggle to control her. (Photograph by J. Alcock.)

ately gather her up and fly away from the spot (Alcock, 1981b,c). Often several males converge on a calling female more or less simultaneously; if a male fails to remove the female before a rival arrives, a fight for her occurs. Thus there is an advantage to the male that can spirit the female away to copulate with her safe from interference. (Depending on the species, copulation may be begun in flight or the male may alight on a shrub to court his partner before mating.)

Even after the start of copulation, there may be a chance for usurpation of the female before insemination is completed. The transfer of gametes and associated materials can require a substantial amount of time (see Table 11.1 in the next chapter), during which a rival may discover and "steal" a mating female from her partner. To the extent that the original male fails to transfer a full complement of sperm, his fertilization opportunities are reduced (especially if the opponent copulates with the female). Again, the probability that this will happen can sometimes be reduced by removing the female from an area of high male density. It is common in the swarming Diptera for a freshly united pair to drift or fly out of the mass of circling males and to complete copulation some distance from the point where the two individuals met. For example, in the empidid fly *Rhamphomyia nigripes* the female first accepts a prey from a male in a swarm and then the pair descends to a perch site where they continue to mate while the female feeds upon the nuptial present (Downes, 1970).

In butterflies postnuptial flights are also common (Fig. 10.2). It is usually the male that flies off with his partner dangling beneath him as they leave the location at which courtship and pairing took place. Typically these flights occur when a pair is harrassed by a male (Shields and Emmel, 1973), but in some danaines and one pierid the postnuptial flight is spontaneous in the sense that no external stimulus appears to trigger the departure of the pair from the copulatory site (Brower, Brower, and Cranston, 1965; Pliske, 1975; Rutowski, 1979). The function of induced postnuptial flights seems clear enough; if the paired male can repulse the courting single male, he may avoid premature separation (as well as reducing his conspicuousness to predators). Because it is usually the male that carries the female away, his moving wings provide signals that identify his gender to potentially interfering rivals. This may deter continued courtship by unpaired opponents. The cases in which a female is the active member of the pair occur in species with sex-linked mimicry; the female may be safer flying and displaying her mimetic color pattern than if she were carried with wings folded (Rutowski, 1978a). In these species departure from the mating site must be to the female's advantage as well as to the male's. However, Rutowski (1979) raises the possibility that in some species a female may prefer to remain in a location with numerous males in order to determine the ability of her mate to repel opponent males, if this in some way helps her assess the suitability of the male as a mating partner. Additional information is required to determine if there are sometimes conflicts of interest associated with postnuptial flights.

Counteracting the Attractant Signals from Discovered Females

A second way for a male to decrease the chance of interference from mate searching rivals is to produce a signal that in some way reduces the attractiveness of cues provided by receptive females. In the mealworm beetle (Happ, 1969), various bark beetles (Rudinsky, 1969; Nijholt, 1970; Vité and Williamson, 1970), the armyworm moth (Hirai, Shorey, and Gaston, 1978), and the bee *Centris adani* (Frankie, Vinson, and Colville, 1980) males that have found a female release a scent that appears to make her less attractive to other males (Fig. 10.3).

The function of the male's contribution has been interpreted in a

COURTSHIP OF THE QUEEN BUTTERFLY

<u>FEMALE BEHAVIOR</u>   <u>MALE BEHAVIOR</u>

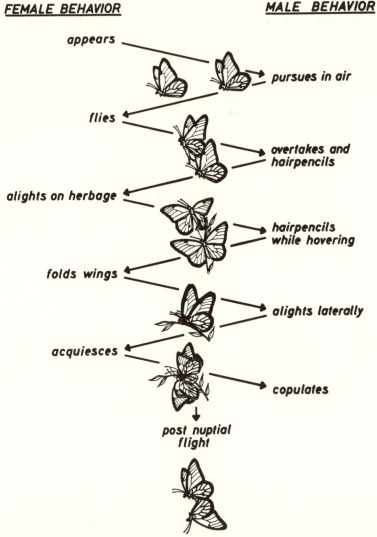

appears

pursues in air

flies

overtakes and
hairpencils

alights on herbage

hairpencils
while hovering

folds wings

alights laterally

acquiesces

copulates

post nuptial
flight

10.2 Postnuptial flight in the queen butterfly occurs after courtship and the initial phase of copulation has been completed. (From Brower, Brower, and Cranston, 1965.)

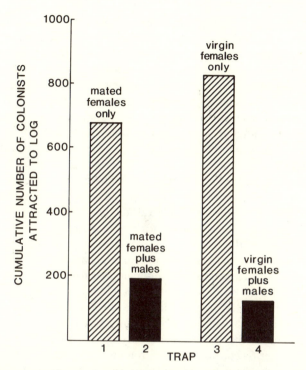

10.3 Log traps with caged males of the ambrosia bark beetle, *Trypodendron lineatum,* attract far fewer additional colonists (primarily males) than traps with females only, regardless of whether the females are virgin or not. Incoming males are repelled by scents associated with established males. (From Nijholt, 1970.)

number of ways. One approach, which is clearly group selectionist, has been to argue that the male's "inhibitory pheromone" helps prevent overpopulation (Shorey, 1973) while promoting "reproductive efficiency" by preventing undue competition for a female and by dispersing the population of searching males (Hirai, Shorey, and Gaston, 1978). This hypothesis fails to consider how both the releasers of the scent and the "inhibited" receivers might gain reproductively by their responses. One possibility is that the pheromone somehow chemically masks the female's odor, freeing the male from the risk of interference. One untested but interesting proximate possibility is that the male scent molecules bond with the female sex pheromone to produce an altered molecule that is no longer volatile and therefore cannot attract rivals from a distance (L. D. Marshall, personal communication). A more widely advanced hypothesis is that the male produces his own volatile signal that is perceived by other males and blocks their approach. But what might an approaching male gain by detecting a rival male's chemical signal and by avoiding the location? The signal could communicate accurately to a searcher that he is unlikely to reproduce by continuing to hunt for the source of the attractant sex pheromone. The male scent might indicate that a competitor has already arrived near the signaling female. Therefore the female may have already mated or be about to copulate, making further effort to locate her nonproductive. This would be true if females generally mate promptly with the first male to reach them, are not susceptible to takeover attempts, and become completely nonreceptive after copulation. In such a case the approaching male would do well to go elsewhere in search of an uncontested female rather than persist in what would usually be a fruitless (or costly) attempt to mate.

In the case of the bark beetles, a relatively high concentration of male scent relative to female sex pheromone coming from an infested tree could inform a mate searching male that the level of competition for females would be high at that location. A male that has found a female and entered her burrow (in those species in which the females are the colonizers) will defend the site vigorously against intruders and is difficult to displace (Ryker, in press). The dispersing male may often do better by continuing to search for a tree in an earlier stage of colonization with a more favorable sex ratio (for him)—that is, many more females than males.

Reducing the Conspicuousness of Courtship Display

Because sexually interfering male insects are frequently attracted to the signals other males produce to draw females to them (see Table 6.2) or are attracted by the activities of courting pairs, a third way in which to lower the risk of takeover is to court an encountered female in a manner that will not draw attention to her presence. The noisy attractant calls of some crickets and katydids cease when a female makes contact with the caller. Cricket males often switch at this time to a much softer courtship chatter (Alexander, 1962), the relative inconspicuousness of which may improve the male's chances of inducing the female to mate without interference from nearby territory owners and satellite males (Fig. 10.4). Less intense signals may also reduce the risk of parasitism and predation.

Similarly, in various katydids males shake or vibrate the plant or leaf on which they have been calling when they detect vibrations from an arriving conspecific (Busnel, Pasquinelly, and Dumortier, 1955;

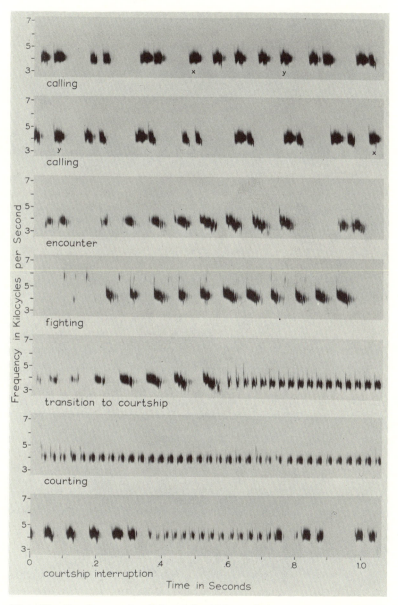

10.4 The signal repertoire of the cricket *Teleogryllus oceanicus*. Note that the courtship signals are less intense than the territory announcement (and mate attraction) call. (From Alexander, 1962.)

P. D. Bell, 1980b; G. K. Morris, 1980). Alexander (1975) suggests that such a signal would communicate solely with the arrivee and not with male neighbors of the caller. If the attracted individual were a female, substrate shaking would help the two make contact and provide information on the weight of the male, a factor of potential interest to a female, but the message would not travel to other plants to alert potential interlopers.

In *Anurogryllus arboreus,* however, several males may be calling from the same tree trunk; a switch to an alternative mode of communication once a female had been attracted might only alert the other males present. Unlike most crickets, a male of this species trills loudly throughout copulation (Fig. 6.3; T. J. Walker, 1980). Although there is no evidence of sexual interference in this species, perhaps mating males draw less attention to their partners by persisting in long-distance calling.

The Physical Prevention of Takeovers

Males that have found a potential mate and are in the midst of courtship or copulation may, despite their efforts to avoid it, find themselves confronted by a competitor. In order to deal with the problem of takeovers, males perform a number of actions that make separation from a mate and theft of the female less likely. One tactic is the maintenance of close contact in the precopulatory phase. In various beetles, including some cicindellids, staphylinids, and cerambycids (Parker, 1970c) as well as the acridid grasshopper *Locusta migratoria* (Parker and Smith, 1975), certain flies (Tarshis, 1958; Banegas and Mourier, 1967; Linley and Adams, 1972), the ambush bug *Phymata fasciata* (Dodson and Marshall, manuscript), the collembolan *Sminthurides aquaticus* (Schaller, 1971), and a veliid water-strider (Kellen, 1959), males grasp potential mates with some portion of their anatomy and do not let go for many hours or even days (Fig. 10.5). In most cases males simply ride on the backs of females without engaging their genitalia, the so-called passive phase of Parker (1970c). In this position the male may court the female and make copulatory attempts while preventing other males from mounting his partner. But in the collembolan *S. aquaticus,* which lives on the water surface, the male uses his specially modified antennae to grip the female's antennae and is transported by her for periods of up to 3 days. When the female becomes receptive, the male lowers his abdomen, deposits a spermatophore, and guides the female over his sperm packet.

In certain cerambycid beetles the passive phase and its function are

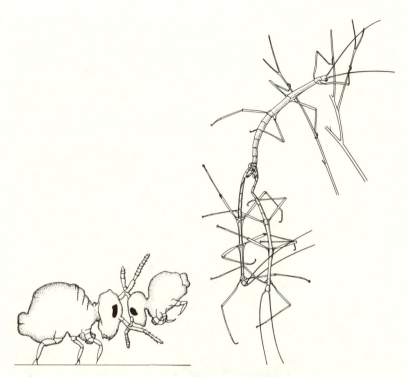

10.5 (*Left*) Passive phase in the collembolan *Sminthurides aquaticus*. The male holds a potential future mate with his specialized clasping antennae. (From Schaller, 1971. Reproduced, with permission, from *Annual Review of Entomology*, vol. 16, © 1971 by Annual Reviews, Inc.) (*Right*) Combat between two male walkingsticks, each of which holds a female with his terminal claspers. (From Sivinski, 1978.)

somewhat different from pure precopulatory guarding. Some species of these beetles have elongate protibiae and protarsi, apparently adapted for maintaining contact with females (Hughes, 1981). A male that encounters a receptive female may grasp her, copulate briefly, then use his fore legs and midlegs to grip the female firmly in a "half-mount" (Fig. 10.6). The male follows the female about for a time while she oviposits and then he copulates again; the female lays another egg (or eggs) followed by still another copulation. By remaining in contact with his mate in the interval between copulations, the male may help prevent takeovers that could lead both to postcopulatory sperm displacement and the loss of opportunities to mate again with the same female and fertilize more of her eggs. In this last sense, the male's behavior constitutes precopulatory mate guarding.

10.6   "Half-mount" in a curculionid beetle in which the male maintains contact with a female with whom he has already mated and with whom he may copulate again in the near future. (Photograph by A. P. Smith.)

Having discussed in detail in Chapter 1 the costs and benefits of precopulatory mate guarding, we need only summarize the key points. Time spent guarding a potential mate is time lost for acquiring additional females. It is not at all unusual for a male to wait by a female that is on the verge of copulating, as the gains often exceed the time costs involved. But cases in which males guard females for a prolonged period are more intriguing. When hours or days are spent with a female, the time investment becomes substantial. If that investment is to yield a greater return than a strategy of searching for immediately receptive mates, one expects certain conditions to prevail. Among the factors that could tend to reduce the cost of mate guarding or raise its benefits are (1) a scarcity of potential mates, (2) the likelihood that a male which copulates will fertilize his partner's egg(s), and (3) the existence of cues that enable males to determine when a female is likely to become receptive. As outlined in Chapter 1, the conditions affecting the fly *Lynchia hirsuta* and the grasshopper *Locusta migratoria* appear to be favorable for the evolution of precopulatory mate guarding.

Repulsion of Interfering Males

A male loses a great deal if a female he has found is stolen at a late stage in courtship. A robbed male loses an opportunity to fertilize eggs, as well as the time and energy invested in a fruitless courtship. As a result, males waiting for a female to become receptive in the midst of courtship are generally far from passive when confronted by another male. The guarding male desert locust uses its body as a shield and its legs as weapons to block a rival from contacting the ovipositing female. If successful, he will be able to fertilize her shortly after she has completed laying her eggs (Parker and Smith, 1975). Likewise, male cerambycids employ their long antennae to detect and lash out at intruders (Hughes, 1981) and males of the tenebrionid *Onymacris rugatipennis* trailing after a potential mate will turn to wrestle with rivals that have picked up the trail (Fig. 10.7; Hamilton, Buskirk, and Buskirk, 1976).

In the pteromalid chalcidoid wasps, males that have discovered a female typically climb onto her thorax and perch there strcking her antennae. The female, if receptive, will signal her readiness to copulate by a change in the position in which the antennae are held. The courting male then backs up to grasp the female's abdomen and insert his aedeagus into the female's genital aperture, which she exposes only at this time (Fig. 10.8). Because of the courting male's forward position on the female, he is vulnerable to sneaker males that slip onto the posterior of the female and wait for the opportunity to fertilize her (van den Assem, Gijswijt, and Nubel, 1980; van den Assem, Jachmann, and Simbolotti, 1980). When the courting male discovers the other male and the female *in flagrante delicto,* it is too late. (In the pteromalid wasps, females typically use sperm from the first mating only.)

There is some variation in the courtship patterns of pteromalids and van den Assem, Gijswijt, and Nubel (1980) suggest that the distinctive behavior of males of *Pteromalus puparum* may have evolved in response to selection created by sneaker males. Courting *P. puparum* stroke the female's antennae briefly and then back up, whether or not she has signaled her receptivity. The cycle is repeated over and over until the female permits copulation to occur. By continually moving to the abdomen of the female, a courting male improves his chances of

10.7 Wrestling match between tagged males of *Onymacris rugatipennis* for the right to follow a female who may later accept a mate. (Photograph by W. J. Hamilton III.)

detecting (and repelling) any rival that has slipped onto the female during a bout of antennal stroking.

More generally, the willingness of males of various insects to interrupt courtship to chase off intruders may have as one function the prevention of exploitation of their courtship by the other males. For example, males of the otitid fly *Physiphora demandata* (Alcock and Pyle, 1979), the tephritid fly *Anastrepha suspensa* (G. Dodson, 1982), and the damselfly *Calopteryx maculata* (Waage, 1973) invariably stop courting if another male approaches. This response has a definite cost, for the female may leave while her suitor deals with his opponent. Or sometimes while the male chases off one intruder, another arrives and steals the female. The reason males give top priority to repelling a rival may have to do with protection of a territory, inasmuch as prolonged territory ownership may enable the male to gain access to more than

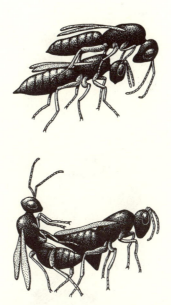

10.8   A female pteromalid wasp of the genus *Muscidifurax* requires courtship signals from a mounted male before changing her abdomen shape to permit copulation. A male in the courtship position runs the risk that a rival male, by copulating from the rear, will steal his mate at the moment she becomes receptive. (Drawing by M. H. Stewart.)

enough females over the long run to compensate for the risk of losing one mate while driving off an invader. A complementary function, however, could be that by repelling another male, the courting male prevents this individual from stealing the female at a precise moment when she becomes receptive.

An ingenious solution to the problem of coping with intruders while at the same time reducing the chance that the female will decamp has evolved in the collembolan *Dicyrtomina minuta* (Schaller, 1971). The male builds a "fence" of spermatophores around the female and defends the area against rival males by jostling them away. This usually prevents his opponents from replacing his spermatophores with their own before the female has had a chance to accept one of the owner's sperm capsules.

Prevention of Takeover during Copulation

Many of the structural features of male insects that enable them to grasp their females firmly in precopulatory maneuvers can also be put to good use during copulation itself. The male that can employ

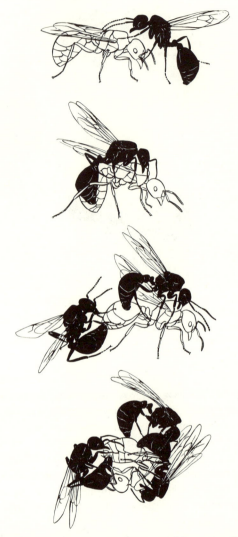

10.9 Competition among male harvester ants of the genus *Pogonomyrmex* for possession of a female. (*Top*) A male (black) approaches a female and attempts to mate. (*Next*) After copulation has begun, the male releases his grip on the female's thorax and falls back. (*Next*) Another male arrives and grasps the female's thorax. (*Bottom*) A third male arrives and holds the abdomen of the second male while the female bites at the abdomen of her mate. This male may be replaced by one of those waiting. (From Hölldobler, 1976.)

10.10   A male firefly, *Pteropteryx malaccae,* uses specialized hooked wing covers, which he places under the elytra of his mate, as an abdominal clamp to hold the female (*right*) firmly during copulation. (Photograph by J. E. Lloyd.)

powerful legs to grip his partner may prevent another male from pulling him from his mate or wedging him off her. Rozen (1977) notes that males of the mellitid bee *Meganomia binghami* have enlarged hind legs, which are modified to permit the male to wrap these appendages about the "waist" of the female. This is a useful ability in a species in which competition for possession of females is intense and in which some males attempt to strip a female of her rider.

Modifications of the legs that promote better grasping of a female are common in male insects, including "sex combs" in *Drosophila* (Spieth, 1952) and the spiny hind legs of certain fleas, which interlock with the tarsal spurs of the female. Copulation in these fleas may be prolonged; but the male, who is underneath the female, does not contribute to locomotion, his only goal being to retain possession of his mate (Mitzmain, 1910). In *Sepsis* flies the male's legs are used to clamp about the base of the wing of the female (Parker, 1972). This may not only make his removal more difficult but may also prevent escape by the female, another important function of male gripping devices in the insects (Spieth, 1952).

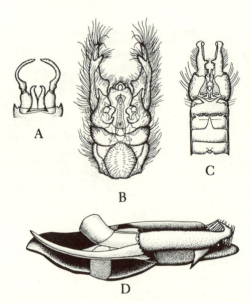

10.11   Diversity in the structure of the terminal segments of the male abdomen of insects, showing the repeated independent evolution of grasping devices and spines that may help a male to remain linked with a mate. (*A*) Terminalia of *Mirawara aapta*, a mayfly with prominent claspers. (From MacKerras, 1970.) (*B*) Terminalia of *Neoaratus hercules*, an asilid fly with hairy claspers. (From McAlpine, 1970.) (*C*) Terminalia of *Notiothauma reedi*, a scorpionfly, with large gripping structures used to hold the female. (From Crampton, 1931.) (*D*) The aedeagus of *Lytta nuttalli*, a meloid beetle, showing the penis spines that catch on the walls of the female's vagina. (From Gerber, Church, and Rempel, 1972.)

Portions of the body other than the legs may be used to hold the female tightly during copulation. The jaws, logically enough, are used by males of many groups, among them some harvester ants. In *Pogonomyrmex desertorum*, for example, males have extremely strong mandibles, which grip their mates at the waist. Two to four individuals commonly compete for possession of a female, each trying to pull the other from her (Fig. 10.9). Once a male has secured a grip on the female, he will hold on with such bulldog determination that he may sever the female's gaster from her thorax, cutting her in two rather than releasing her (Hölldobler, 1976). Similarly, the lycid beetle, *Calopteran discrepans,* uses his sickle-shaped mandibles to bite through the female's wing covers. Although this injures the female, it enables the male to cling to his mate in the face of assaults from additional suitors (Sivinski, 1981).

Other structures used to prevent takeover include the antennae, which are specially modified in male fleas for use as claspers (Holland, 1955). We have already mentioned the notal organ on the abdomen of male *Panorpa,* which is used by rapists and nonrapists alike to hold the female during copulation (Thornhill, 1980e, forthcoming). Male *Pteropteryx* fireflies use the hardened tips of their elytral wing covers to grip the female's abdomen in a clamp (Fig. 10.10). Finally, male insects often employ their genitalia to hold a mate so firmly that the pair cannot be easily separated. If the only function of male genitalia were to serve as a conduit for the sperm (and allied substances) from male to female, it seems unlikely that the insect "penis" and the structures associated with it would exhibit the spectacular diversity and complexity that have proved to be such a convenient means of identification for the taxonomist (J. E. Lloyd, 1979a). O. W. Richards (1927) long ago recognized that sexual selection may have contributed greatly to elaborate penis morphology and strong genitalic claspers. The importance of the structural uniqueness of genitalia as a reproductive isolating mechanism has probably been highly overrated relative to the importance of this feature in enabling those males selected by females to transfer a complete supply of sperm. The length of the male's aedeagus (Evans, 1969), the presence of spines on the penis that catch in the wall of the female's vagina (Gerber, Church, and Rempel, 1971), the inflatable components (Waage, forthcoming), and the interlocking features of some insect genitalia all may contribute to the prevention of takeovers (Fig. 10.11).

One wonders whether the diversity of copulatory positions of insects may not also be related to different kinds of risks of takeover within the group. The "typical" male-above, female-below position is only one of many employed by insects. Even within a single order the variety may be impressive (Fig. 10.12). If a male must clamber onto a female to insert his aedeagus, he may remain in this position to prevent other males from having a chance to mate (but see Fig. 10.13). Moreover, the riding male may maneuver in response to assaults from others attempting a takeover. In the bee *Nomadopsis puellae* copulation may be a lengthy business, with the male mounted on the back of the female as she forages from flower to flower (see Fig. 11.21). If touched by a mate searching rival that darts down to grasp the female, the rider flips sharply backward, using his body to catapult the incoming male outward away from the female (Rutowski and Alcock, 1980). Males that use other copulatory positions may utilize specific rejection responses to combat takeovers that are especially effective because of the position of their body relative to that of their mate; however, there seems to be little information in the literature on this point.

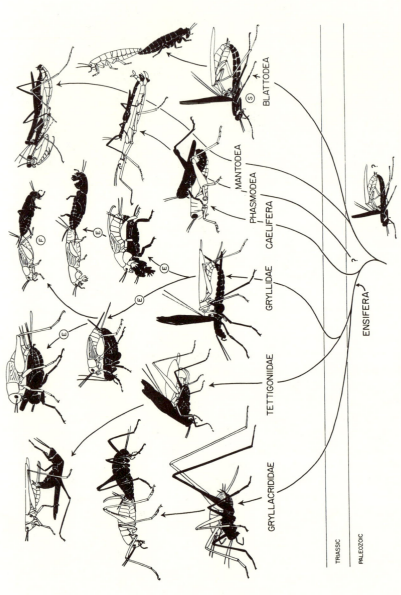

BLATTODEA

MANTODEA

PHASMODEA

CAELIFERA

GRYLLIDAE

TETTIGONIIDAE

GRYLLACRIDIDAE

ENSIFERA

TRIASSIC

PALEOZOIC

10.12 Diversity in the copulatory positions of the Orthoptera. The arrows show possible evolutionary path-
ways leading from an ancestral orthopteran's mating position (*below*) to more recently evolved patterns. In
species in which takeovers are possible, males may be under selection to assume a position that permits effective defense against sexually interfering rivals. (From Alexander, 1964.)

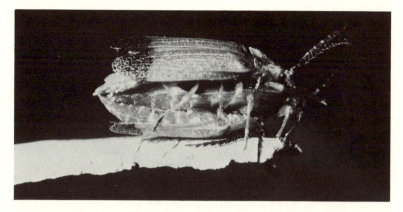

10.13   Copulating trio of lycid beetles. The female is sandwiched between two males, each of which has inserted his aedeagus into her. (Photograph by J. Alcock.)

We have shown that a great variety of structural and behavioral characters may contribute to the ability of a male to thwart interfering rivals. These adaptations improve the odds that a discoverer of a female will actually inseminate her. But the contest to fertilize eggs does not end with the transfer of gametes. In the next chapter we document an even more impressive array of tactics and countertactics that males use in sexual competition after insemination has occurred.

11

Sperm Competition and the Fertilization of Eggs

As WE NOTED in earlier chapters, sperm competition is a dominant feature of the reproductive contest in many insects whose females mate multiply. If females may accept more than one partner, selection will favor males that inseminate already mated females, provided that their sperm can to some extent supersede those of earlier donors. On the other hand, selection will also favor males that can prevent their rivals from gaining access to their recent mates if there is risk of sperm displacement (Parker, 1970a,c,d, 1974b).

This chapter examines the postinsemination phase of intrasexual selection. We first demonstrate that multiple mating by female insects occurs and that it leads to competition among the sperm a female receives. We examine the typical outcome of sperm competition, which is sperm precedence (the last-male fertilization advantage mentioned previously). In the final section we consider the diversity of sperm competition tactics that males employ in species with multiple-mating females. These traits increase the probability that an individual male will be the last to copulate with a female, at least during one reproductive cycle, and so will fertilize all or most of her mature eggs.

We begin with the case of the damselfly *Calopteryx maculata* to illustrate the point that copulation does not assure a male of exclusive fertilization rights, and to show how the male's behavior in this species has been shaped by that reality.

Sperm Competition in a Damselfly

Males of the damselfly *Calopteryx maculata* are conspicuous and elegant insects, with jet-black wings and shiny iridescent green bodies, found along small streams and brooks in the eastern United States. The males compete for possession of oviposition sites attractive to females, with the population at any one time consisting of some individuals that hold valuable territories and some that are currently unable to claim and defend a suitable location (Waage, 1973).

Typically, an egg-laden female that flies to a stream to oviposit is detected first by a male territory owner near the point at which she intends to lay her eggs. After a brief courtship the female is seized in tandem by the male, prior to copulation and sperm transfer. In *C. maculata*, as in other odonates, copulation involves indirect sperm transfer (Fig. 11.1). While in tandem the male transfers a quantity of sperm from his testes, located near the tip of his abdomen, to a sperm vesicle found on the underside of the second segment of his abdomen. Associated with the sperm vesicle is a penis with a sperm channel on its dorsal surface; sperm travel along this groove into the female when she orients her genitalia over the male's penis (Fig. 11.2). The penis is inserted into the bursa copulatrix, where sperm are

11.1   A copulating pair of *Calopteryx maculata*. The female, on the left, has twisted her abdomen around so that her genitalia are over the male's penis, the sperm transferring structure located on the underside of his abdomen. (Photograph by J. Alcock.)

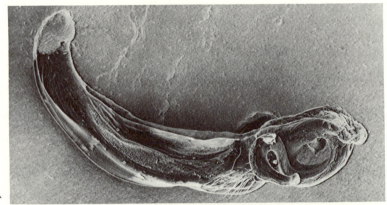

A

B

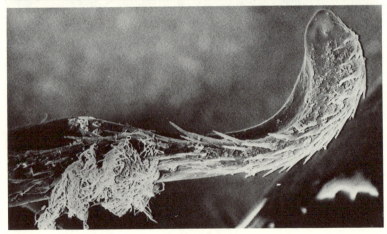

C

received prior to storage in a T-shaped spermatheca (Fig. 11.3). Thus females have the capacity to store sperm from a mating for use in fertilizing their eggs over a period of time (Waage, 1979a).

After mating, the male releases his partner and resumes surveillance of his territory. His mate usually flies down to the water in the area under his control. While she is on the stream ovipositing with just the tip of her abdomen underwater, or while perched in weeds near the water's edge in the midst of traveling from one oviposition site to another, the female remains receptive, no matter how many times she has copulated during the day. There are numerous males eager to take advantage of this fact. Subordinate nonterritory holders lurk in streamside vegetation waiting for an opportunity to approach a gravid female; territorial males attempt to steal one another's ovipositing mates at times when a neighbor is busy defending his site against another intruder. Both classes of males sometimes succeed in mating with a female that has copulated with another male a few minutes to an hour or so earlier.

The behavior of *C. maculata* poses several puzzles. First, why are egg-laden females willing to mate with almost any male that approaches them even if they are in the middle of an egg laying bout? Perhaps by retaining her willingness to mate during the entire oviposition time, a female can test her original mate's ability to fend off other males for lengthy periods. If he is to prevent a takeover, a male of this species must perch by his mate(s) and repel all competitors interested in stealing her. Presumably only physiologically competent, strongly territorial males are likely to succeed in meeting these demands; as a result, a female is likely to lay most of her eggs after having mated with a dominant male. Weakened individuals, or those low in social status, will usually be quickly supplanted by more powerful males that from a female's standpoint may represent superior mates.

11.2 Scanning electron micrographs of the penis of *Calopteryx maculata*, showing the spines that entangle and help remove rival sperm stored in the female. (*A*) A side view, showing the horns projecting from the head of the penis. (*B*) A closer view of the penis head of a copulating male, showing white sperm masses caught on the hairs on the lower side of the penis and on a lateral horn shown in closeup in (*C*). When collected, the male was in the process of removing a competitor's gametes. (From Waage, 1979a. Copyright 1979 by the American Association for the Advancement of Science.)

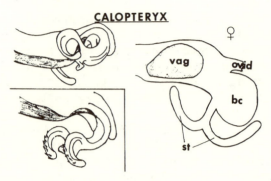

11.3   The bursa copulatrix (*bc*) and spermatheca (*st*) of the female of *Calopteryx maculata*. The male's penis (*left, above*) enters the vagina (*vag*): the two horns of the penis (*inset*) fit in the spermathecal horns and scrape out sperm stored by the female. (From Waage, forthcoming.)

But if females mate with a series of partners, which male actually fertilizes her eggs? Waage (1979a) answered this question by collecting random samples of females in various stages of mating and examining their sperm storage organs. Sixteen females collected while in tandem (prior to mating) proved to have nearly full complements of sperm stored in their bursa copulatrix and spermatheca. These females did not have to remate in order to replenish depleted sperm supplies. A sample of 24 females collected in the midst of copulating had almost no stored sperm at all. This could only happen if the females expelled older sperm during mating or if their mates had removed any stored sperm before transferring their own. (Females collected after copulating had full supplies of sperm once again.)

Waage demonstrated that sperm precedence occurs because males are able to use their penis to remove any sperm stored within a female. To this end, the damselfly penis is endowed with stiff hairs that point backward along the shaft of the organ. In addition, there is a pair of prominent horns that project from the head of the penis out to the side (Figs. 11.2 and 11.3). During the first phase of copulation the male rhythmically undulates his abdomen, moving the penis back and forth within the female. The penis penetrates the spermatheca, the hairs entrap the stored sperm, and the male sweeps the sperm out of the female's genital tract. When 88 to 100 percent of the "old" sperm have been removed, the male releases his own sperm to travel down the penis into his mate.

If the same female copulates again, the sperm just injected will be speedily removed. Because of the risk of sperm displacement, territorial males guard their recent partners by perching a short distance from the female while she oviposits (Fig. 11.4). If another male approaches, he is assaulted and chased from the territory. Waage (1979b) found that a guarded female oviposited without interruption for about 13 minutes, during which time she laid 90 to 140 eggs, all (or most) of which were presumably fertilized by her most recent mate, the guarding male. Females that chose to lay their eggs without a guard male present averaged less than 2 minutes of undisturbed oviposition. These individuals laid only about 15 eggs before being chased from the water by another male; they would often mate again quickly, with a consequent loss of egg fertilizations for a previous partner.

The behavior of this damselfly dramatically demonstrates the conflicting nature of the selection pressures acting on males that must deal with multiple-mating females (Parker, 1970c,d, 1974b,c). On the one hand, selection favors mechanisms (in this case, a remarkable penis) that facilitate sperm precedence, because they enable a male to benefit from copulations with nonvirgin females. On the other hand, selection favors tactics (in this case, guarding behavior) by males that

11.4   A male of *Calopteryx maculata* (second from left, with wings spread) guarding three ovipositing females on his territory. (Photograph by J. Alcock.)

prevent opponents from benefiting from sperm precedence. We shall now explore these two kinds of interacting adaptations in a broader context.

Sperm Competition and Internal Fertilization in Other Insects

Because females of C. *maculata* and many other insects mate with more than one male and store sperm internally, sperm competition is inevitable. This is not to say that sperm competition cannot occur in species that practice external fertilization. For example, in an aquatic invertebrate like a sea urchin, males and females release sperm and eggs into their environment and fertilization takes place externally. Here the competition among males for access to eggs will result in selection affecting the timing of the release of sperm, the locomotory properties of male gametes, their egg "detection" abilities, and their capacity to penetrate the membrane of the ovum and achieve fertilization. Males that release their sperm at times when the probability of egg fertilization is greatest should enjoy an advantage, as should males whose sperm are unusually mobile, quick to adhere to unfertilized eggs, and prompt in entering the egg.

In vertebrates with external fertilization, such as many fishes, competition among males for fertilization opportunities is often pronounced and has resulted in a complex array of egg defending behaviors on the one hand and surreptitious fertilization tactics on the other (see, for example, Dominey, 1980; Gross and Charnov, 1980). Internal fertilization is not necessary for sperm competition to occur, but its evolution alters the framework in which the phenomenon takes place because it gives the female better control of the fertilization process.

Parker (1970c) has proposed that the first form of internal fertilization originated when a female found a spermatophore (that a male had deposited on the ground) and stored it internally. The change from external to internal fertilization may have been associated with the invasion of terrestrial habitats; the deposition of spermatophores that are acquired by females without copulation is characteristic of a number of "primitive" insects (Fig. 11.5). In some of these species the males scatter their sperm packets and rely upon receptive females to perceive and collect them. Male-male competition for access to eggs expresses itself in the tendency of males to consume the spermatophores of their rivals when they encounter them (Christiansen, 1964; Schaller, 1971). (Because the consumer then replaces the eaten spermatophore with one of his own, the behavior has been interpreted

11.5 The spermatophore of a collembolan. The male deposits this structure on the leaf litter and the female must locate it in order to fertilize her eggs. (From Schaller, 1971. Reproduced, with permission, from *Annual Review of Entomology*, vol. 16, © 1971 by Annual Reviews, Inc.)

as a device to ensure that females are offered only "fresh" sperm, thus maximizing the production of offspring by the population as a whole—an unnecessary group selectionist argument in light of the obvious gains for the spermatophore consuming male).

In other species of primitive terrestrial insects males guide their mates to spermatophores, which they attach to the substrate or to silken threads (Fig. 11.6). This method perhaps increases the likelihood that a deposited spermatophore will actually be used by a female before it is eaten or destroyed by a competitor male.

A further potential improvement in sperm placement is offered by copulation, and Parker (1970c) suggests that it evolved under the pressure of intrasexual selection in species that at one time practiced external placement of sperm. Males able to insert their sperm directly into the female may enjoy a reproductive advantage over those that rely more heavily on female cooperation to achieve this end. Through copulation a male can inject his sperm close to the eggs that are available to be fertilized.

Still, in most female insects sperm are prevented from traveling *directly* to the eggs and instead are shunted into one, two, or three spermathecae that store the sperm received from copulatory partners (Fig. 11.7; Engelmann, 1970; R. F. Chapman, 1971). The exceptions, primarily the mayflies, are extremely short-lived as adults and therefore presumably derive no benefit from the ability to store sperm for some time. But for a female with a longer adult phase, a spermatheca has at least two major virtues. First, the female need not mate immediately prior to each oviposition bout, if these occur at intervals, but can use stored sperm at the precise moment it is needed. This frees her from the need to encounter mating partners just before each period of egg laying and saves her the costs (time, energy, and risk) associated with copulations. Second, a sperm storage organ heightens her control

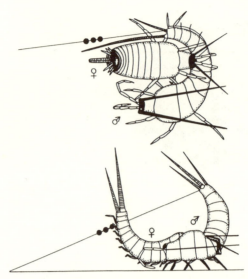

11.6   Mating in the thysanuran *Machilis germanica*, in which the male first finds a female and then deposits his sperm on a silken line, after which he guides the female to the sperm, improving the chances that she will accept his gametes. (From Schaller, 1971. Reproduced, with permission, from *Annual Review of Entomology*, vol. 16, © 1971 by Annual Reviews, Inc.)

over the fertilization process. Typically, when a female is about to oviposit, the egg passes down the oviduct to a special pouch near the exit from the spermatheca. The female can control the position of the egg and can release exactly the number of sperm guaranteed to fertilize it. This enables her to use sperm highly efficiently so that she need not store, transport, and maintain quantities of sperm in great excess of the number of eggs she is likely to produce.

There is a special advantage to sperm storage and controlled egg fertilization for hymenopterans and other insects that employ a haplodiploid method of sex determination. A female bee, ant, or wasp is able to produce females by fertilizing her eggs and produce males by laying unfertilized haploid eggs. This potentially permits the female to control the sex ratio of her progeny (Gerber and Klostermeyer, 1970; Werren, 1980), an especially adaptive feature in those species that typically are highly inbred. In addition, if a female can determine the sex of each egg she lays, she can match levels of investment to the sex of her offspring. Thus if it is adaptive to produce relatively large males under some conditions, a female can lay an unfertilized egg on a relatively large prey item or on a large quantity of stored brood provisions, with the result that she will have a larger-than-average son.

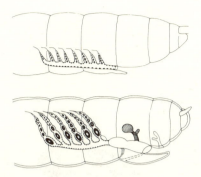

11.7   Diversity in the internal reproductive tract of female insects. (*Above*) A mayfly female's reproductive system, showing the absence of a spermatheca or other sperm storage organs. (*Below*) The reproductive tract of a butterfly female with a spherical spermatheca for maintenance of a supply of sperm from spermatophores formed within the bursa copulatrix by males. (From Weber, 1954.)

Much more generally, if a female stores sperm before she uses it, she improves her control over the paternity of her progeny. If a mating partner is unsatisfactory for some reason, the female can potentially accept or even solicit another mating with consequent displacement of the unacceptable male's sperm. Through sperm storage female insects gain better control over a spectrum of reproductive decisions ranging from when to fertilize their eggs to which of several males' sperm will fertilize their gametes. These advantages may help offset the metabolic expenses of sperm storage (Davey and Webster, 1967) and the costly aspects of copulating more than once.

The full spectrum of possible advantages of polyandry is examined in detail in Chapter 14 (see in particular Table 14.4). Here we need only summarize a key point, which is that the acceptance of multiple mates is *not* necessarily related to the depletion of stored sperm or a loss of their viability. In many insects sperm are stored for periods of up to a year or more without ill effect (Bequaert, 1953; Reid, 1958; Gordon and Gordon, 1971). Moreover, the entomological literature is replete with observations of polyandrous species in which a single mating is sufficient to fertilize all the eggs of a female (Dick, 1937; DeLong, 1938; Bequaert, 1953; Gerber and Church, 1976). Norris (1954) consequently suggested for the desert locust that the function of multiple mating is to stimulate oviposition because secondary matings are not necessary for fertilization of the eggs.

This explanation is most unlikely for several reasons, including its implication that some males altruistically serve as "oviposition stimulators." If males regularly copulate with already mated females, our

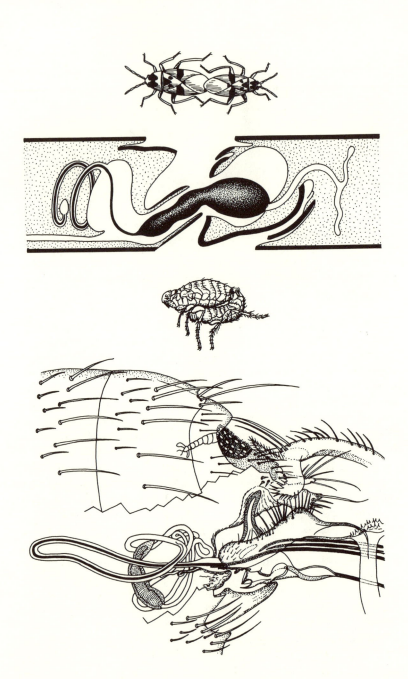

working hypothesis must be that they do so because there are genetic gains to be made. Otherwise males would discriminate against nonvirgins. Moreover, if females accept another male when they already have sufficient sperm reserves to fertilize their lifetime production of eggs, they probably gain something by using the new sperm. If not, females that become completely nonreceptive after securing a large sperm supply would gain an advantage. In fact, in those groups in which females mate multiply (and there are many representatives among the Odonata, Lepidoptera, Orthoptera, Heteroptera, Mecoptera, Coleoptera, Hymenoptera, and Diptera) there is a marked tendency for sperm precedence to occur (see reviews by Parker, 1970c; Boorman and Parker, 1976).

Structural and Behavioral Mechanisms of Sperm Precedence

Many structural and behavioral attributes have evolved in male insects that can be interpreted as devices to facilitate sperm precedence. The example of *Calopteryx maculata* is a dramatic illustration that the penis of male insects may be more than a simple sperm conducting tube. This conclusion was reached long ago by the anatomist Snodgrass (1946:67–68), who upon examining the genitalia of fleas wrote, "It is hard to imagine why a male insect should need, for the mere transfer of sperm to the female, an organ so elaborate in structure and so complex in its mechanism as the intromittent apparatus of the fleas." Selection for complexity of design and function arises if the penis can serve as a locking device to prevent the disruption of copulation *or* if it can be employed to assist in successful sperm competition, a possibility that we shall explore here.

We have already mentioned Parker's (1970c) suggestion that copulation itself arose as a result of sperm competition, with selective advantage to males that could insert their sperm inside a female, as this

---

11.8   Males with very long sperm transferring organs. (*Top*) A copulating pair of lygaeid bugs shown from above. (*Next*) A diagrammatic lateral view of the terminal segments of the male and female bugs in cross-section. The male's aedeagus is the long, thin, black structure shown penetrating right into the female's long, coiled spermatheca. (From Weber, 1930.) (*Next*) A copulating pair of ceratophylline fleas shown from the side. (*Bottom*) A diagrammatic lateral view of the terminal segments of the female (*above*) and male (*below*). The male's aedaegus is shown in black as it lies within the female's long, coiled spermatheca. (From Holland, 1955.)

would place their gametes closer to the eggs than externally deposited sperm. Selection along these lines might favor males with an aedeagus (= penis) of a length and design to facilitate placement of the sperm in the location where it was most likely to be stored and used. If the female of the species has a sperm storage organ located at some distance from the opening to her genital tract, then selection may result in the evolution of an unusually elongate penis, examples of which are found in lygaeid bugs (Weber, 1930; Bonhag and Wick, 1953), chrysomelid beetles and tipulid flies (Gerber, 1970), tephritid and otitid flies (G. Dodson, 1978a,b; Alcock and Pyle, 1979), and fleas (Holland, 1955). Fully extended, the remarkable coiled aedeagus of the otitid fly *P. demandata* is as long as the body of the fly. In those species in which the match between female genitalia and male penis has been studied, it has been shown that the penis reaches well into the spermathecal ducts, in some cases even into the spermatheca (Fig. 11.8). As a result, a male may be able to flush his rival's sperm from the spermatheca, although W. F. Walker (1980) questions whether this could occur in species with narrow spermathecal ducts. Alternatively, by placing the tip of the aedeagus near the proximal end of the spermatheca, a male may be able to position his sperm close to the exit of the storage organ while pushing competitor sperm distally. His sperm might, therefore, be first out of the duct and down to the eggs when the female oviposits. "Last in, first out" may be a common mechanism of sperm precedence, and the structure of the aedeagus may be related to this goal. These speculations could be tested by determining whether possession of a relatively long intromittent organ is characteristic of insects with multiple-mating females that have lengthy internal reproductive tracts.

Still other morphological aspects of the insect aedeagus may promote strategic sperm placement. J. E. Lloyd (1979a:22) argues that "it is not inconceivable that males will have evolved little openers, snippers, levers and syringes that put sperm in the places females have evolved ('intended') for sperm with priority usage—collectively a veritable Swiss Army Knife of gadgetry!" He notes that in the scutellerid bug *Hotea* the male penis is a huge bulky device that cuts its way through the vagina and body cavity of the female to deposit sperm directly in the spermatheca. The evolution of "traumatic insemination" (in which males use their aedeagus to pierce the body wall of the female) in the Strepsiptera and bed bugs (Hinton, 1964) may also have evolved as a result of the advantage some males gained by short-circuiting the customary pathway within the female for sperm transport (Parker, 1970c). In some members of the bed bug group, the aedeagus enters the female in the typical fashion but then penetrates the roof of the vagina (in a manner reminiscent of *Hotea*) with sperm placed in the haemoceol. This may represent a first step toward complete abandonment of a vaginal route for sperm, culminating in use of the male's genitals to penetrate the external wall of the female's abdomen (Fig. 11.9).

Traumatic insemination of this sort may have led to counterevolution in females, with the development of a highly specialized tissue mass (the spermalege) beneath the area of cuticle penetrated by the male. These tissues receive the male's sperm and direct it to the circulatory channels that eventually carry the gametes to the female's eggs (if the sperm are not digested as a nutritional supplement instead). If this Byzantine scenario is correct, it provides a classic illustration of the results of a coevolutionary race between the sexes with respect to which will control the fertilization process.

The penis may be used to remove a rival's sperm, as well as to place

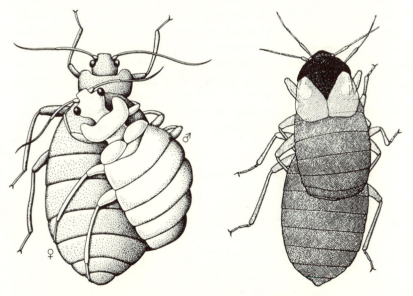

11.9 Traumatic insemination. (*Left*) The male bedbug *Cimex lectularius* employs his dagger-like aedeagus to puncture the female's cuticle on the underside of the middle of her abdomen, injecting his sperm into a specialized tissue mass there. He bypasses the female's genital opening altogether. (From Weber, 1930.) (*Right*) In the anthocorid bug *Xylocoris flavipes* the male pierces the female through an intersegmental membrane on her back, under which the sperm receiving tissues lie. (Drawing courtesy of J. Carayon.)

the male's own gametes in strategic position. In *C. maculata,* as we have seen, the male's aedeagus scrapes and pulls out rival ejaculate from the bursa copulatrix. Waage (forthcoming) has found that in other damselflies there is also a close correspondence between the shape of the male intromittent organ and the structure of the female genitalia (Fig. 11.10), which is strongly suggestive of sperm removal. In several species the morphological evidence has been supplemented by data in support of sperm precedence derived from examination of sperm volumes in females dissected prior to, during, and following copulation.

In the dragonflies, however, the situation is less clear because the male penis often has an erectile component that is not fully inflated in insects that have been killed or are not in the act of mating. This makes it difficult to analyze the correlation of male and female genitalia during copulation. Nevertheless, there are indications for several species that males may use the penis less for sperm removal than for repositioning of rival ejaculates. For example, in *Sympetrum rubicundilum* the male's penis terminates in two coiled inflatable processes that appear to fit into the large, paired spermathecae. The penis lobes may push stored sperm deeper into the spermathecae, enabling the copulating male to place his sperm next to the exit close to the oviduct, to gain a "last in, first out" advantage (Fig. 11.11).

Finally, it is possible that males of some insects are able to achieve sperm precedence through injection into the female's genital tract of substances that cause release of any sperm stored within the spermatheca(e). Such substances may mimic or exaggerate the chemical

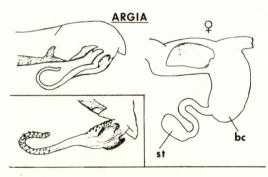

11.10   A diagram of the male penis and female reproductive tract in a species of *Argia,* a damselfly. The penis (*left*) has a hood-like extension (*inset*) that neatly matches the design of the female's spermatheca (*st*) and bursa copulatrix (*bc*). (From Waage, forthcoming.)

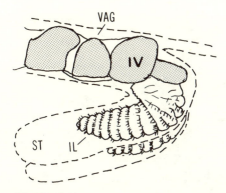

11.11 The penis of the dragonfly *Sympetrum rubicundilum* has two inflatable lobes (*IL*) extending from the IV abdominal segment. When the penis is inserted in the vagina (*VAG*) of the female, the lobes may inflate to push the sperm of rivals deeper into the female's spermatheca (*ST*), where they are unlikely to be used to fertilize her eggs. (From Waage, forthcoming.)

trigger that females employ when releasing sperm for the purposes of fertilization. Gilbert, Richmond, and Sheehan (1981) originally thought that the enzyme esterase 6, transferred in quantity to the female in the ejaculate of male *Drosophila melanogaster*, might serve this function. More recent evidence indicates that this hypothesis is unlikely; the precise mechanism of the second-male advantage in this species is still to be discovered, although esterase 6 does in some way affect the rate of sperm lost from the female's sperm storage organs (D. G. Gilbert, 1981; Gromko, Gilbert, and Richmond, forthcoming).

Insurance Matings and the Last-Male Advantage

The last-male advantage can be achieved in other ways than the removal or repositioning of the sperm stored within a female. In species in which the precise moment of oviposition cannot be determined by the male, he may try to prevent his mate from copulating again until she has laid some eggs. Alternatively, he may copulate more than once with the same female, which may improve his fertilization success in several ways.

Consider the male belostomatid waterbug who carries the eggs of his mates on his back and provides various forms of paternal care that improve the hatching rate of the eggs. He has a special incentive to increase the likelihood that he will actually fertilize the eggs of his mate, for if he fails to do so he will be providing costly parental care for

11.12   A copulating pair of belostomatid waterbugs. The male, on the right, is carrying a mass of eggs laid by his mate(s). (Photograph by R. L. Smith.)

a competitor's offspring. (The assumption that females can store sperm from previous partners is correct. R. L. Smith, 1979b, severed the ejaculatory duct of several males and paired the "vasectomized" bugs with females that had mated before with intact males. These females subsequently laid fertile eggs even though they received no sperm from the surgically altered partner.)

A normal, intact male avoids the risk of "cuckoldry" by never permitting a female to place her eggs on his back until he has had a chance to copulate with her (Fig. 11.12). Even then he allows her to glue no more than three eggs in place before insisting on another copulation, repeating this cycle about five times before the pair separates (Fig. 11.13). If in belostomatid females there is a mechanism that permits the last sperm in to be the first sperm out to make contact with an egg at the time of oviposition, a male would through repeat matings regularly replenish the number of his gametes in the optimal location for egg fertilization.

The effectiveness of the repeat-mating strategy was examined by Smith in a series of sperm precedence experiments. Male waterbugs

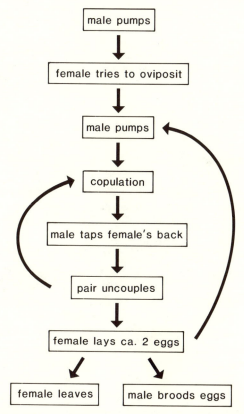

11.13   Flow chart of the activities of copulating waterbugs (*Abedus herberti*). Males *always* require at least one copulation before permitting the female to oviposit on their back. Egg laying is interrupted at regular intervals for one or more additional copulations. (From R. L. Smith, 1979a.)

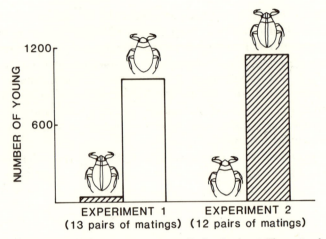

11.14   Sperm precedence experiments in *Abedus herberti*. The striped morph of *A. herberti* is caused by a single dominant allele. Females that mate first with a homozygous striped male and then with an unstriped individual almost always produce unstriped offspring. If the sequence of pairing is reversed, the offspring will be striped. Sperm precedence therefore is nearly 100 percent. (From R. L. Smith, 1979b. Copyright 1979 by the American Association for the Advancement of Science.)

with a dominant genetic marker (an allele that leads to the development of a distinctive pale stripe down the back of the bug) were mated with females that had earlier copulated with unstriped males. The reciprocal sequence of matings was also performed, with the result shown in Fig. 11.14. The sperm that the females used came, almost without exception, from their last partner.

There are other insects whose males may mate with the same female at fairly short intervals while releasing her between copulations. These could serve as "insurance matings" if the female may have had another partner during the intervals between encounters with her original mate. The damselfly *C. maculata* experiences conditions that should favor insurance matings. Territorial males release their partners after sperm transfer and resume their defense of the oviposition territory. The female joins the male in his territory in order to lay her eggs there. Although she is guarded by her mate at a short distance, there is some chance that she will be chased off the site by an intruder or will voluntarily depart to oviposit elsewhere for a time. She may copulate with a new male, then return to her previous mate's territory in the same afternoon. Territorial males appear not to respond sexually to their recent mates (but perhaps only in some populations—see Alcock, 1979e, and Waage, 1980). This ability enables a male to avoid copulating again too quickly with the same female. He saves time that can be used to defend his territory and respond to additional females that arrive near his site to mate and lay their eggs. Moreover, it gives his mate a chance to lay her eggs (most of which he has probably fertilized) rather than having to engage in superfluous additional matings with him. On the other hand, the probability that a female will have copulated with an intruder, despite the guarding efforts of her territorial partner, increases with the passage of time. At some point an additional mating with the same female may be adaptive for a resident male, in that it enables him to displace sperm of any rival that she may have accepted since his last mating with her. In one Virginian population of *C. maculata* repeat copulations did occur—with an average interval of about 40 minutes between the two matings (Alcock, 1979e). (See also Sakagami et al., 1974, on repeat copulations in the dragonfly, *Hemicordulia ogasawarensis*.)

A more clear-cut example of insurance mating is provided by the bee *Anthidium maculosum* (Alcock, Eickwort, and Eickwort, 1977). Here too females copulate more or less on demand. A territorial male that is defending a patch of flowering food plant mates for 30 seconds with each female that visits his territory. After copulating she may leave or continue to forage on the flowers under the male's control. If she remains, the territory owner usually modifies his patrolling behavior so that he keeps flying back to his recent mate; although he may come close to her, he does not at once copulate again with her. Should a new female arrive in his territory, he will immediately try to grasp her and will inseminate her even if this is within a minute or two of completion of his previous mating. Males, then, do not enter a sexual refractory period after mating but in some way can identify recent mates and avoid recoupling with them for a time. But if a female remains within a male's territory for about 6 minutes, the male will eventually pounce upon her and copulate once more (Fig. 11.15). Because territory owners compromise their postcopulatory guarding behavior, presumably in order to detect and respond to new females that arrive while an earlier mate is present, nonterritorial intruders regularly secure matings within their territories. By mating at intervals with the same female, a territory owner probably improves the chance that his sperm will be the last received by a female before she returns to her nest to lay an egg.

A somewhat different pattern of repeat matings has been described

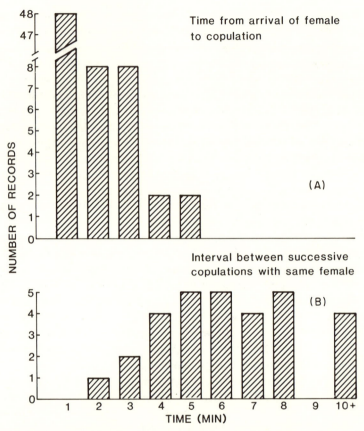

11.15   Mate recognition by the megachilid bee *Anthidium maculosum*. (*A*) Males that succeed in mating with a female generally do so within 2 minutes of her arrival on the territory (mean = 1.6 minutes). (*B*) Males that copulate a second time with a previous mate during a prolonged foraging visit by her wait an average of 6.5 minutes before reinseminating the female. (From Alcock, unpublished.)

by Reid (1958) for the mountain pine beetle (*Dendroctonus montico-lae*). He found that when a male joined a female in her incipient burrow, copulations were frequent the first day, declining to once per day thereafter until the female began to oviposit. The initial burst of matings is conceivably related to sperm competition. A male that finds an unpaired female has no way of knowing whether or not she has already acquired sperm from another male. Nothing is known about sperm precedence in this species. We shall assume that it occurs but that one follow-up mating does not lead to complete removal of sperm stored from previous matings. (Sperm precedence is often not 100 percent; see Boorman and Parker, 1976). Let us say that one copulation results in 70 percent sperm displacement. Two matings, therefore, might yield 91 percent displacement (0.70, plus 0.70 × 0.30) and three matings, 97 percent (0.70, plus 0.70 × 0.30, plus 0.70 × 0.09). After the first day's orgy of mating, selection might favor daily insurance matings if there was some chance of stolen matings in these beetles. Such matings would be rare when the male remains in a burrow with his female and guards the nest entrance against intruders, but they are possible because the male does not stay in direct physical contact with his mate at all times (Goeden and Norris, 1964; Ryker and Rudinsky, 1976).

Repeat matings also occur with some frequency in various Orthoptera, especially the crickets (Alexander, 1961; Loher and Rence, 1978). In these cases too, females may mate with more than one male during a period of receptivity, and although a male attempts to remain close to his mate, she may wander off or be stolen by a satellite. This possibility may favor males that mate again at intervals with the same female. In the giant weta, for example, receptive females are willing to copulate repeatedly. A. M. Richards (1973:201) notes that "the female is passive, and is kept in place by the male. If a mating pair are disturbed, the female tends to wander away, but the male immediately searches for her and they remate."

Preventing a Female from Accepting a New Partner

The adaptations we have discussed thus far enable a male to displace rival sperm and thereby increase the probability that his sperm will fertilize his mate's eggs. An alternative method of achieving the last-male advantage is to copulate with a female and then prevent her from accepting any other male until she has oviposited. In order to lower the probability that the female will mate again, a male may provide her with substances that make her nonreceptive, in much the way drosophilid fruit flies transfer an enzyme that may activate sperm release within their mates. Materials present in the ejaculates of some insects are known to trigger mechanisms that cause females to become reluctant to mate for some time. In their classic study of receptivity in the house fly, Riemann, Moen, and Thorson (1967) showed that by the end of the first 10 minutes of an hour-long copulation, a male had transferred the bulk of his gametes. Copulation continued for an additional 50 minutes in order to permit the male to transfer spermless seminal fluids to his mate. If the female did not receive these seminal substances, she retained her receptivity; if she did get them, she became nonreceptive for most or all of the remainder of her life.

We shall discuss the relationship between male-donated fluids, spermatophores, and nuptial gifts and changes in female receptivity in detail in Chapter 14. The key point for our discussion here is that in most of these cases males appear to offer metabolically valuable substances to their mates. The female that has received nutritious "gifts" during copulation becomes nonreceptive while she produces and lays some eggs, which are likely to be fertilized by gametes received from her most recent partner. By giving a mate useful resources, a male may make it possible (and advantageous) for the female to avoid mating for a time that she can then devote to egg laying.

Many male insects, rather than changing female receptivity with a nuptial gift, guard their females to prevent them from accepting a new partner. One possible tactic involves application of a chemical signal to a female that announces that she is not receptive, even though she might actually be capable of copulating again. In the bee *Centris adani* (Frankie, Vinson, and Colville, 1980) and the butterfly *Heliconius erato* (L. E. Gilbert, 1976) females that have mated carry away with them not only the sperm of their partner but also a distinctive male scent. In the bee, citral from mandibular glands, which males use to mark grass stems in their territories, is also applied to the bodies of females during copulation. This presumably makes the female smell like a male to members of her species. In the butterfly, males also transfer to the female a scent identified with males. The material is first produced by male pupae to discourage adult males from perching on the pupal case and fighting for access to the individual that will emerge from it (which happens if the pupa is a female). The freshly emerged male could be injured in the fights; adult males are repelled by the odor because they have nothing to gain genetically from fighting for access to a virgin male. Once a male has copulated, he donates the same substance to his mate, passing it into storage organs in the tip of her abdomen. Gilbert has suggested that the chemical is a pheromone that blocks other suitors from gaining access to the female, forcing monogamy upon her. The same argument could apply to the bee.

However, the notion that the antiaphrodisiacal scent *deceives* other males into missing opportunities to mate has the logical difficulty that we discussed in Chapter 10 in the context of male "inhibiting" pheromones. If it were advantageous for the male to ignore the antiaphrodisiac provided by a rival, selection would favor mutant individuals that lacked the capacity to detect these deceitful messages. Selection might even favor males that found the odor attractive and so were able to use it as a cue to locate receptive females that competitors were trying to hide. Furthermore, the hypothesis that males of *H. erato* can force their mates to be monogamous is weakened by the finding that females have specialized abdominal glands ("stink clubs") for storing the repellent substance (Fig. 11.16). This must mean that it is to a female's advantage to be made unattractive after mating, at least for a time, otherwise females without such glands would enjoy greater success. Females probably release the repellent only when they gain by doing so. After receiving a spermatophore, female butterflies typically lose their receptivity for a period of days. During the time when ample supplies of sperm and allied materials remain within her, a female can readily advertise her lack of receptivity by releasing the male-donated signal when harassed by courting males. In turn, a sexually motivated male may gain by leaving the female when he detects the "male" odor if the odds are that she will never respond to his courtship. Finally, the original mating partner gains by helping his mate reduce the time she must spend evading unwanted suitors, thereby increasing the time available for activities that more directly contribute to her reproductive output and his fitness.

This argument also applies to females of the bee *Centris adani*. The male odor, while it persists, probably counteracts any virgin scent that lingers on the female. In this way the male helps his mate avoid other

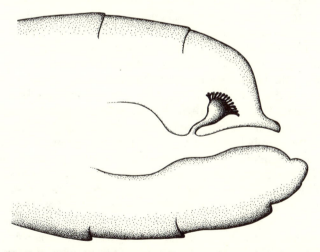

11.16   The fringed "stink clubs" of female *Heliconius erato* occur in the tip of the abdomen and are used to accept a secretion of male-repelling material from a copulatory partner. (From Eltringham, 1925.)

males that might be attracted to her for a short time after mating. Potential suitors presumably are not accepted by mated females and therefore gain by ignoring male-scented bees. In both the butterfly and the bee the antiaphrodisiac appeals to a competitor's genetic self-interest, as deflected rivals can probably gain more by searching for receptive females than by trying to copulate with a recently mated individual unlikely to accept his gametes.

Mating Plugs

The same type of conflict of sexual interest arises in an analysis of mating plugs donated by a male to his copulatory partner. These substances, which seal the female's genital opening, provide a physical rather than a pheromonal barrier to other males. We have already discussed how drone honey bees and male ceratopogonid midges may leave their severed aedeagus in the female's genital opening as a suicidal finale to mating. While this sacrifice may decrease remating by a female, it does not prevent it completely.

The use of body parts as mating plugs also occurs in some ants in which there is intense competition among males for receptive females in dense mating aggregations. In some harvester ants of the genus *Pogonomyrmex* mating occurs at landmark sites where vast numbers of males gather to await the arrival of receptive females. In several of these species females have been seen with the male copulatory apparatus attached to their genitalia. Females that wish to terminate copulation bite at the male's gaster, with the result that the male may disengage abruptly leaving the tip of his abdomen in place. Just who is responsible for the male's sacrifice is not completely clear. Either the female is able to sever her mate's abdomen with her mandibles or else the male is able to break off his genitalia in response to a signal from the female (Hölldobler, 1976; Markl, Hölldobler, and Hölldobler, 1977). It is not known whether the detached male genitalia prevent or reduce the number of copulations by females, although queen harvester ants frequently do mate more than once. On the one hand, it is possible that a female may cut through her mate's abdomen (or induce him to engage in genital autotomy) when it is in her interest to prevent additional matings; on the other hand, perhaps females require several copulations to fill the sperm storage organ completely (Hölldobler, 1976). If a male that mates with a virgin is able to prevent her from mating again, he may benefit from reduced competition for access to her eggs but she may suffer reduced fecundity. Thus conflicting interests may lead to prolonged copulation by some males and a lethal reaction by females.

Less drastic solutions to the problem of how to physically plug a mate's genitalia include the deposition, either internally or externally, of a secretion that performs this function. In fact, the male honey bee may apply both an external seal in the form of his genitalia and an internal block of mucopolysaccharides that coagulate within the female (Blum, Glowska, and Taber, 1962). In a diversity of other insects the male also transfers to his mate substances that coalesce or harden once inside the female. Frequently in these cases the result is a spermatophore, sometimes with a tube running to the spermatheca for the sperm to travel along and with a distal end blocking or filling the female's reproductive tract so that she cannot accept additional spermatophores (Fig. 11.17). The materials within the spermatophores are sometimes metabolized, perhaps serving a nutritional function once dissolved, but acting as a mating plug until then (Landa, 1960; Gerber, Church, and Rempel, 1971). In a variety of Diptera including drosophilid fruit flies and mosquitoes, the male's ejaculate contains

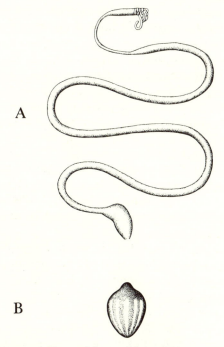

11.17   (*A*) The spermátophore of the grasshopper *Locusta migratoria,* which is about 2 cm long and is shown here uncoiled. The spermatophore fills the female's spermatheca. See also Fig. 13.2. (From Gregory, 1965.) (*B*) The spherical spermatophore of the cockroach, *Periplaneta americana,* is only 1 to 2 mm long. (From Gupta, 1947.)

substances that become a gelatinous or waxy mass that plugs the female's sperm receiving organs (Gillies, 1956; E. T. Nielsen, 1959; Lum, 1961; Linley and Adams, 1972; Fowler, 1973).

An external plug or sphragis is applied to the female's genital opening in a water beetle (O. W. Richards, 1927) and a few butterflies. In the lepidopterans males first transfer a spermatophore and then secrete viscous substances derived from the male's accessory gland over the genital area. The sphragis hardens upon contact with the air and obstructs other males' genital claspers. (Originally the sphragis was thought to be an oviposition guide of some sort.) In the genus *Acrea* pairing takes place without courtship, so perhaps a mated female would be vulnerable to forced copulation if she lacked the sphragis (G. A. K. Marshall, 1901). Mating pairs of *Euphydryas editha* are sometimes jostled by "supernumerary" males; by leaving an external plug in place, a male prevents rivals of this sort from seizing the female as soon as the first mating is completed (Labine, 1964).

Males that inject their mates with inducers of nonreceptivity or apply mating plugs are free to leave one partner after inseminating her to search for others. This benefit, however, is offset if other males (or the female herself) are able to remove, dissolve, or otherwise overcome or counteract the copulatory block. Perhaps the most effective method of mate guarding, therefore, is for the male to remain near his partner after inseminating her, to physically thwart other males that would copulate with her. This can be done in three ways: (1) by extending copulation past the time when *insemination* had been completed so as to prevent other males from assuming the copulatory position, (2) by disengaging from the female but retaining contact with her (the male might stay mounted on the female), or (3) by disengaging from the female completely but remaining close enough to her (in the manner of *Calopteryx* damselflies) to detect approaching rivals, which could be chased away. We shall discuss each of these possibilities in turn.

Copulatory Mate Guarding. In order to demonstrate that males may remain *in copula* as a form of mate guarding, one must first eliminate other reasons why copulation might be relatively lengthy. From a male's perspective genital linkage with a female can serve a variety of functions, one of which is the transfer of sperm. Table 11.1 presents a sample of copulatory times for the Diptera to illustrate the great diversity within even one group. Because in some instances sperm transfer is accomplished in a matter of seconds, the process of gamete exchange obviously does not have to be prolonged. We assume that if the sole function of copulation were sperm donation, then matings would

Table 11.1   Duration of copulation in various flies, showing the great interspecific (and substantial intraspecific) variation in mating times.

Family/species	Copulation time	Reference
Chironomidae		
Glyptotendipes paripes	3 to 5 sec	E. T. Nielsen, 1959
Drosophilidae		
Drosophila spp.	30 sec to 1½ hr	Spieth, 1952
D. suboccidentalis	7 to 28 min	Spieth, 1952
Tabanidae		
Chrysops fuliginosus	3 to 59 min	Catts and Olkowski, 1972
Hippoboscidae		
Pseudolynchia maura	Few min to 2 hr	Prouty and Coatney, 1934
Syrphidae		
Microdon fuscipennis	Few min to 2 hr	Duffield, 1981
Limoniidae		
Erioptera gemina	5 min to 2 hr	Savolainen and Syrajämäki, 1971
Scatophagidae		
Scatophaga stercoraria	30 to 35 min	Parker, 1974c,d
Tephritidae		
Rioxa pornia	Up to 35 min	Pritchard, 1967
Euarestoides acutangulus	1 to 3 hr or more	Piper, 1976
Muscidae		
Musca autumnalis	1 to 1½ hr	Teskey, 1969
Sarcophagidae		
Sarcophaga spp.	Up to 4 hr	Thomas, 1950
Bibionidae		
Plecia nearctica	Up to 50 hr	Thornhill, 1976b

tend to evolve toward brevity so that males and females would have more time to invest in other critical activities.

Copulation, however, can involve more than just the donation of gametes. Males may invest added time in one mating above and beyond that needed for insemination of the female (1) to transfer materials such as seminal fluids, spermatophores, mating plugs, and nuptial gifts, which may sometimes prevent sperm displacement, and (2) to displace the sperm of competitor males contained within the female. (It is also possible that males may prolong copulation if a mated pair is safer from predators or competitors than an unpaired individual; see McLain, 1980, 1981.)

On the first point, it takes time to mobilize the nongenetic materials to be given to a female (which may equal 25 percent of the male's body weight) and still more time for them to travel through the penis into the female. We noted earlier that while sperm passage in the housefly consumes only 10 minutes, the pair remains together for another 50 minutes, during which the female receives an injection of glandular secretions from her partner. In the Orthoptera copulation is often lengthy (up to 18 hours in *Locusta migratoria*) and is associated with the passage of a complex spermatophore from male to female (Gregory, 1965), although some species pass large spermatophores in a matter of minutes (G. K. Morris, 1980; Gwynne, 1981). Matings of several hours are common in butterflies, as the male forms a spermatophore within his mate (Scott, 1972). The same is true for several beetles that require 2 to 4 hours to complete the transfer of a spermatophore (Landa, 1960; Lew and Ball, 1980). In the pentatomid *Nezara viridula* males that mate for a relatively short period are more vulnerable to sperm displacement than individuals that copulate for a long while (McLain, 1980), although the mechanism is not known.

On the other side of the coin, sperm displacement may require time. We have already discussed how males of *C. maculata* devote a portion of copulation to the removal of stored sperm, after which the male transfers his own gametes. In other insects the mechanism of sperm competition is far less well documented but in the dung fly, *Scatophaga stercoraria*, the duration of copulation is definitely correlated with displacement of sperm given the female by her earlier partners. Females of this species mate each time they come to dung to lay a clutch of eggs. Parker (1970c) found that a female receives sufficient sperm from a single mating to fertilize four batches of eggs with no decline in fertility. Therefore when a male encounters a nonvirgin female, she will probably already have sperm within her even if she has not mated since her last oviposition episode. The male's task therefore is to replace the female's stored sperm with his own gametes, and he is capable of doing this (Parker, 1970c). The degree of sperm precedence, however, is a function of the time spent in copula.

The demonstration of sperm precedence involves pairing each female first with either a normal or a sterile, irradiated male and then mating her with a second male of the other type. A female that copulates first with a normal and then with a sterile male will lay only about

20 percent of the number of fertile eggs that she would have laid had she mated only with a normal male. This shows that about 80 percent of her eggs were "fertilized" by sterile sperm from the second male.

The 80 percent figure is achieved, however, only if the second male mates for 30 to 35 minutes, the time normally devoted to copulation in this situation. By separating the flies prematurely, Parker derived the graph shown in Fig. 11.18. By extrapolation from these data, a male that copulated for an hour, instead of 35 minutes, would displace all his rival's sperm and fertilize all of his mate's clutch, not just 80 percent. But as Parker (1978a) shows in an optimality analysis, there is a conflict between achieving maximum egg gain from one female and securing additional copulations. Matings of 35 minutes are optimal because extension of copulation past this point yields a relatively low and declining yield in eggs fertilized per minute which can be exceeded by searching for and finding additional partners.

Contact Guarding through Prolonged Copulation. Thus prolonged copulations may be favored for a variety of reasons other than mate guarding. But there are cases in which sperm transfer time and sperm displacement seem not to be responsible for lengthy copulation. In some of these cases it is plausible that males retain the copulating position primarily in order to prevent their mates from copulating again.

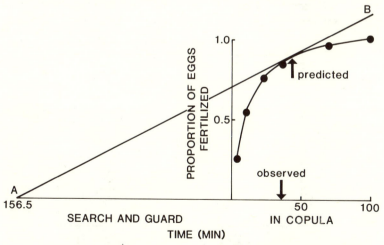

11.18 The theoretical optimum duration of copulation in the dung fly *Scatophaga stercoraria,* contrasted with the actual average duration of mating in this fly. (From Parker and Stuart, 1976. Reprinted by permission of the University of Chicago Press.)

The evidence comes from species in which the duration of mating is variable, with unusually lengthy matings associated with male-male competition. For example, consider the case of the glowworm fly, *Arachnocampa luminosa* (A. M. Richards, 1960). Typically, copulation lasts about an hour, but in situations in which there are fights for possession of a female, a male may remain linked to the mate for up to 7 hours. Similarly, Uéda (1979) and Hassan (1981) found that males of libellulid dragonflies prolong copulation when the frequency of sexual interference from rivals is high. A more speculative case involves a limoniid fly whose copulations last from 5 minutes to 2 hours. In this species a pair may fly out of a swarm but be accompanied by another male who perches nearby. The supernumerary male may copulate with the female when she is released (Savolainen and Syrajämäki, 1971). Perhaps males that detect a lurking rival lengthen copulation in an effort to induce him to abandon his wait, although this hypothesis remains to be tested.

A clearer case of the effect of a male's social environment on the length of copulation comes from a study of the stink bug *Nezara viridula* (McLain, 1980). In laboratory experiments in which four males were placed with two females, the average duration of copulation was significantly longer than when the sex ratio was 2:4 (Fig. 11.19). In this bug the degree to which a first male prevents sperm displacement is positively correlated with the length of copulation (which varies from 5 minutes to 14 days!). When the density of competitors is high—and thus the probability of a repeat mating by a partner is

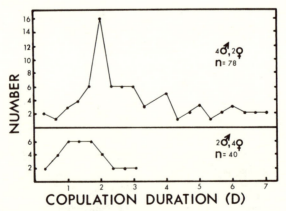

COPULATION DURATION (D)

11.19   The operational sex ratio and the duration of mating in the stink bug *Nezara viridula*. When males outnumber females in a caged population, copulation is prolonged. (From McLain, 1980.)

also high—males lengthen the time spent in copula and thereby increase the percentage of their mate's eggs that they are likely to fertilize (see also Sillén-Tullberg, 1981).

Some male insects may also mate with one female for a long while in order to monopolize her until such time as she is highly likely to oviposit rather than mate again. For example, the male lovebug, by copulating for 50 hours and more, may make it advantageous for the female to lay her eggs immediately after she is finally released (Thornhill, 1976b). Females of this species typically mate just once, but might mate more often if their first male failed to devote a long time to copulation. If after a lengthy mating a female were to accept another male, she might have to endure still another protracted union before getting a chance to lay her eggs, a system favoring females that oviposit promptly when freed.

The strategic timing of even a moderately long mating may make it adaptive for a female to lay her eggs after copulating with one male and not accept another, at least for a time. As a case in point, females of the otitid fly, *P. demandata,* sometimes mate on consecutive days, but copulation occurs only in the afternoon after 1300 or 1400 hours and before 1700 hours (Alcock and Pyle, 1979). Mornings are probably spent feeding, resting, and egg laying. When a male does secure a receptive female, he will not release her for several hours; this ties the female up for much or all of the daily mating period. When a female is finally released, it is late afternoon and she routinely leaves the area. As a result, any eggs she lays the next morning will probably be fertilized by her mating partner of the previous afternoon (assuming that sperm precedence occurs in the fly). The same hypothesis has been applied to the katydid *Copiphora rhinoceros,* which mates for 4 hours, largely consuming one entire nocturnal mating period (G. K. Morris, 1980).

Copulatory Mate Guarding and the Operational Sex Ratio. A male that remains with a female either to guard her against takeover or to displace an opponent's sperm cannot transfer gametes to other females for the duration of the association. Our prediction, therefore, is that mate guarding by postinsemination copulation should tend to occur when the probability of acquiring multiple mates is relatively low. The prediction can be tested by observing mating behavior in phasmid walkingsticks and through a study of intraspecific differences in the duration of copulation in a bee.

The insect record for prolonged copulation (79 days) is held by a walkingstick. (We assume that in this and other species with lengthy matings, sperm is transferred to the female early in the association

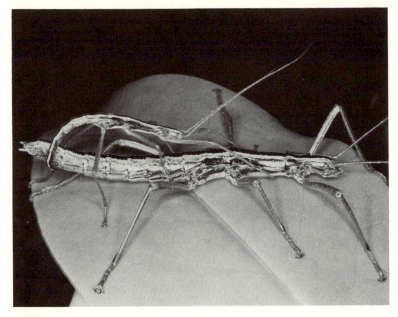

11.20   Copulating walkingsticks. The small male may act as a living chastity belt for the female. (Photograph by A. P. Smith.)

and copulation persists long after insemination has been completed.) In some other members of the group, mating is measured in hours rather than days, providing a wide range of diversity in mating times. Sivinski (1978) discovered a correlation between the length of mating and the degree of sexual dimorphism in walkingsticks. Males range from one-quarter the length of females to the same size, depending on the species (Fig. 11.20).When adult males are considerably smaller than adult females, one assumes that the time required to reach sexual maturity is much less for males. If mortality rates are reasonably constant over time and equivalent for both sexes, there will be fewer mature females than mature males and thus an operational sex ratio with a male bias. The positive relation between sexual dimorphism and copulation time in phasmids matches the predicted trend toward protracted copulation in species with the greatest male bias in the operational sex ratio.

The nature of intraspecific variation in copulation length in the andrenid bee *Nomadopsis puellae* also supports the prediction (Rutowski and Alcock, 1980). Males of this species adjust the duration of each copulation in accordance with the relative abundance of competitor males and available females. Females begin collecting pollen and nectar in the early morning and continue until sometime between 1100 and 1300 hours, when the flowers of their major food plant, the desert dandelion, close up for the day. Females are thought to be able to gather sufficient materials in one morning to produce a complete brood ball, which is constructed in a cell in an underground nest burrow (Rozen, 1958). An egg is laid on each brood ball sometime after it is completed, most often in the early afternoon after the period of foraging is finished.

Females mate with any male that pounces on them while they are foraging on the flat surface of a dandelion flower (Fig. 11.21). Sperm precedence probably occurs, judging from the failure of males to search for mates early in the foraging period, whereas male activity is very high and the intensity for competition for females great in the latter part of the morning. These observations suggest that males are competing for access to females that are likely to lay an egg after completion of a final foraging trip.

Correlated with the increase in male searching and fights for possession of a female is a change in the average duration of copulation. Early in the day males invariably copulate for less than a minute and then release a mate to search for another. As the day progresses, males

11.21 Copulating pair of *Nomadopsis puellae*. The male rides on the back of the female as she gathers pollen for her brood. (Photograph by J. Alcock.)

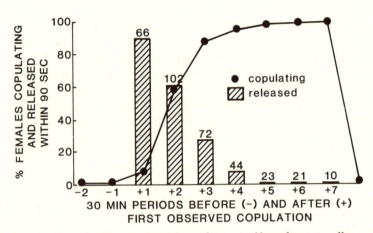

11.22 Variation in the duration of copulation in *Nomadopsis puellae* as a function of changes in the operational sex ratio over the course of the morning. The numbers above the bars represent sample sizes. When females outnumber males (*line graph*) most matings last less than 90 seconds (*bar graph*). Copulations lengthen when single foraging females are scarce. (From Rutowski and Alcock, 1980.)

begin to remain with their partners for longer periods of time, riding on the female's back as she travels from flower to flower (Fig. 11.22). In this position the male actively repels incoming males searching for females. The change in the duration of mating from 30 to 60 seconds to 30 or more minutes probably does not reflect a 30 to 60-fold increase in the quantity of sperm or other materials transferred to females. Instead, prolonged mating may help males guard females that are reasonably likely to be on their last collecting trip of the day and thus on the verge of ovipositing. By monopolizing a female in the late morning, a male may dramatically improve his chances of fertilizing an egg (again, the assumption is that males and female would not mate repeatedly unless there were genetic gains for them both).

By contrast, early in the morning a male may fertilize an egg only if he is fortunate enough to contact a rare out-of-phase female who failed to complete a brood ball the preceding day and who will finish her provisioning of the cell after an early-morning trip. Most of the in-phase females he encounters will mate with other males on other foraging excursions before laying an egg, so that the expected gain from remaining with any one female during an early foraging trip is very low. Therefore males at first dash quickly from one female to the next in an attempt to copulate with as many individuals as possible, increasing their chance of mating with an out-of-phase female who may be missed by other males before she returns to her nest. As the day wears on, more and more males begin to search for mates and then more and more begin to remain in prolonged copulation; single females become scarcer and the operational sex ratio becomes male biased. Thus the genetic payoff from a "copulate-briefly-with-many-females" strategy falls. A protracted mating-with-one-female strategy is favored until, by the end of the morning, each foraging female is mounted by a guarding male who accompanies her until she returns to her nest and burrows through the sand away from other males.

The case of *Nomadopsis* provides a fine illustration of the trade-off between guarding and nonguarding. The male that remains with one mate improves his chances that this female will lay an egg fertilized by his sperm, but at the cost of being unable to locate and copulate with additional females during the guarding period. The nonguarder is free to search for and mate with as many single females as possible during a morning, but the probability that he will actually fertilize any one female's egg is low. Over the course of the daily foraging time, the change in the availability of unguarded females and the change in the likelihood that a female will oviposit after completing her current foraging trip alter the average genetic gains derived from the two competing strategies, leading to a complete switch to copulatory mate guarding by males.

Contact Guarding without Prolonged Copulation. A male need not remain in genital contact with a female in order to guard her closely. In the dung fly, *Scatophaga stercoraria*, a male retracts his genitalia after copulation but does not dismount while his partner oviposits (for about 15 minutes). Instead, he remains above the female ready to repel male harassers (Fig. 11.23). If an attacker does succeed in displacing the first male, he will copulate promptly with the female and fertilize about 80 percent of those eggs remaining to be laid (provided that he too can prevent another male from taking his mate from him). Parker (1974b) has shown that a mutant male that did not exhibit contact guarding would be one-quarter to one-third as successful in gene transmission as normal males over a wide range of male densities. Because there are usually large numbers of males searching for mates on dung pats, the chance that an unguarded female will be overlooked after her mate departs is small indeed.

The primary advantage of nongenital contact guarding may be to permit the female to oviposit while the male guards her. Otherwise the costs and benefits of this form of mate protection are nearly the same as those associated with copulatory mate guarding. Although there are intermediate patterns between the two options (in which the male al-

11.23   Mate guarding by a male dung fly. The male stands above his oviposit-ing mate and uses his body as a shield to repel an intruder. (From Parker, 1970d.)

ternates between copulation and a passive phase—see for example, Wright, 1960; Hagley, 1965; Alcock, 1976; Sivinski, 1978), here we will consider an example of purely nongenital contact guarding in which the copulating male withdraws his genitalia but stays in contact with his mate for a period of time before releasing her, in the manner of *S. stercoraria.*

The example is the cricket *Teleogryllus commodus,* the subject of a comprehensive study by Loher and Rence (1978). Male crickets typi-cally attract mates with a loud calling song, followed by a courtship signal that may induce an attracted female to climb onto the back of the male. While in the female-above, male-below position the male presses a spermatophore upward into his mate's genital opening. Although some pairs separate promptly after a brief mating, other pairs copulate for a long period (presumably as a mate guarding device). In another group the male feeds his mate an offering from his dorsal glands so that she remains on his back for periods of more than a hour after the pair have uncoupled their genitalia (Khalifa, 1950; Alexander, 1961; Loher and Rence, 1978).

Teleogryllus commodus is an example of yet another type. Transfer of the spermatophore requires only 3 minutes on average. During this time the male guides the long (1 cm), thin tube of his spermatophore into the duct leading to the female's spermatheca. After the tube is in place, the remainder of the complex spermatophore is pushed into the female's genital chamber, where it is held in place by an attachment plate. After the male disengages from the female, sperm begin to flow from the spermatophore into the spermatheca by way of the thin tube, a process that requires about 70 minutes before the spermatophore is completely drained of its contents.

After uncoupling, the male attempts to retain contact with his female by placing his antennae across her body (Fig. 11.24). Initially she is restless and may make several attempts over the first 20 minutes or so to break away from her mate. When she moves off, the male vibrates his antennae and may chirp aggressively before pursuing his mate. Should he achieve antennal contact once again, she will remain motionless for a time until her next move. If the male can retain contact during the initial period of restlessness, then the female becomes inactive for another hour or so (Loher and Rence, 1978).

The early escape attempts of females may be designed to test the dominance of a mate. In nature the crickets are evidently territorial, as judged from the wide spacing of males (8 to 12 m apart) and their avoidance of nearby singing conspecifics (Campbell and Shipp, 1979; Evans, manuscript). On the other hand, there is considerable movement and burrow changing in populations of the cricket, particularly at certain times of the year (Evans, manuscript). This suggests that females moving toward a territory owner may come in contact with mobile subdominants of untested quality. If her mate does not keep close to her, she may inadvertently copulate with such a subordinate male in a dominant male's territory (although there is no evidence currently that satellite males exist in this species). Assuming that it is advantageous for her to mate with a dominant male, she will benefit by moving about, thereby forcing her mate to follow and to chirp if he is to retain control of her. Even after mating, a female is receptive and if she escapes from one partner and is met by another, she will mate again. Her new partner may succeed in pushing the original sperma-

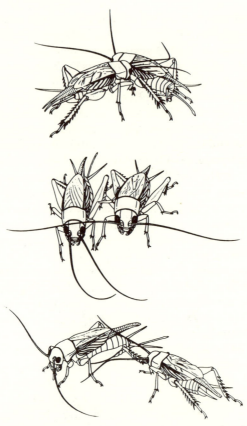

11.24 Contact guarding by the cricket *Teleogryllus commodus*. The male attempts to maintain physical contact with his recent mate, often by placing his antennae across her body. (From Loher and Rence, 1978.)

tophore out onto the ground, to be replaced by his own (Loher and Rence, 1978).

Even if no other male is immediately available for a second copulation, the female takes action to reduce the fertilization rate of any partner who does not stay by her side. An *unguarded* female reaches back with her mouthparts and removes the spermatophore from her body about 30 minutes after mating. Many, perhaps a majority, of the sperm will not have left the spermatophore by this time. They are consumed along with their container, and in nature the female probably then seeks out another male to fill her sperm storage reservoir. But if she is guarded successfully for a total period of about 80 minutes, the female does not remove and consume the spermatophore until 105 minutes have passed from the moment of copulation. At this stage the spermatophore is empty and the male has realized the maximum genetic gain possible from his donation of the spermatophore to the female.

We interpret the value to a male of guarding as twofold: first, to prevent his partner from mating again with consequent loss of the spermatophore and its contents and, second, to cause his mate to delay removal of the spermatophore until all the sperm have moved into the spermatheca. The male may, as Loher and Rence suggest, physically prevent the female from reaching back to grasp and eat the spermatophore, but it seems equally likely that the female does not even try to do so in the presence of a competent guarder, as it is presumably in her interest to fertilize her eggs with sperm from a dominant, territorial male.

Although the average period of guarding exceeds by 15 minutes the average time needed for the spermatophore to be emptied, males may benefit by erring on the safe side rather than permitting a female to wander off prematurely. Moreover, the intensity of guarding by the male does decline at just about the time (50 to 60 minutes after mating) when the spermatophore is likely to be drained. At this point he begins to compromise his guarding activities by going out on brief patrols, each time returning to rest near his mate.

Contact and Noncontact Guarding in the Odonata

In damselflies and dragonflies, as well as in crickets, there is a full spectrum of guarding activities, ranging from species in which there is no interaction between the sexes after mating to species in which a brief copulation is followed by a long period in which the male escorts the female to one or more oviposition sites, all the while retaining a grip on her with his terminal claspers (Fig. 11.25). For the odonates, to a greater degree than for the crickets, ecological data are available to test the prediction that the time investment in guarding is related to the likelihood that an unguarded female will mate again (with subsequent sperm displacement). A second testable prediction is that contact guarding should occur in species in which receptive females are scarce, because a male will lose few mating opportunities while guarding one partner.

Typically, among the odonates, the failure to mate guard is correlated with low risk that a female will mate again for some time after copulating. For example, males rarely guard their mates in species

11.25   Contact guarding by a damselfly male, which remains in tandem with his mate following copulation and stays with her while she oviposits. (Photograph by J. Alcock.)

whose females oviposit underwater where they are safe from rivals, although in these species pairs often fly in tandem to suitable oviposition sites. Another group of species in which there is no mate guarding is characterized by females that oviposit in concealed locations. In the dragonfly *Tanypteryx hageni* females are released after mating and later slip unguarded into the thick emergent vegetation above spring-fed bogs and *walk* through the plant growth to oviposit in the spongy bog material (Clement and Meyer, 1980). Likewise, if there happen to be separate mating and oviposition periods either daily or seasonally, there is no guarding. In all these situations there is little risk to a male that his partner will copulate before laying the eggs his sperm will fertilize (Sakagami et al., 1974; Schmidt, 1975; Waage, 1979b).

At the opposite end of the spectrum are the species whose males do not release their mates until they have completed laying their entire clutch of eggs. Contact guarding through retention of the tandem grip appears to be a highly effective manner of assuring the fidelity of one's mate. It is evidently extremely difficult for a male to force an opponent to release a female because takeovers in contact-guarding odonates are rarely seen.

The contact guarder, however, pays for effective monopolization of one mate: (1) he is unable to mate with other females while in tandem and (2) he is unable to defend his territory, should he have one, during the guarding time. Intraspecific variation in mate guarding offers the chance to determine if these factors are correlated with differences in male behavior toward females. In the dragonfly *Sympe-trum parvulum* Uéda (1979) found that the degree to which males guard their mates depends on both the social status of the male and the potential risk of takeovers. In high-density populations in which there was considerable interference with mating pairs, subordinate males employed tandem (contact) guarding, whereas in low-density situations they did not. Dominant males *never* contact guarded, but instead perched (hovered) by their mate and therefore were free to defend the territory and respond to any additional females that might arrive while one was ovipositing. Nonterritorial, low-ranking males had, one assumes, a much lower probability of securing a series of mates in short order and therefore enjoyed a greater genetic gain through intensive guarding of one individual at times when she and her eggs might be quickly lost if she were not defended.

A Comparative Analysis of Mate Guarding in Two Damselflies

We can also look for the ecological correlates of contact and noncon-tact mate guarding by comparing two related species of damselflies that differ in their manner of mate guarding. We have already de-scribed how males of *C. maculata* release their mates immediately after copulation and are content to employ noncontact guarding, de-spite the fact that their females are still receptive and oviposit above water in areas where many other males may be searching for mates. While perched near his female, a territory owner can repel most in-truders intent on mate thievery, although he sometimes fails. An un-encumbered guarding male gains because (1) he can defend the in-tegrity of his territory and (2) he can respond sexually to females newly arrived in his section of stream.

If a male did not defend his territory continuously, it would be oc-cupied promptly by a newcomer and this male would vigorously attack the original resident upon his return, especially if the new occupant had mated during his rival's absence. These battles are not only costly to a male in terms of time and energy but they may also result in the loss of his territory, and with it the loss of many additional opportuni-ties to mate with females drawn to the area.

Opportunities for frequent mating are relatively good in *C. macu-lata* because the operational sex ratio is not as strongly male biased as in many other odonates. Territory owners averaged nearly five copula-tions per day in one study (Waage, 1979b). Because most matings occur in a 3- to 5-hour period each day and because any one female may oviposit for periods in excess of 2 hours (Alcock, 1979e), a male that remained in tandem with one mate for the duration of her oviposition bout might well miss a number of other matings. Thus both in the short term and the long term, the cost of contact guarding for *C. maculata* is relatively high.

The significance of this factor can be further examined by contrasting the behavior of *C. maculata* with that of another calopterygid damselfly, *Hetaerina vulnerata* (Alcock, 1982b). In many respects males of *H. vulnerata* behave similarly to *C. maculata*. Some individuals defend territories at stream locations that contain oviposition material attractive to females. Other males are nonaggressive satellites. Males in both categories attempt to capture gravid females near the water and copulate with them. At this point some important differences emerge. Males of *H. vulnerata* do not release their mates after a brief copulation, but instead remain in tandem after completion of sperm transfer. Pairs fly along the stream until they locate a likely oviposition site, usually but not always within the male's territory (if the copulating male owns a territory). Females are released at these sites and submerge almost instantly. A male gains nothing by following his female underwater; she is safe from takeover there. But the male does perch near the spot where he released his mate, even if it is outside his territory. He remains on guard, though there would appear to be little advantage in providing surveillance for a female that cannot mate while underwater (Fig. 11.26).

Thus there are two peculiarities in the behavior of *H. vulnerata*: (1) a male may abandon his territory to accompany a female to an egg laying location outside his area and (2) a male will guard an ovipositing female despite the general rule that submerged females are not guarded because they are presumably safe from competitors.

Noncontact guarding of a submerged female becomes more understandable when one discovers that females often reject one oviposition site, come to the surface, and fly off. If their mate is present, he quickly recaptures the reemerged female, resumes the tandem position (without a preceding copulation), and transports her to a new location. (Much the same thing happens in unrelated species of *Enallagma*, whose females also regularly lay only a portion of their clutch of mature eggs in any one submerged site and then reemerge to go elsewhere to lay more eggs; see Eriksen, 1960; Bick and Bick, 1963). In *H. vulnerata* if the female's original partner is not present (often because he has been driven off by the owner of the steam section in which he has released his mate), she is likely to be grasped by another male. If so, she will mate again readily, presumably using this last male's sperm to fertilize the unlaid portion of her clutch.

11.26   Noncontact guarding by the damselfly *Hetaerina vulnerata*. The male perches on a rock at the point where he released his mate. She submerges to oviposit underwater. If she reemerges, the male may be in position to recapture her and prevent her from mating with another male. (Drawing by M. H. Stewart.)

These observations partially explain why males may leave their territories in tandem with a female. Because females often refuse to lay some or all of their eggs in a mate's territory, males that fail to escort their mate off-territory (if she insists on leaving) would surely lose her to a rival. In order for temporary abandonment of the territory to be profitable for the resident damselfly, he must gain a sufficient number of egg fertilizations through his control of one female to offset any egg losses through missed mating opportunities with other females. A male that leaves his territory cannot copulate with females that arrive there while he is gone. Moreover, when a territory is left unguarded, another male often claims it and could potentially prevent the absentee landlord from reclaiming the site on his return. In actuality, these potential costs of escort and guarding behavior are slight in the case of *H. vulnerata.* In contrast with *C. maculata,* receptive females are very scarce, leading to a highly unbalanced operational sex ratio and an average copulation rate of one mating per 3.6 days for a territorial male. The chance that a receptive female will be missed while the male guards a current mate for an hour or so is extremely low. In addition, territory owners have only one chance in ten that they will not be able to recover the territory from a newly established occupant on their return from a bout of guarding. The distinctive features of male behavior in *H. vulnerata* are biologically sensible, given the relative risks of mate thievery and missed matings due to guarding.

With this chapter we conclude our discussion of the possible effects of sexual selection on competition among male insects for mates and their eggs. It is appropriate to end with the analysis of sperm competition and male behavior because this topic exemplifies the impact of the revived adaptationist approach following the group selection era. If one views reproductive behavior as a cooperative endeavor designed to help perpetuate the species, one is unlikely to look for mating plugs, competition among rival ejaculates within females, or male genitalia that remove sperm. Indeed, prior to Parker's pioneering paper on sperm competition and insect behavior (1970c) there was little or no awareness of the possible effects on individual reproductive success of the many characteristics discussed in this chapter. In the years since publication of Parker's article, there has been great interest in reinterpreting male postcopulatory behavior from the perspective of individual selection. Although much remains to be learned about male competition for egg fertilizations, the development of this research topic demonstrates the positive contribution of a modern evolutionary approach to an understanding of insect reproductive behavior.

12

Selective Mate Choice by Females

THE MATERIAL presented in Chapters 4 through 11 shows that intra-sexual selection is a potent force for evolutionary change and has been responsible for a remarkably wide range of attributes in male insects. We turn now to the second component of sexual selection, epigamic or intersexual selection, which we introduced in Chapter 3. Epigamic selection occurs primarily when choosy females make discriminating choices among males as mating partners. Although currently it is widely accepted that intrasexual selection has played a dominant role in shaping male reproductive behavior, the importance of female choice in the evolution of male characters remains controversial.

In this chapter we examine why it has been so difficult to document the occurrence of female choice despite the theoretical argument that it should be a key pressure on male behavior. To illustrate that female choice actually does occur, we describe the behavior of a scorpionfly, *Hylobittacus apicalis*, one of the few species for which adaptive dis-crimination among potential mates by females has been unequivocally demonstrated. We then discuss the diversity of potential benefits, ma-terial and genetic, that males of some species appear to offer to their mates. If males vary with respect to the quantity or quality of these benefits, then we predict that female choice will evolve and we point to some evidence in support of this proposition.

The Limited Evidence for Female Choice

Earlier we developed the argument that the sex which makes the greater parental investment per offspring will be a limited resource for members of the opposite sex. Usually female animals allocate more to parental effort than males. As expected, males typically compete among themselves for possession of females because male reproductive success is limited by the number of eggs an individual fertilizes. Females are usually courted by many males, so the potential exists for females to choose among would-be partners. If females consistently preferred males with certain characteristics, they could greatly influence the evolution of male reproductive adaptations. But has the potential for female choice been realized? Has epigamic selection been an important factor in the evolution of male behavior? There are three major reasons why this question has not yet been fully resolved (Maynard Smith, 1978b; Borgia, 1979; Thornhill, 1980a).

First, there is considerable indirect but almost no unambiguously direct evidence that females actually do choose certain conspecific males over others under natural conditions. This is because of the difficulty of separating the effects of intrasexual from epigamic selection. Second, many of the traits of males that females might use to discriminate among potential partners (such as bizarre courtship dances) seem to be irrelevant to female fitness. Thus it is unclear how female preferences, if they exist, could promote female reproductive success. Third, one of the attributes of males that females might prefer would be superior genetic material. But if female preferences are consistent, any genetic variation underlying the key male traits that females could use to choose among males should be eliminated by sexual selection. If the male population were genetically uniform with respect to these gene quality markers, there would be no reason for female choice (if gene quality is the only factor that influences mate selection by females, as it often seems to be). We discuss these three points in the sections ahead.

Wallace (1889) criticized Darwin's female choice theory on the grounds that there are few observations supporting the claim that females discriminated among males (Thornhill 1980a). Poulton (1890) came to Darwin's defense, arguing that the lack of evidence stemmed from failure to carry out a sufficient number of field studies. Over the years ample evidence was collected to show that "females are very fickle indeed and usually remain for a long time unimpressed by the displays of large numbers of suitors before finally accepting one of them" (Mayr, 1972:92). Mayr and others have argued that the demonstration of female "coyness" is evidence of the important role of female choice in animal reproduction. But coyness on the part of a female might also occur if her eggs were not mature (Woodhead, 1981) or if she determined that the environmental conditions were unsuitable for reproduction, rather than because she had assessed a potential mate and found him unworthy. West-Eberhard (1979) has also suggested that females might be reluctant to mate because they were "afraid" of males that were larger or stronger and more aggressive and thus potentially dangerous. Coyness in itself is not unequivocal evidence for female choice, although it does suggest it.

Most other lines of support for mate selection by females are also more suggestive than certain. Female choice is often inferred from the discovery that some males with particular attributes, such as possession of a territory or a distinctive courtship behavior, are associated with more females or copulate more often than other individuals (see for instance Mason, 1964; Scheiring, 1977; McCauley and Wade, 1978; reviews in E. O. Wilson, 1975; Halliday, 1978; Wittenberger, 1981a). In these cases, however, female choice may not determine the relative reproductive success of males. Instead, males may compete among themselves for key positions in the environment, such as oviposition or emergence sites, in which receptive females occur. The male that controls one of these locations gains access to mates; his reproductive success will be settled by his interactions with competitor males rather than through female preferences.

Laboratory studies of female choice have produced much the same kind of ambiguous support for mate discrimination. For example, the studies of Maynard Smith (1956), showing that females of *Drosophila pseudoobscura* mate more with outbred than with inbred males, can be interpreted in two ways: (1) females actively prefer males with certain genotypes or (2) outbred males are better competitors or more active courters. In this and other similar work, the success of some males may be due either to the passive acceptance by females of winners of male-male competition *or* may be the result of active choice of certain individuals by females.

It is arguable that females can choose where and when to be receptive and that they do not have to copulate with a particular male because they always have the option of leaving and looking for another male (Maynard Smith, 1978b). But if time and energy are limited commodities for a female, there will be constraints on her freedom to postpone mating, constraints that males may exploit. Moreover, in some circumstances a female may be forced to remain with certain males until they mate if the males are larger and more powerful.

Borgia (1979) has clarified this problem somewhat by categorizing male-female associations in polygynous populations in relation to the probable freedom that females have in mate choice. He distinguished four types of associations:

(1) Females acquire resources on their own because their mates provide nothing, and thus females are free to choose among all available males on the basis of phenotypic differences reflecting relative genetic quality. This association corresponds to male dominance polygyny and types of scramble competition polygyny.

(2) Males collect resources and provide them to females in exchange for sexual access, but females may collect resources; females can choose among all males in terms of resources and/or genetic benefit to offspring.

(3) Males control resources, and females can *only* obtain resources by mating with the male(s) controlling them.

(4) Males capture or otherwise directly control females.

In type (3) (a kind of resource defense polygyny) and type (4) (female defense polygyny) female choice may be severely constrained by male behavior.

Given the restrictions on active female choice and the confounding influence of intrasexual competition on male reproductive success, it is perhaps not surprising that there are so few clear-cut demonstrations of female choice of mating partners. Table 12.1 gives some of the more convincing examples of this phenomenon among insects. We shall discuss the case of *Hylobittacus apicalis* in detail. Generally, however, more observations of male-female interactions are needed to determine whether female choice contributes to variation in male mating success.

The Adaptive Value of Female Choice

In many species males have in their possession a resource that females apparently assess. This resource may be a territory, a nest site, or a food item. If females choose males on the basis of resources that the females themselves will use or that will be provided to offspring, this presents no problem for female choice theory; an immediate reproductive benefit can be envisioned for a discriminating female. However, there are few studies showing that females actually increase their own reproductive success by choosing certain males over others, which is a requirement if one wishes to show that female choice in a

Table 12.1 Some insect species in which female choice may be operating.

Taxon	Nature of study	Criterion of choice	Reference
Mecoptera			
Hylobittacus apicalis	Field	Size of nuptial prey	Thornhill, 1976c, 1980c,d
Other bittacids	Field	Size of nuptial prey	Thornhill, 1977, 1978b; Alcock, 1979b; Byers and Thornhill, 1983
Panorpa spp.	Lab and field	Size of nuptial dead arthropod; also salivary pillar	Thornhill, 1979a, 1980e, 1981, forthcoming
Blattoidea			
Nauphoeta cinerea	Lab	Male dominance	Breed, Smith, and Gall, 1980
Diptera			
Scatophaga stercoraria	Field	Oviposition site	Borgia, 1979, 1981
Drosophila spp.	Lab	Rare male phenotype	Ehrman and Probber, 1978 and references therein
Drosophila melanogaster	Lab	Sperm level	Markow, Quaid, and Kerr, 1978
Coleoptera			
Tetraopes tetraophthalmus	Field	Body size	Mason, 1964; Scheiring, 1977
Chauliognathus pennsylvanicus	Field	Body size	McCauley and Wade, 1978
Orthoptera			
Conocephalus nigropleurum	Lab	Body weight	Gwynne, 1982

resource-based mating system is adaptive (see Maynard Smith, 1966; Thornhill, 1976c, 1980a).

There are many other species in which males do not offer resources to females or parental care to their offspring, but at best provide an elaborate courtship that comprises complex movements, flights, or ritual chases. It is in species of this sort that female choice theory is most problematical, because the only benefit females can possibly receive is the genes of a male. How can the courtship antics of male insects and the sometimes bizarre morphological features associated with courtship provide an indication of male genetic quality?

While this question has yet to be fully resolved, at least one worker has tried to test whether female choice enhances reproductive success in a species in which the only apparent male contribution is his genes. Partridge (1980) separated adult virgin female *Drosophila melanogaster* into two groups. Each female in one group was mated with a male selected at random by the investigator. The females in the other group were placed in containers with a number of males and therefore had the opportunity to choose among several potential mates. The larvae produced by females that mated randomly with respect to male genotype were allowed to compete throughout larval development with larvae of a genetically different strain. The offspring of females that had a choice of mates were also placed in competition with larvae of the other strain. Because the offspring of the experimental females could be identified, Partridge was able to show that individuals given the chance to choose their mates produced significantly more offspring that survived competition with rival larvae.

These results suggest that females are able to select males with genes that improve the ability of their larvae to compete for resources. Unfortunately, it is not certain that female choice actually occurred in this particular experiment. The results might have been the product of male-male interactions rather than female choice if some males were able to outcompete others in the struggle for access to females and if their progeny were also superior under conditions of competition for larval food. Thus the degree to which female choice occurs in *Drosophila* and the extent to which it elevates female reproductive success is not certain at the moment.

Female Choice and Male Genetic Variation

There is still another problem associated with the theory of female choice, this one related to population genetics. If male characteristics are the product of epigamic selection, the assumption is that there are fitness-related genetic differences among males that have been detected by females and used to discriminate among potential mates. However, if females do select mates on the basis of their genetic properties (as expressed in the male phenotype), they would theoretically deplete any additive genetic variance among males, especially if certain individuals were much more attractive partners than others. In a population without genetic variation any phenotypic differences among males with respect to the key characters will not reflect underlying genetic differences in males relevant to female reproductive success (Williams, 1975; Halliday, 1978; Maynard Smith, 1978b). This is not a purely theoretical concern; there are data showing that those attributes most directly related to fitness have the lowest heritabilities (Harpending, 1979), and sexual selection on males is known to be intense in some species (see for example Wiley, 1973; Bradbury, 1977, 1981; Borgia, 1979). In addition, genetic variation underlying a trait subjected to strong artificial selection is often rapidly exhausted (Falconer, 1960; Mukai, Schaffer, and Cockeram, 1972; Harpending, 1979).

Why should females expend calories and incur risks to choose among genetically similar males in species in which males provide only sperm to their mates? Perhaps the answer is that genetic variation may be more persistent than one would think, given the argument just outlined above. Maynard Smith (1978b) suggests that even in male populations under severe and consistent selection for certain characters, there will be a small but significant amount of genetic variation introduced into the population by harmful mutations and by the genetic shuffling associated with establishment of newly introduced favorable mutations. A second reason why all genetic variation might not be eliminated by selection is that selection pressures are not always consistent in space and time within a generation (Falconer, 1960; Mayr, 1963; Borgia, 1979). For example, Cade's studies (1979b, 1980, 1981b) on genetic variation in cricket calling behavior among males (Chapter 9) suggest that the adaptive value of calling may shift radically through time or from place to place depending on the abundance of acoustically orienting parasites.

Population genetic models that predict the quick removal of additive genetic variance from a population assume that the trait under selection is related to one or a few genetic loci. If many loci are involved, however, genetic variation will persist far longer in the face of selection (Lande, 1976). Many traits are polygenic, especially behavioral attributes, and most species that have been studied with respect to genetic variation have been found to be genetically variable at a great many loci. For example, the highly polygynous orangutan is one of the most genetically variable primates known (see Antonovics, 1976).

However, let us accept for the moment that any differences among males with respect to a female-preferred trait are not the result of genetic variation in a current population. Although evolution (changes in gene frequency) due to selection will cease without genetic variation, selection may continue. In other words, female preferences once evolved may persist in the temporary absence of heritable variation in the preferred characters (Alexander, 1977; West-Eberhard, 1979). Females that make choices based on supposed male genetic quality are faced with the same problem that confronts the human animal breeder who wishes to choose superior breeding stock. After a period of artificial selection for a desired trait, most of the variation remaining will be nonheritable, but the breeder's best strategy may be to choose the variants that would, if they were heritable, produce superior offspring. The animal breeder cannot tell for sure if the residual variation among individuals is caused by genetic differences among them or is entirely the product of environmental differences. If the variation is linked to genetics, the breeding program may advance. At least by choosing the phenotypically "superior" individuals, the breeder is likely to create a new generation whose members are not inferior to those in the preceding generation.

Likewise females of polygynous species have little to lose by selecting a mate with a desired property. They may win a good deal if the phenotypically superior individual can transmit his attrributes to their mutual progeny. Imagine a case in which females prefer large males because in the past this preference was associated with enhanced offspring success. Assume that currently there is no genetic variation associated with differences among male body sizes, but occasionally a mutant arises that contributes to the production of larger body size. Females might gain by preferring larger males, because even if large size is usually nonheritable the female will at least produce offspring of average fitness. If the female happens to choose the rare large mutant, she may gain considerably in terms of genetic propagation. A nonselective female would be unlikely to pick the unusual favorable mutation and might even accept a male with a deleterious mutation that, among other things, would lead to reduced body size. Thus there may be persistent selection in favor of female preferences based on male genetic "quality" even if genetic variation underlying male differences is only irregularly present.

Even though female choice has rarely been demonstrated in an unequivocal fashion and even though the adaptive value of female choice may not be obvious, especially when the preferred male attribute is supposed to be "superior genes," there are reasons for thinking that the potential for epigamic selection exists. We shall examine a clear-cut example of active choice by females of certain males with attributes that elevate female reproductive success. We shall then consider in detail the range of possible differences among the males of a species that females *might* use to choose superior mating partners. It is our conviction that in the years ahead many additional cases of adaptive female choice based on subtle differences in the material or genetic benefits offered by males will be discovered if researchers are alert to the possibility of epigamic selection.

Female Choice in the Scorpionfly, *Hylobittacus apicalis*

The mecopteran *Hylobittacus apicalis* (Bittacidae) is a medium-sized predaceous insect about 2 cm long that inhabits the woodlands of much of the eastern United States. During July in southeastern Michigan the adult population numbers in the thousands per hectare. The sexes are similar in size and color, but are distinguishable from a distance on the basis of differences in flight pattern and genitalia. Individuals are easily marked and followed in the field by an investigator as they capture prey and interact with other scorpionflies. Both sexes are slow fliers and make only short flights among the herbs. Between flights they hang from a leaf or twig by their forelegs (Fig. 12.1).

H. apicalis supposedly is a relatively primitive insect. Yet females have the ability to evaluate males on the basis of the quality of a nuptial gift (a prey arthropod) that the male offers the female during courtship and upon which she feeds throughout copulation. Females prefer males with large prey as mating partners over males with small prey. This preference increases the number of eggs laid by a female, and probably her survivorship as well (Thornhill, 1976c, 1979a, 1980c,d).

Both sexes of *H. apicalis* prey on arthropods, with insects constituting 90 percent of their diets. Males and females can capture prey by pouncing upon a perched victim or by grabbing insects in flight, but females only hunt prey when male density is low. Otherwise they depend entirely on prey provided by their partners.

The sequence of sexual behavior begins when a male catches a prey for himself (or steals one from another male) and begins to feed on it. The male may consume his victim entirely (particularly if it is a small insect), discard its husk, and then obtain another. On the other

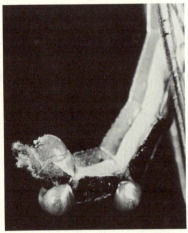

12.1 Male *Harpobittacus australis* hanging from vegetation while holding a dead moth as a nuptial present for a female. Note the slightly exposed pheromonal glands on the male's dorsal abdomen, which are shown in detail at right. Males of *Hylobittacus apicalis* behave similarly. (Photographs by J. Alcock and G. F. Bornemissza.)

hand, if a male catches a relatively large prey he is likely to sample it briefly, then carry it about with his hind legs, flying a few meters at a time from one plant to the next. After each flight the male, while hanging by his forelegs from a perch, releases a long-distance attractant pheromone from glands in his abdomen. The odor may attract a female who will alight in the foliage and hang there facing the male, who then presents his nuptial prey to her (Fig. 12.2). The female, now also holding the prey, feeds as the male attempts to copulate with her. If the prey is at least 16 mm^2 (the length of the insect × its width or height—a housefly is about 22 mm^2) and palatable, the female will allow the male to copulate for 20 minutes or longer. The average copulation lasts 23 minutes, at which time the male disengages his genitalia and pulls the prey from the female's grasp.

If the prey is smaller than 16 mm^2 or unpalatable (a ladybird beetle, for instance) the female either terminates the interaction prior to copulation (50 percent of observations) or copulates briefly (about 5 minutes). These short copulations are terminated by the female, not the male, as she uncouples her genitalia from the male and releases her grasp on the prey despite the male's attempt to maintain genital contact and to again present the food item to the female.

Studies conducted to determine when sperm transfer occurs in a

12.2   A copulating pair of *Hylobittacus apicalis*. The female is feeding on the prey offered her by the male. (Photograph by R. Thornhill.)

mating by *H. apicalis* reveal that (1) up to 5 minutes of mating is required for *any* sperm to be transferred to the female, (2) the duration of mating over the range from 5 to 20 minutes is positively correlated with the number of sperm transferred, and (3) after 20 minutes there is no increase in sperm transferred with increased time in copulation, although male sperm reserves are not exhausted at this point (Fig. 12.3). These findings help us understand why males and females behave as they do during mating. Females that are offered small or unpalatable prey reject the male outright or else mate but accept no sperm from their partners. If a female receives a large nuptial gift, the male gets a chance to transfer a complete quantity of sperm, after which he ends the mating in order to use the prey as a gift for another female.

A male that mates for 20 minutes or longer, in addition to completely inseminating a female, induces her to reject other males and to begin egg laying. A substance produced in the male's accessory glands and passed via his penis into the female's reproductive tract in the last few minutes of a male-terminated copulation is the stimulus that regulates female sexual receptivity and oviposition. Females mated for 20 minutes or longer remain sexually nonreceptive for approximately 4 hours, during which time they lay about three eggs. But females that

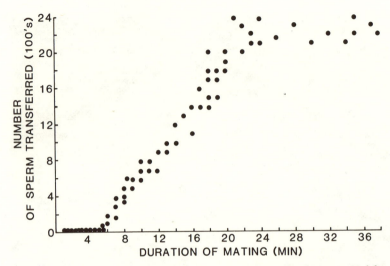

12.3   Total sperm transferred in relation to duration of copulation in *Hylobittacus apicalis*. During the first 5 to 6 minutes, no sperm are passed to the female. Then gametes are accepted for 15 minutes, at which time sperm transfer is complete. (From Thornhill, 1980d. Copyright © 1980 by Scientific American, Inc. All rights reserved.)

mate for less than 20 minutes continue to be receptive and search out males until they receive a suitable large prey item. In areas with dense populations a female can find a second male quickly after rejecting a male that has offered a small nuptial prey. If the second male too fails to offer an acceptable food item, the female will continue to seek a male with a large prey (with whom she will copulate for 20 minutes or more) regardless of the amount of nutrients she has received from males with small nuptial gifts. At the end of each period of sexual nonreceptivity and egg laying, females copulate again—each time preferring a male with a large prey item. This sequence of mating, oviposition, and nonreceptivity continues throughout the day.

Females of *H. apicalis* therefore discriminate against males with small prey by rejecting them outright or by mating with them for a such a brief period that few or no sperm are transferred. Even when some sperm are passed to a female in an incomplete copulation, it is unlikely that they fertilize many eggs. The female is likely to mate again and use the sperm of her second partner, the typical pattern in species in which males contribute an important resource such as a nuptial offering to females (Gwynne, forthcoming a).

Female choice in *H. apicalis* is apparently unrelated to male size or male age, because the size of the nuptial prey possessed by males is not correlated with either attribute. The major criterion females use in evaluation of mates is prey size and, therefore, the nutrient content of the prey. A number of other species of scorpionflies of the family Bittacidae also appear to exhibit female choice on a similar basis (Thornhill, 1977, 1978b; Alcock, 1979b; Byers and Thornhill, 1983).

Evolutionary Effects of Female Choice in *H. apicalis*

Having shown that females prefer males with nutrient-rich large prey rather than those with nutrient-poor small prey, we may ask how the preference affects the reproductive success of males and females of *H. apicalis*.

Epigamic selection by females seems certain to have shaped some components of male behavior in scorpionflies. Theoretically there are two ways in which sexual competition can contribute to variance in the reproductive success of a sex. There may be differences among individuals in their ability to win intrasexual conflicts and differences in their attractiveness to members of the opposite sex. Both factors operate in *H. apicalis* and are separable to some extent. First, there is direct and indirect competition among males for nuptial gifts. Only a small fraction (2 to 10 percent) of the males in a population will be carrying prey at any one time. These are the winners of male-male competition for access to prey offerings. They have survived risk taking associated with obtaining prey and have competed successfully with other males in finding food items. Few males at any one moment have gifts, because suitable prey are limited in number (this leads to indirect competition among individuals, in that those that secure a rare prey reduce the availability of victims for other males). In addition, there is direct competition for nuptial gifts when a male attempts to steal another male's offering while he calls for females.

Males that succeed in securing a gift are not guaranteed to reproduce, for females discriminate against the 10 percent of the prey-bearing males that offer an unsuitably small prey. Males with no prey at all have no chance of reproducing because of female choice; individuals unable to secure large prey rapidly and to avoid predators while doing so also experience reduced reproductive success because of epigamic selection.

The risk of mortality from predation induced by the female requirement for a large nuptial gift may be an important factor affecting male reproductive success. Spiders take many more males of *H. apicalis* than females (although the adult sex ratio is initially 1:1), probably

Table 12.2   Male and female activities between copulations in *Hylo-bittacus apicalis*. Males travel more widely than females and are more often killed in spiderwebs. (From Thornhill, 1977, 1979c, 1980c.)

Sex	Activity	Mean proportion of time spent in activity	Mean distance traveled/ 60 min	Number observed dead in webs
Male	Foraging for prey	0.5	34 m	347
	Feeding on prey	0.3		
	Attracting females	0.2		
Female	Traveling to mate	0.2	16 m	148
	Laying eggs	0.8		

because females rarely have to hunt for food on their own and thus are less likely to collide with a spiderweb than males (Table 12.2).

Female choice not only results in increased male mortality but in unexpected ways indirectly affects the evolution of male behavior. Because hunting is hazardous, males of *H. apicalis* appear to have evolved the ability to adopt an alternative tactic of prey thievery. Some males under certain circumstances fly to calling rivals and adopt female-like behavior, perching close to the prey-laden male and waiting for the opportunity to steal the gift when it is presented to them. If successful, the thief will use the prey to feed a female he attracts, without having to invest in a lengthy and risky search for a prey of his own (although the savings are offset to some extent by the expenses associated with locating males and trying to steal their prey; Thornhill, 1979c). Female choice is therefore responsible not only for the evolution of gift-giving males but also, indirectly, for the evolution of female mimicry and prey stealing by males.

Female Choice and Female Genetic Success

Thus far we have only considered the effect of female choice on male reproductive success in *H. apicalis*. Do female preferences raise female fitness? The answer is yes. One can determine in the field how many eggs a female lays after interacting with males that have offered different kinds of prey. The large, visible eggs adhere to the female's genitalia for a few minutes before dropping to the litter on the forest floor. Females that discriminate against males with small prey lay more eggs per unit time than females that fail to be choosy. In addition, discriminating females probably also enjoy increased survivorship. By selecting a male with a large prey item, a female will not have to hunt on her own, thereby avoiding prey-searching flights and the associated risk of spider predation. At least in high-density populations the costs of discrimination (energy expended and risks taken) are probably small; females can locate a second mate quickly under these circumstances and without much traveling. A female preference for males with large prey could also improve her chance of producing offspring that are outstanding hunters, to the extent that her mate's foraging abilities are heritable.

The many advantages that females may derive from nuptial feeding not only account for the maintenance of the trait currently, but also have a bearing on the question of how the behavior originated and its value to male reproductive success. We can imagine that predatory males that were in the midst of consuming a prey sometimes encountered a receptive female. Some of these individuals might have permitted their mates to share the food or were unable to prevent theft of the prey by their partner. Perhaps those original "present givers" experienced heightened reproductive success because their mates, which had been fed, would not have to forage (and run the risk of being eaten by a spider) during the time they were laying eggs fertilized by the males that gave them food. (We assume that initially as well as currently the last male to complete a long mating with a female would fertilize the majority of her eggs until she mated again.) The original nuptial gift givers might also have gained if by feeding their partners they were able to prolong mating, with consequent advantages in sperm displacement and other aspects of sperm competition.

Male Insects Offer Their Mates Diverse Benefits

The potential benefits received by females from their mates can be categorized as material or genetic (Table 12.3). The case of the scorpionfly illustrates how a discriminating female can gain resources from her mate that raise her fecundity and probably improve her survivorship. We have briefly mentioned that a selective female may derive some genetic benefits as well. If a male's talents in securing prey are inherited to some degree, a discriminating female's reproduction would be enhanced by the relatively great foraging skills of her sons, who should do well in a social environment in which males compete for nuptial gifts. A mother's reproductive success might be improved

Table 12.3   Effects on female fitness of male-provided benefits.

Material benefits from mate
  Increased female survivorship
  Increased female fecundity
  Increased female competitive ability
  Increased egg quality and therefore increased offspring survivorship or competitive ability
  Increased survival of immatures
    Paternal care
    Male-provided oviposition sites

Genetic benefits from paternal genes
  Improved survivorship of female's offspring
  Improved attractiveness of son to mates

---

further if her daughters also received "superior genes" that promoted hunting efficiency, a useful ability if prey-bearing males were scarce during her daughter's generation.

In the next several sections we examine the range of potential material benefits that male insects other than *H. apicalis* offer to females and the possible influence of their resources on female fitness. We then elaborate on how paternal genes alone may affect female fitness. This requires presentation of competing hypotheses on why females might choose males with certain "attractive" attributes and an assessment of their relative plausibility.

The Nature of Material Benefits Provided by Male Insects

Male insects present their mates with a diversity of potential material benefits (Table 12.4), some of which we have already mentioned. Here we collate the various possibilities and make the argument that if males vary in the quality of benefits they can offer to a female and if the female can discriminate among potential mates, then female choice based on some criteria related to material benefits is adaptive and can be expected to occur.

The first and perhaps most familiar category of goods and services males may provide is paternal care for a female's progeny. For reasons discussed in Chapter 3, male animals (insects included) usually attempt to be as polygynous as possible, with the result that it is not advantageous for them to divert mating effort into parental effort. Still, there are some exceptions. Males of certain sphecid wasps and several

Table 12.4   Potential material benefits provided to females by male insects.

Parental care: defense of a female's nest or offspring, feeding the young, brooding eggs, etc.
Male protection of female
Oviposition or brooding sites
Nutritional resources collected and/or controlled by males
    Pollen and nectar
    Arthropods
Nutritional resources synthesized by males
    Passed in ejaculate
        Sperm itself
        Accessory-gland material
            Spermatophores
            Other
    Products of external glands
    Products of salivary glands
    Specialized exoskeletal (nonglandular) structures

bark beetles station themselves in or near the nest of their mate (see Fig. 7.6). The sphecid males are present to repel parasites that enter burrows and whose parasitic offspring would destroy the wasp's progeny (Peckham, 1977; Colville and Colville, 1980; Hook and Matthews, 1980). Bark beetle males may prevent usurpation of their mate's burrow by a rival female and they remove frass from the nest galleries, thereby freeing the female to lay more eggs (Goeden and Norris, 1964; Ryker and Rudinsky, 1976). Ashraf and Berryman (1969) have shown for one species that in galleries in which the male is present with his mate, the females lay about 60 eggs, whereas females that lack an assisting male have about half the number in their tunnels. In all these cases males are in effect protecting the eggs laid within the nest they defend, albeit probably as an incidental result of the male's efforts to drive off all intruders, some of which are rival males who would steal their mates. Thus male defense is largely mating effort. From the female's perspective, however, the male's help may be valuable whether or not it is incidental to the male's goals. To the extent that males vary in their ability to help guard the nest, females may gain by choosing males with above-average helping potential. There is no direct evidence that they do, nor is there evidence on selection for superior egg brooding or larval feeding ability in the few male insects that engage in these paternal activities.

Exclusive postcopulatory care by males is known in only about a hundred species of insects, all in a few families of hemipterans (Ridley, 1978; R. L. Smith, 1980). The best-studied examples are in the belostomatid waterbugs, whose males carry eggs deposited on their backs by their mates (see Fig. 11.12). For many days the egg bearing males engage in paternal activities that aerate the eggs, prevent fungal growth on the eggs, and assist in nymphal emergence as well as in protecting the offspring from predators prior to hatching (Voelker, 1968; Cullen, 1969; see especially R. L. Smith, 1976a,b, 1980). In addition, males of a few carrion beetles (Milne and Milne, 1976) and dung beetles (Halffter and Matthews, 1966) may join their mates in provisioning brood cells or feeding their larval offspring (see Fig. 8.6).

Closely allied with male protection of a nest site that a female has constructed is direct defense of the female herself. We have already suggested that a nest guarding male sphecid or scolytid is probably trying to maintain exclusive egg fertilization rights with the nest owner. The same ultimate goal characterizes the effort of male dung flies (Parker, 1970d, 1974b,d), damselflies (Waage, 1979a,b), and other insects (Gwynne, forthcoming a), which guard their mates after copulation while the female oviposits. Here too the female may gain certain benefits by having as her partner a competent protector, if such an individual can reduce the degree of harassment from other males intent on copulation. Interference of this sort can, at the very least, force the unguarded or weakly guarded female to expend time and energy in an unnecessary (for her) copulation or force her to flee from or repel the superfluous males. In some dense populations of the damselfly *Calopteryx maculata,* females that are unguarded oviposit for an average of about 2 minutes before being chased off by a male, whereas guarded females enjoy an average of 15 minutes of uninterrupted oviposition (Waage, 1980).

Harassing males may also damage the female, reducing her life span and reproductive success substantially, as sometimes happens in dung flies. In struggles among males for possession of a female, the unfortunate female may be trampled into the dung (Fig. 12.4). Besmeared with moist excrement, the female may be unable to fly or may even drown. It therefore behooves a female to have as her mate a male that can best protect her from injury. Large males are superior in this respect because they are less likely to be challenged by a rival intent upon takeover (Sigurjónsdóttir and Parker, 1981); and if a struggle does occur, large males can usually repel their opponents (Borgia, 1979, 1980). Borgia (1981) has demonstrated that female dung flies do select relatively large males. (Variation among males is pro-

12.4   Females of *Scatophaga stercoraria* sometimes drown in dung as a result of fights among males for possession of mates. (From Parker, 1970b.)

nounced, with wing lengths ranging from 5 to 12 mm.) In part this is accomplished by movement of the female to the prime oviposition area on a dung mound, the pinnacle of the pat, where the dung is thickest and where egg desiccation and food depletion are least likely to occur. Large males claim these sites in competition with their rivals, so that if a female can reach the high point of a pat she is likely to be mated and guarded by an individual who can protect her effectively.

A third category of material benefits provided to the female by some male insects consists in territorial sites suitable for egg laying or the rearing of young, which are transferred to females in return for copulations. The dung fly and certain territorial damselflies and dragonflies might be placed within this group as well as in the "female protector" category, because males do control access to oviposition materials. Mating enables a female to use the resources her partner has secured. In many of the examples of male territorial defense of oviposition sites we have discussed, the male either relinquishes his territory to his mate or, more often, permits the female on his territory to do as she

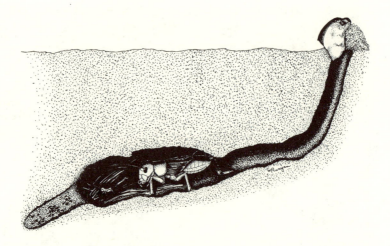

12.5   A female of the cricket *Anurogryllus muticus* in a burrow originally constructed by her mate. (From West and Alexander, 1963.)

likes. Here we shall merely provide another similar example. Males of several species of crickets call from burrows they have constructed; when a female arrives and mates with the resident male, he turns his burrow over to the female, who then uses the location as a safe spot in which to place her eggs and nurture her young (see Fig. 12.5; West and Alexander, 1963.)

The resources that males control and offer to a female may be nutritional as well as ovipositional. We have discussed a case history at length earlier in this chapter. There are many other examples, including species in which a male defends a patch of food that he makes available only to his mates (the flower-defending *Anthidium* bees of Chapter 8 and the carrion-defending *Panorpa* scorpionflies of Chapter 9). There is also a diversity of species whose males may collect a nuptial present to offer a female, including some predatory empidid flies (Downes, 1969; Alcock, 1973), nectar collecting thynnine wasps (Given, 1954; Alcock, 1981b,c), and the seed offering lygaeid bug *Stilbocoris natalensis* (Carayon, 1964). The bug not only gathers food (a fig seed) for his mate, but he injects it with saliva, apparently predigesting it for her (Fig. 12.6).

It would be most unlikely that all males controlled resources of equal value to females; therefore, as with *Hylobittacus*, female choice based on resource quality should be widespread in this category of helpful males. Individuals that do not have a present to offer are ap-

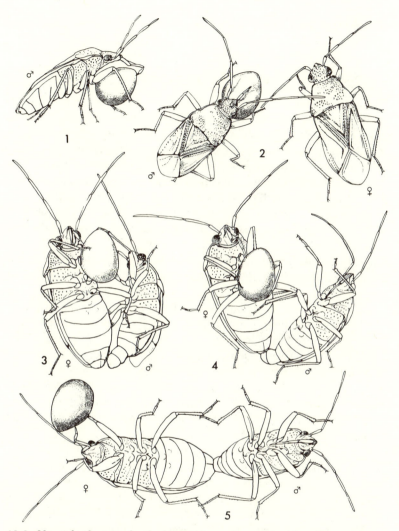

12.6   Nuptial gift giving by the lygaeid bug whose male (*1*) collects, (*2*) pre-digests, and (*3*) transfers an edible seed to his partner. In return, (*4*) and (*5*), the female permits the male to copulate while she consumes the gift. (From Carayon, 1964.)

parently unsuccessful in the lygaeid bug and in some empidids (Carayon, 1964; Downes, 1969). Males of *Anthidium* with flower-poor territories and male *Panorpa* with small carrion items do not attract mates as often as competitors that control superior territories or large chunks of carrion (Thornhill, 1979a, 1981; Severinghaus, Kurtak, and Eickwort, 1981).

The Synthesis of Nutritional Gifts by Males

We have already described cases in which males provide nuptial gifts that they have manufactured internally, for example the salivary secretions of *Panorpa* scorpionflies and some tephritid fruit flies (Stoltzfus and Foote, 1965; Fletcher, 1968). We have also suggested that materials in the male's ejaculatory fluid or spermatophores may have nutritional value for their mates. Because adult females of many insects must acquire limited resources such as protein, lipids, or even water if they are to maximize egg production, the acquisition of these materials from their mates may have considerable impact on their reproductive success. Any variation among males in the ability or willingness to make these offerings could become a basis for mate choice.

We begin by considering the possible nutritional value of sperm. Although sperm design may have evolved primarily in the context of intrasexual selection, it is conceivable that some aspects of sperm morphology and ejaculate composition have taken on a nutritional role (Fig. 12.7). For example, in polyspermy, a common feature of insect reproduction (Davey, 1965), numbers of sperm penetrate the egg, although only one contributes chromosomes to the nucleus of the zygote. Nonfertilizing sperm within the fertilized egg may contribute resources for its development (Sivinski, 1980a). In addition, insect sperm often contain a large crystal derived from mitochondria, which is completely absorbed by the zygote (Periotti, 1973). Very large sperm are produced in some *Drosophila* (Beatty and Burgoyne, 1971), featherwing beetles (Dybas and Dybas, 1981), and the backswimmer *Notonecta glauca* (Afzelius, Baccetti, and Dallai, 1976). Although possibly related to sperm competition, these features are suggestive of a nutritional role for the sperm (Sivinski, 1980a).

There are also records of male insects that provide more than one sperm morph; those that are not employed in fertilization may provide nutrition for the female (Sivinski, 1980a). The commonplace observation that males transfer far more sperm than a female needs for egg fertilization can be interpreted as the product of sperm competition among rival males *or* perhaps as a means by which the female acquires resources ("excess" sperm) that she can metabolize and use for her reproductive gain. In this context it is interesting that adult males of some moths produce anucleate sperm, which they include within the spermatophore along with a complement of nucleate sperm

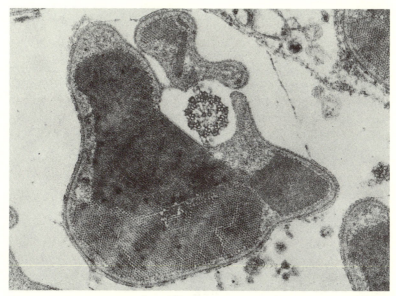

12.7   Photomicrograph of a cross-section through a giant sperm of the back-swimmer *Notonecta glauca*. The sperm with its large mitochondrial crystals may provide nutritional substances for the zygote. (Photograph by B. A. Afzelius.)

(Friedlander and Benz, 1981). Only the nucleate sperm, however, leave the spermatophore and travel to the spermatheca (Friedlander and Gitay, 1972). Because the anucleate sperm remain within the bursa copulatrix away from the fertilization site, it is hard to imagine that they assist in fertilization or sperm competition. They may, however, play a role in female nutrition.

The use of sperm by females as a nutritional supplement is still essentially speculative. Far more convincing evidence exists that the accessory-gland products of males transferred to the female during mating (Alexander, 1964; Leopold, 1976) contain substances that may elevate the egg production, or survival chances, of the female (Fig. 12.8). For example, Bentur and Mathad (1975) found that female crickets survived longer during periods of starvation if they had access to males and received spermatophores from them. In the scarab beetle *Melontha melontha* the male forms a spermatophore within the female's bursa copulatrix that is subsequently digested enzymatically. The presence of the spermatophore is related to the development of the female's ovaries and fat body (Landa, 1960).

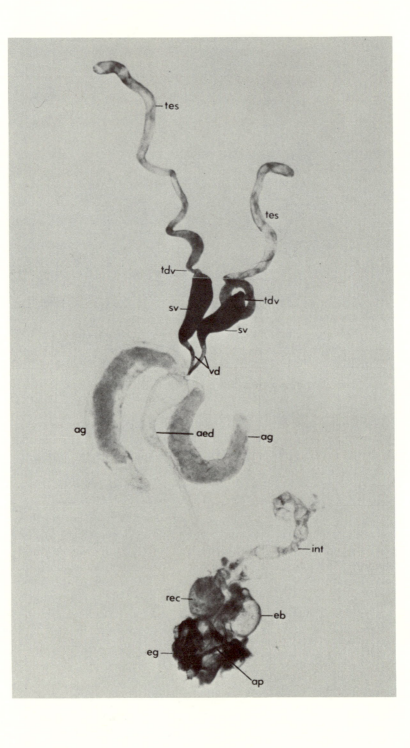

The two groups for which the nutritional role of the spermatophore have been best established are the Lepidoptera and Orthoptera. In the Lepidoptera, as in *Melontha,* the male secretes materials into the female's bursa where they form a discrete spermatophore (Ehrlich and Ehrlich, 1978). Amino acids in these secretions may be used by females in egg production (Goss, 1977; Boggs and Gilbert, 1979; Boggs, 1981a,b; Boggs and Watt, 1981) and may be critical for maximum egg production (Dunlap-Pianka, Boggs, and Gilbert, 1977). A single spermatophore produced by a male *Heliconius* contains the nitrogen equivalent of 15 to 30 eggs (Boggs and Gilbert, 1979). Experiments by Boggs and Gilbert indicate that in *Heliconius* and *Danaus* butterflies the spermatophore protein is especially important in enhancing early egg production before females locate pollen sources (Fig. 12.9). Amino acids are not the only nutrients provided by lepidopteran males. Marshall (1980) has found that spermatophores of *Colias philodice-eurytheme* are composed of proteins, hydrocarbons, cholesterol, glycerides, and phospholipids. He proposes that what male Lepidoptera may offer mates depends upon the key nutrients that limit female reproduction, the availability of those nutrients, the cost of their acquisition, and the timing of egg development relative to copulation (Marshall, 1982).

Studies of various orthopterans have also provided strong evidence for the participation of spermatophore nutrients in female reproduction (see the reviews by Sakaluk and Cade, 1980; Gwynne, forthcoming b). In acridid grasshoppers, spermatophores are placed inside the female's reproductive tract where they are digested (Farrow, 1963; Loher and Huber, 1966). Males of *Melanoplus sanguinipes* pass an average of seven spermatophores into a female per copulation (Pickford and Gillott, 1971). Friedel and Gillott (1977) used immunoelectrophoretic methods to show that proteins produced by male *M. sanguinipes* were taken up unchanged (that is, not digested) by oocytes within 24 hours, and that three-quarters of the male-produced protein in the female's hemolymph is transferred to developing eggs within 72 hours after mating. The empty spermatophores are ejected after most of the soluble protein has been removed.

12.8 The accessory glands and other internal organs of a male fruit fly, *Drosophila melanogaster. aed* = anterior ejaculatory duct, *ag* = accessory gland, *ap* = basal apodeme of the aedeagus, *eb* = ejaculatory aedeagus, *eg* = external genitalia, *int* = intestine, *rec* = rectum, *sv* = seminal vesicle, *tdv* = testicular duct, *tes* = testes, *vd* = vas deferens. (From Gromko, Gilbert, and Richmond, forthcoming.)

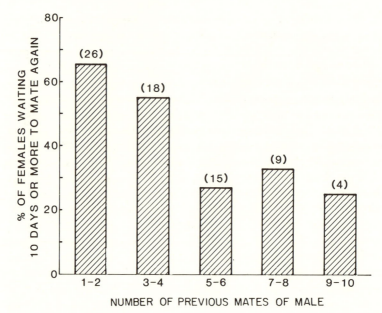

12.9 In the butterfly *Heliconius cydno* females whose partners had mated only a few times (and therefore provided larger spermatophores) generally waited longer (and laid more eggs) before mating again than females whose partners had mated many times. The median interval between matings by females was 9 days. The numbers in parentheses are sample sizes. (Data courtesy of C. L. Boggs.)

Riegert (1965) showed that multiple mating by female *M. sanguinipes* increases fecundity. Individual females housed with many males had increased numbers of eggs, better hatching rates, and improved offspring survival rates when compared with females housed with a single male. Because females of this species obtain sufficient sperm in 10 minutes of mating to fertilize eggs for 3 weeks (Pickford and Gillott, 1971), the long (5-hour) copulation probably occurs so that females can secure more protein donations from their partners (Gwynne, forthcoming b).

Laboratory experiments with *Orchelimum* katydids reveal that some nutrients from the spermatophore are also used in this group for egg production (Gwynne, 1982). Radiolabeled amino acids fed to immature males were traced later to their mates' eggs, although some of the label was also present in somatic tissues. Both fertilized and unfertilized eggs dissected from females were labeled; therefore the label was not received solely from male sperm.

In addition to the genital route of entry of male spermatophores and other accessory-gland products, female insects may eat the spermatophore in cases where it is deposited externally (Toschi, 1965; Leopold, 1976; Thornhill, 1976a; G. K. Morris, 1979; Gwynne, 1982, forthcoming b). Most species in the orthopteran suborder Ensifera (crickets and katydids) employ external spermatophores that are applied to a female's genital opening and are later consumed by her (see Fig. 13.22 in the next chapter). Several lines of evidence suggest that consumed spermatophores serve as a source of female nutrition as well as operating in sperm transfer. In some species of crickets males typically produce two spermatophores during mating, and the first spermatophore is eaten by the female (Gabbutt, 1954; Mays, 1971; T. J. Walker, 1978 and references therein). In certain Ensifera a part of the spermatophore, termed the spermatophylax (after Boldyrev, 1915), appears to be a specialized food source that the female eats after copulation. Some crickets exhibit this structure (Alexander and Otte, 1967), but it is in the tettigoniids (katydids) that the spermatophylax is most highly developed, reaching 30 percent of the male's body weight in the genus *Ephippiger* (Busnel and Dumortier, 1955; Gwynne, 1982, forthcoming b)

Male cockroaches are known to introduce nitrogenous nutrients into females via urates that coat the spermatophore (Roth, 1967). Mullins and Keil (1980) conducted a radiotracer study to show that nutrients from the urates of male German cockroaches (*Blatella germanica*) are partially incorporated into eggs and therefore may represent a nutritional contribution by the male. Gwynne (forthcoming b) has suggested that because urates are waste products, they may not be very costly and do not constitute a major contribution. However, male-provided urates (via tergal gland feeding) in the cockroach *Xestoblatta hamata* (Fig. 12.10) do increase female fecundity (Fig. 12.11; Schal and Bell, 1982), perhaps because the diets of these cockroaches are nitrogen poor. The urate-laden tergal glands comprise about 20 percent of the male body weight and are emptied by the feeding female.

There is evidence that males of different ages and mating histories produce spermatophores of different sizes. In *Orchelimum* katydids a mated male requires a week before it can produce another full-sized spermatophore (Feaver, 1977, forthcoming). Likewise, Gwynne (1982, forthcoming b) has found that male katydids generate smaller spermatophores after having mated recently. This pattern applies also to certain Trichoptera (Khalifa, 1949), Diptera (Linley and Adams, 1972), and Lepidoptera (Boggs and Gilbert, 1979; Rutowski, 1979;

12.10   *Above,* a copulating pair of *Xestoblatta* cockroaches. *Below,* After coupling, the female feeds on urate crystals provided by the male. (Photograph by C. Schal.)

L. D. Marshall, 1980). Not only is the male donation smaller, but copulation time (with its associated costs) is generally longer when a male's accessory glands have been depleted. Finally, young males of the cockroach *Diploptera punctata* provide smaller spermatophores than older roaches (Stay and Roth, 1958). Females mated with 4-day-old males averaged far fewer litters (0.4) than females given the opportunity to mate with 4-week-old males (2.2 litters). There seem to be advantages for the females that can discriminate against young or recently mated males so as to avoid mating with an inferior resource donor.

Other Material Benefits Derived from the Male

Males of some insects have tissues other than the accessory glands that synthesize nutritional substances for females (Roth and Willis, 1952; Roth and Stahl, 1956; Alexander and Brown, 1963; Alexander and Otte, 1967; Roth, 1969; Thornhill, 1976a; T. J. Walker, 1978; Gwynne, forthcoming b). For example, female tree crickets (*Oecanthus*) before, during, and after mating feed on male external metanotal glands, located posterior to the base of the male's front wings

(Fig. 12.12). Experiments by P. D. Bell (1979b) with *O. nigricornis* showed that females that fed on glandular products for longer periods of time laid significantly more eggs. Similarly, as just noted, males of some cockroaches have tergal glands located on the dorsum of the abdomen (Brossut et al., 1975), and females feed upon the exudate from these glands at the time of mating (see Fig. 14.3; Roth, 1969).

Another group whose males feed their mates with glandular secretions are the malachiid beetles (Matthes, 1962). Within the group there are a wide variety of secretory organs located on the head or elytra. Females "nibble" on tufts of hairs associated with these glands, probably imbibing fluids from them. Nothing is known about the quantity of materials transferred, their value to females, or the varia-

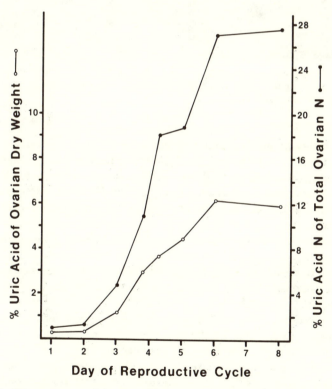

Day of Reproductive Cycle

12.11 Two measures of the contribution of uric acid to ovarian development in the cockroach *Xestoblatta hamata*. Uric acid is obtained by the female when she feeds on urates after mating at about day 4. When the female lays an egg case on day 8, the uric acid content of her ovaries falls almost to zero. The male's contribution of urates represents a significant input to her nitrogen pool. (From Schal and Bell, 1982.)

12.12   A female *Oecanthus* tree cricket feeding from her mate's metanotal glands after mating. Note small round spermatophore at base of female's ovipositor. (Photograph by M. Zorn.)

tion in the quantity or quality of substances offered. It would not be unreasonable to suspect that, for example, male body size might be related to the amount stored in the glands (as has been shown for the salivary secretions of *Panorpa* scorpionflies; Thornhill, 1981). If so, females may use size variation among males in the selection of mating partners.

The same potential for adaptive mate choice exists in those species whose females eat the entire body of their mates, as has been reported for mantids (Roeder, 1967; Edmunds, 1975), a carabid beetle (Fabre, 1910), and some ceratopogonid midges (Goetghebuer, 1914; Edwards, 1920; Downes, 1978). Cannibalism of a mate may be less drastic, as in those insects whose males sacrifice only a portion of their body, such as the spurs on the male's legs in some Orthoptera (Mays, 1971) or the hind wings of male *Cyphoderris* (Haglidae). Although male (and female) haglids are flightless, the males have both fore and hind wings. The latter are unsclerotized, fleshy structures resembling husked sunflower seeds. A female feeds on the hind wings prior to and during copulation until the male pulls away from his mate at the moment the spermatophore is transferred (Fig. 12.13) (G. K. Morris, 1979; Dodson, Morris, and Gwynne, forthcoming). Females may

12.13 The fleshy hind wings of haglid katydid males are eaten by their mates during copulation. (*Above*) The edge of an edible hind wing appears beneath the raised fore wings of the male, near the fore leg of the copulating female. (Photograph by D. Gwynne.) (*Below*) The hind wings of a restrained male haglid. (Photograph by G. K. Morris.)

prefer virgin males, with unchewed wings, to mated males, with wings largely consumed by previous partners.

This description of the variety of ways in which male insects may assist their mates nutritionally will undoubtedly prove to be incomplete as, in the years ahead, more workers look for and discover subtle mechanisms of resource transfer between mates. The existence of male attributes of this sort has a bearing on the maintenance of sexuality, inasmuch as "helpful" males partially compensate their females for the various costs of sexuality. For our present discussion the existence of resource donating males demonstrates that the potential is great for adaptive female choice based on differences among males in their ability to offer useful goods and services. For the most part, however, the challenge of showing that females do make adaptive discriminations based on male offerings remains to be met.

Genetic Differences among Males and Female Choice

An even sharper challenge exists with respect to showing that females can select individuals with "superior" genes to be their mates. Yet in many insect species males provide their mates only with sperm, and therefore the only possible ultimate criterion for female choice is the value of a male's genes for female reproductive success. It is worth considering, then, whether male genetic quality may be a basis for selective mating by females.

Just as in the case of material benefits, there is the *potential* for reproductive gain among females able to pick males that will provide them with sperm whose genes carry the hereditary basis for certain desirable traits. For example, it is conceivable that a female can receive male genes that raise the fitness of her offspring (and thus her own genetic success) by increasing their attractiveness to mates, their disease resistance, their size, their skill in foraging for food or avoiding predators or parasites, and so on. There is no reason to believe that these characteristics cannot be inherited in some cases, but the question is, how would a female know that a male was a carrier of useful genes? Her "judgment" could not be based directly on an examination of a male's genotype, but would have to depend on an assessment of his phenotype. Only if there was a correlation between phenotypic cues and genetic quality would a female gain by selecting males with those cues.

When females interact with males prior to mating, male insects and other animals often engage in striking behavior and employ unusual,

and sometimes extraordinarily exaggerated, morphological traits—
apparently in an attempt to induce receptivity in a potential partner.
As we noted in Chapter 3, Darwin (1859, 1874) felt compelled to develop the theory of sexual selection with its female choice component
partly because he needed a special explanation to account for the evolution of puzzling male characteristics. His argument was that female
preferences for certain attributes could override natural selection
against what were survival liabilities. He did not, however, attempt to
explain how such female preferences originated and he said little
about the selective maintenance of these traits, other than that females preferred them.

R. A. Fisher (1958) argued that male traits could arise and be
maintained by female choice by means of a so-called runaway process.
The first step in Fisher's scheme requires that a female be attracted by
a heritable male trait that enhances survival of her offspring. Any female with a preference for the male attribute would gain reproductively because she would produce more fit offspring. This would lead
to a spread in the population of the male trait, as well as female preference for the trait. When females with the preference became common, the second step of the runaway process would come into play.
Females that happened to produce males with a trait that exaggerated
the attractive components would bear unusually attractive sons, and
females that chose such males as mates would in turn gain by having
superattractive male progeny. This could lead to ever more exaggerated manifestations of the sexually desirable property until finally
the high costs of the extreme trait (in terms of reduced survival
chances) outweighed the benefits gained through sexual selection.
Once an original preference is established, female choice based on
purely aesthetic grounds can lead to the evolution of extreme courtship antics and freakish morphological attributes in males (Fig.
12.14).

O'Donald (1962, 1980) and Lande (1981) have developed Fisher's
argument in greater detail, demonstrating the mathematical feasibil-
ity of his hypothesis. Lande argues that a male character need not
even *initially* confer a direct positive effect on female fitness, because
an arbitrary preference may arise by drift and set off the runaway
selection.

Fisher's model can be illustrated with a hypothetical example taken
from the Hawaiian *Drosophila* (Spieth, 1966, 1968). Males of some
species have large and strikingly patterned wings, which they employ
in courtship of females (Fig. 12.15). The original mutant with slightly
larger than average wings might have been a bit better at flying and at

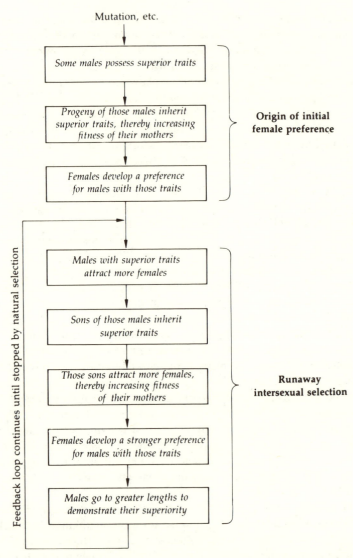

12.14 The hypothetical chain of events that leads to runaway selection. (From Wittenberger, 1981a.)

escaping from predators. Females preferring such males could gain an advantage, inasmuch as their offspring would also enjoy the benefits of flying efficiency and escape from predation. The proportion of males with larger wings and of females that preferred large-winged males

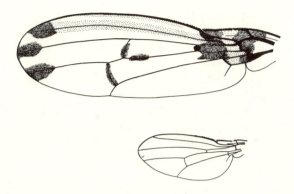

12.15   The relative sizes and wing patterns of males of the Hawaiian *Drosophila heteroneura* (*above*) versus the cosmopolitan *D. pseudoobscura* (*below*), which is similar to *D. melanogaster*. (Drawing by L. S. Kimsey.)

would begin to increase in frequency. As the preference for certain males spread, at some point females that selected those large-winged males would begin to benefit *solely* because they would produce sons that other females would find attractive. This would initiate runaway selection, which would promote the evolution of ever-larger wings, even when wing size passed the optimal point for flight efficiency and predator avoidance. The elaboration of wings would continue via runaway selection until the mating success of *Drosophila* males with unusually large wings was offset by their decreased survival chance (and thus, ultimately, decreased reproduction).

Is Female Choice Arbitrary?

In species whose males offer genes alone, female choice can be based only on aesthetic grounds if the Fisherian model is accurate. But are female preferences made without respect to aspects of their progeny's fitness other than the sexual desirability of their sons? An alternative to the runaway selection hypothesis holds that although certain components of male behavior and appearance used in persuading females to mate may superficially appear to be unrelated to anything other than courtship itself, they actually convey useful information about male genetic quality. The traits females prefer may be reliable phenotypic markers of survival ability, resource gathering, and so on. One proponent of this view is Zahavi (1975, 1977a,b), whose handicap principle makes the following argument.

Zahavi believes that elaborate male characters preferred by females represent a true handicap to the male that identifies his survival ability. A female prefers a handicapped male because he improves her chance of mating with a male who will supply her offspring with superior survival genes. If we interpret the oversized wings of Hawaiian *Drosophila* from the perspective of the handicap principle, they represent a real impediment to survival. Females prefer males able to survive with larger wings because these individuals have managed to live despite their greater potential vulnerability to predators. Males that can do this are demonstrating unequivocally that they possess superior genes for survival ability. A female mating with such a male would confer her partner's superior genes on her offspring, which would enjoy heightened reproductive success through improved survival. A male that did not possess the flight-hindering wings would lack unambiguous phenotypic evidence that a female could use to determine whether a short-winged male was or was not a better-than-average survivor. In the absence of a clear-cut indicator of survival ability, a female serves herself best by picking a handicapped male, according to the argument.

Zahavi's principle has received considerable attention from theorists. Both verbal and mathematical genetic models have been used to analyze the conditions under which the handicap hypothesis might apply (Davis and O'Donald, 1976; Dawkins, 1976; Maynard Smith, 1976b, 1978b; Borgia, 1979; West-Eberhard, 1979; O'Donald, 1980). In general, these analyses indicate that the hypothesis will only operate under restricted circumstances (see Zahavi, 1977a; G. Bell, 1978; Halliday, 1978; Maynard Smith, 1978b).

The major difficulty with the idea is that females with a preference for handicapped males will produce offspring with the handicap. According to Maynard Smith (1978b), Zahavi's form of sexual selection would only be effective if Fisher's runaway selection were also operating. But before the handicap principle is laid to rest, it would be desirable to test the hypothesis using real animals rather than computer-simulated genetic models or verbal arguments. Ideally the animal would be one in which the males do not transfer anything except sperm to their mates. Otherwise, it is difficult to separate the effects of resource quality from male genetic quality on female choice in species where males provide more than their genes (see Weatherhead and Robertson, 1979; Wittenberger, 1981b; Searcy and Yasukawa, 1981, for an example of the problems of interpreting the relative importance of resources and genes to female choice in the red-winged blackbird). Unfortunately no such example exists from the insects. We shall test Zahavi's theory as best we can with an examination of the behavior of the scorpionfly *H. apicalis* (Thornhill, 1980c).

It is possible to argue that aspects of the behavior of males of *H. apicalis* are survival handicaps that, if overcome, will reliably demonstrate the ability of a male to live a long time. Females can then use this information to select a male that has experienced considerable risks and survived. We know that movement in the habitat exposes male scorpionflies to predation from spiders. The collection of nuptial prey by a male can therefore be viewed as his handicap. Zahavi's hypothesis leads to the following predictions about prey collection by males of the scorpionfly: (1) males should prefer prey in a range of sizes that would demonstrate to females that the males had taken risks to obtain the nuptial present and (2) females should prefer males possessing prey that can only be secured with associated risk taking.

Both of these predictions are met. Males usually choose prey that are 16 mm^2 or larger, and females prefer males that offer gifts of this sort. Prey of attractive size are much less abundant than are smaller insects. But do the females choose males with large prey because of the resource value of the gift *or* because of the survival demonstration of the male? Observations of scorpionflies do not permit us to differentiate between these two possibilities; we need a prediction whose test will enable us to make the necessary distinction.

If the handicap principle applies to scorpionflies, we can predict that males with *unusually* large prey are preferred by females. Relatively large items require longer to find and are more dangerous to obtain, because greater movement in a predator-rich habitat is required. Males collecting prey at the upper end of the spectrum of acceptable prey sizes run a greater risk of predation, a survival handicap. Any male that can overcome this obstacle demonstrates exceptional survival capacity and so should be more attractive to females. But females do not show a preference for prey that are relatively difficult and risky to secure. Instead, they have a minimum size requirement only; the prey must exceed 16 mm^2 if the female is to mate for the full 20 minutes. Males take prey 16 mm^2 and larger strictly in proportion to their relative abundance in the environment. This result suggests that the handicap principle is an inappropriate explanation for mate choice in *H. apicalis,* but does not rule out the possibility that the principle is operative in other species.

Zahavi's argument is only one of a cluster of related hypotheses which state that females may gain something over and beyond the production of sexually attractive sons if they mate with males of certain genetic constitutions. As we suggested at the outset, there are many potentially heritable attributes of males that females might use to make adaptive mating decisions. In particular, indicators of a male's ability to gather resources efficiently or of his resistance to disease or parasites would be highly relevant to adaptive mate choice by females. The ability to transmit these attributes to offspring would benefit both sons and daughters (Trivers, 1976; Borgia, 1979; Thornhill, 1979a, 1980a). In contrast, traits that confer only a sexual attractiveness advantage on sons might well have a negative effect on the fitness of daughters (Searcy, 1979). It seems worthwhile to consider whether courtship and other traits of male insects might not convey information about the genetic quality of a male relative to the likely competence of his offspring in a broad array of activities, not solely the ability of sons to induce females to mate. Despite the mathematical soundness of Fisher's runaway selection model, we suspect that the opportunities for nonarbitrary mate choice by females are so varied and significant that purely aesthetic features rarely play a central role in female discrimination of mates. The major goal of the next chapter will be to examine the courtship behavior of males and other aspects of male-female interactions to determine if females use the information they secure from these interactions to choose males that enable them to produce more or better offspring.

13

Mechanisms of Female Choice

IN THE PRECEDING CHAPTER we established that the potential for adaptive female choice is high in many insects because males vary in the quality or quantity of the material or genetic benefits that they can supply to females at mating. In this chapter we establish that most female insects discriminate in favor of those individuals that have the most to offer. But do females exercise their control of the reproductive process in such a way as to reject the sperm of inferior males and use the sperm of superior ones? We outline the diversity of possible mechanisms of mate assessment, paying primary attention to the role of courtship interactions as a basis for adaptive female choice. The argument that courtship enables females to assess the fitness differences among *conspecific* males is distinct from the traditional hypothesis that courtship functions primarily to prevent hybridization and its disadvantages. Although courtship patterns are often species specific, it does not follow that the only (or the major) function of courtship is to help females pick males of their own species in preference to males of other species. We argue that the evolution of courtship has been heavily influenced by epigamic selection for traits in conspecific males that females can use to determine the likelihood of receiving valuable genetic attributes or useful goods and services from would-be partners.

The chapter concludes with an analysis of cases in which males, rather than females, appear to be the choosy sex. As we noted in Chapter 3, females are usually a limited resource for males, with the customary result that males are undiscriminating and females selective about mating. Yet there are species in which males invest so much parentally or through material donations that they become the limiting sex; if so, sex role reversals are predicted—and they do occur, as we shall see.

Female Control of the Fertilization of Her Eggs

Insects are typical animals in that females have more control over the reproductive process than males. They retain their eggs within their bodies and receive sperm, which they generally store and use at their discretion. Moreover, they do not copulate at every opportunity, often failing to respond to the advances of males. Even when they do copulate they may control the amount of sperm they receive during mating and may also decide to copulate again quickly, thereby displacing the sperm of the prior partner. Thus rejection of a male (in the sense of preventing him from fertilizing some or all of her eggs) may occur at any of the following times:

(1) Prior to copulation, if a female prevents a male from inseminating her;

(2) During copulation, if a female refuses to accept sperm or to store the gametes injected in her genital tract;

(3) After copulation, if a female displaces or metabolizes or never uses the sperm her partner has transferred to her.

Sexual reluctance, or "coyness," is typical of female insects (O. W. Richards, 1927). In general, a female cannot be forced to copulate (although there are apparent exceptions); if she chooses not to cooperate (as often happens) and keeps her genital aperture closed (see Fig. 10.8) or fails to stop moving (Tompkins and Hall, 1981) or refuses to permit the male to mount her by keeping her wings closed (Tauber and Toschi, 1965), she can reject an unwanted male. There are other more active rejection responses as well. A nonreceptive *Drosophila primaeva* female raises her abdomen until she appears to be doing a headstand (Spieth, 1981b). Captured dragonfly and damselfly females that are not receptive fly up instead of alighting (Corbet, 1980); butterfly and wasp females may twist the abdomen away from the male's genitalia (Evans, 1966; Scott, 1973a). Alternatively, a female may use force (see Fig. 5.6), kicking rejected suitors (Linley and

Adams, 1974) or biting them, as in some ants (Marikovsky, 1961; Hölldobler, 1976) and orthopterans (Dodson, Morris, and Gwynne, forthcoming). House fly females have a spine on their midlegs, which they jab into the wings of unwanted males. A male may be "encouraged" to dismount because if he does not the female may tear his wings. Under cage conditions with frequent contacts between males and females, some unfortunate males have had their wings totally shredded by reluctant females (Sacca, 1964).

Among the most dramatic of rejection responses is the dousing of a persistent suitor with a disabling spray, as practiced by the tenebrionid beetle, *Pterostichus lucublandus* (Kirk and DuPraz, 1972). The spray, for the most part used against predators, may also be directed against an unwanted male who, after cleaning himself, falls into a "coma" for several hours, giving the female ample time to disappear.

The tenebrionid example makes the point that rejection of males carries a cost for a female. In this case she must resynthesize the fluids that were sprayed upon the male, and until her spray reservoirs are completely filled again she is presumably more vulnerable to predators. Other rejection reactions may be less costly but still require an expenditure of time and energy. The gain for a female is the savings in time and energy derived from avoidance of an unnecessary copulation and perhaps better control over who will father her offspring. We can predict therefore that the evolution of distinctive rejection responses will occur in species in which at least some males are highly persistent courters.

Presumably in species with persistent males, prolonged courtship is advantageous to the male because some females in the population will mate if the male keeps trying long enough. Weakly receptive females are known to occur in species whose females mate a few times in their lives at intervals of several days or weeks. In these insects females become completely nonreceptive in the period immediately after mating but may gradually regain their receptivity. At some point the female may be mildly motivated to mate again, but require considerable priming before becoming receptive to copulation. This pattern of variable receptivity is known to apply to some butterflies (Scott, 1973a) and probably some lampyrid beetles (J. E. Lloyd, 1980b).

Among many butterflies some females will permit mating if courted at length; as a result, all females run some risk of attracting a persistent courter, presumably because males cannot easily discriminate between weakly receptive and completely nonreceptive females. In order to deal with unwanted males, female butterflies commonly em-

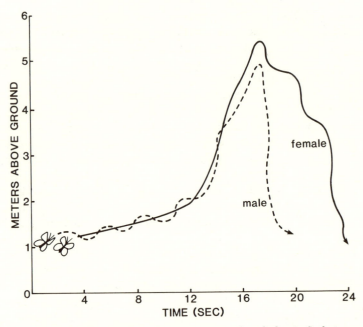

13.1   Ascending flight pattern of a nonreceptive female butterfly being pursued by a persistent male. (Drawing by M. H. Stewart.)

ploy a specific rejection response, the "ascending flight" (Fig. 13.1). Rutowski (1978a,b) has shown that in *Colias* butterflies nonreceptive females pursued by a male for an average of 12 seconds will begin to fly upward, usually with the male trailing behind and then dropping out of the chase after 5 seconds of ascent. The female loops back to the ground (requiring an average of 6 additional seconds), where she resumes her search for nectar or oviposition sites. Because males drop down quickly after leaving the ascending female, it is rare for a pair to resume their interaction after they have separated.

Does the female actually save time and energy by initiating an ascending flight that consumes about 11 seconds of her time (and the energy needed to fly vertically some distance)? Although ascending flights appear to discourage pursuit by males (perhaps because ascent makes pursuit expensive for males and is an unambiguous signal of a lack of receptivity on the female's part), one needs to know how persistent males would be if the objects of their attention failed to engage in an ascending flight. Rutowski (1978a) offered tethered females to passing males and recorded how long they courted restrained females that were unable to fly upward. He found that if a male remained with a female for 12 seconds, there was a better than 50 percent chance that he would remain by her for at least an additional 12 seconds. In general, the longer a male courted a female, the more likely he was to keep at it for an additional 12 seconds. Thus, a female that engages in an ascending flight, after having determined that she is being pursued by a persistent male, has a 50 percent or better chance of saving some time. This increases the time she can spend in feeding or oviposition without interference from a male that will buffet her with his wings in an unwanted courtship. More data for other species are required to determine whether the investment in a specific rejection response yields net time or energy gains for females. But the point remains that these responses enable females to control the number of their partners.

Copulatory and Postcopulatory Mechanisms of Paternity Selection

Even if a female insect does permit a male to mate with her, she need not receive any sperm from him at all (Lea and Evans, 1972). It takes time to transfer sperm; if the female shortens the duration of copulation, as in the scorpionfly *Hylobittacus apicalis*, she reduces the number of sperm she accepts from her mate (see Fig. 12.3). Because there is considerable variation in many insect species in the time spent in copulation (see Table 11.1), there may also be a corresponding variation among males in the quantity of gametes injected into a female. (It is frequently difficult to determine which sex is responsible for controlling the length of copulation, although sometimes female control seems probable; see Allen, 1955; Jordan, 1958; Solomon and Neel, 1973; Thornhill, manuscript b.) In the organ-pipe mud-dauber wasp, *Trypoxylon politum*, some males reside at nests during all provisioning by females (Cross, Stith, and Bauman, 1975). A nesting female may copulate both with the territorial nest defender and with males away from the nest (Colville and Colville, 1980). But matings are longer between the resident male and female; perhaps nest defending males transfer more sperm and fertilize the bulk of the female's eggs. This may be the female's way of choosing a male who will provide her progeny with traits that promote success in nest defense.

Finally, a third level of paternity control may involve postcopulatory manipulations of competing ejaculates by a multiple mating female. The ability of many female insects (Parker, 1970c; R. L. Smith, forthcoming) to store sperm and the size and morphology of sperm storage organs have been interpreted as adaptations to ensure that male gametes will always be available for use by the female. Given the abundance of receptive males, however, female reproductive success cannot often be limited by the availability of sperm. Instead, sperm storage and multiple mating by females may enable them to select stored sperm from "superior" partners for fertilizing their eggs (J. E. Lloyd, 1979a; W. F. Walker, 1980; Sivinski, forthcoming).

In the boll weevil (*Anthonomus grandis*) females have direct control over the amount of sperm displaced from their sperm storage organs. Excision of the external musculature of the spermatheca in this species does not reduce sperm movement to the structure, but does prevent sperm movement *from* the organ and reduces sperm displacement from 66 percent to 22 percent (Villavaso, 1975).

W. F. Walker (1980) has summarized evidence from a variety of insects indicating that the female reproductive tract often plays an important role in the movement of sperm to and from the sperm storage system. His analysis indicates a trend toward low sperm displacement by the last male to mate in insect species with spheroid female sperm storage organs and high displacement in insect species with elongate ones (Fig. 13.2). Walker hypothesized that elongate sperm storage organs may allow displacement of prior sperm into a region of the

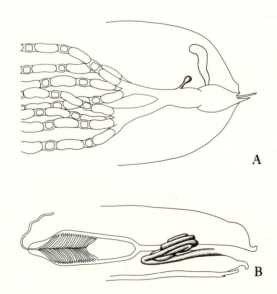

13.2   (A) Spheroid spermatheca of a bumble bee. (From Alford, 1975.) (B) Elongate spermatheca from the migratory locust. (From Gregory, 1965.) Elongate sperm storage organs may facilitate sperm precedence.

organ where they cannot easily reach the orifice of the duct that runs from the spermatheca to the portion of the female reproductive tract where fertilization occurs. Moreover, in some insect species sperm are nourished to an important extent by nutrients produced by the female's sperm storage organs (Davey, 1965; Davey and Webster, 1967). This provides the female with a possible mechanism for influencing the survival or competitiveness of ejaculates from different males (W. F. Walker, 1980).

Even if it is the male rather than the female that directly controls sperm displacement (by, for example, physically removing a rival's sperm—Waage, 1979a) a female can exercise control over the paternity of her progeny simply by permitting a new partner to displace the sperm of a less suitable male with whom she had mated earlier. We have already suggested that this happens in the scorpionfly *H. apicalis;* females that have not received an adequate nuptial meal mate again promptly before laying their clutch of eggs (also Thornhill, manuscript b). The second male's sperm presumably have an advantage over any received from the male that failed to offer a full meal. Likewise, in the house fly *Musca domestica* females rarely mate twice in succession but when they do, they choose a larger male than their previous mate (Baldwin and Bryant, 1981). This enables them to receive the presumably larger accessory gland donation from their second partner and to use his sperm as well (Leopold, Terranova, and Swilley, 1971).

Possible Exceptions to Female Control

In the conflict between males and females with respect to control of reproduction, male behavior may reduce female control of egg fertilization in three ways. First, certain males may make it impossible for females to move freely among rival males by using force or by monopolizing resources or other materials needed by females. Second, males may reduce freedom of female choice by controlling female receptivity or mating frequency through injection of antireceptivity substances or application of mating plugs and the like. (In most of these cases, however, females may activate nonreceptivity mechanisms upon receipt of materials transferred during mating because it is to *their* advantage to avoid copulating for some time.) Third, males may circumvent female choice mechanisms by forcefully inseminating an unwilling female. Although we discussed this phenomenon as it occurs in *Panorpa* scorpionflies in Chapter 9, we shall summarize the key points here with respect to conflict between the sexes.

Under certain conditions male panorpas may find themselves unable to offer carrion or a salivary mass to females and instead may attempt forced copulation, which probably negates female choice (see Fig. 9.3). Females prefer resource providing males; they flee from males that do not offer nuptial gifts, and they struggle (often successfully) to escape from "rapists" that have caught them. Moreover, even if a rapist copulates, in half the matings females are somehow able to prevent sperm transfer. By contrast, females approach, mate with, and accept sperm from gift providing males. The preference for resource donating males enhances female fitness directly by increasing the number of eggs laid per unit time and by reducing the necessity of the female to forage for herself, a risky activity (see Fig. 9.9). Nevertheless, males with low dominance abilities may gain by trying to inseminate rejecting females because in this way they probably do fertilize a few eggs, whereas they would fail completely if they tried to be a resource provider and defender.

Forced copulation probably occurs in other insects (Linley and Adams, 1972; Oh, 1979; Parker, 1979; Smith and Prokopy, 1980; Thornhill, 1980e, forthcoming; Brockmann, manuscript). There is the problem, however, of distinguishing between aggressive courtship and forced copulation. Females may in some cases test the aggressive competence of a male and his motivation to mate by trying to escape from him. This could be one way to secure a physiologically sound, strong male whose heritable traits might benefit the female's progeny. The possibility seems unlikely in *Panorpa* (given the clear disadvantages of not receiving a food present in return for mating) but may apply to other species in which male genetic quality has priority in mate choice decisions.

In any event, the likelihood that females have complete freedom to choose whatever male they prefer may be compromised and constrained by attempts of males to maximize their genetic success. Generally, however, the conclusion that females can exert primary influence over who will father their offspring seems valid (see Thornhill, manuscript b). We can, therefore, proceed to ask how females choose superior mates for the production of descendants.

Female Choice and Information about Male Quality

To exercise mate choice before, during, or after mating, females require information about the value of their would-be or actual partners. For example, females of *H. apicalis* consume a nuptial gift during copulation; the time needed to deplete the prey of its contents provides the critical information a female needs to determine the length of copulation and thus the amount of sperm received from her mate. Postcopulatory assessment of males is also possible and could come from the operation of internal physiological mechanisms (such as stretch receptors in the walls of the sperm storage system) that measure the quantity of sperm or other materials received from a male. Males capable of ejaculating a relatively large amount of material into a female may be either unusually vigorous males and therefore desirable from a gene quality perspective *or* individuals that have provided relatively large amounts of nutritionally valuable substances—to the extent that the components of the ejaculate or spermatophore are digested (J. E. Lloyd, 1979a; Sivinski, 1980a; see Willson, 1979, for an analogous argument relative to plants). In the first case, a female might gain by fertilizing her eggs with sperm from such a male for the genetic benefits her progeny would receive. In the second case, a female might gain both genetically (her partner has attributes that make him a superior resource gatherer) and materially (from increased fecundity due to the nutrition assimilated from the ejaculate). If so, the female can afford to wait longer before mating again to secure a new complement of sperm.

Similarly, postcopulatory analysis of the material benefits offered by some male crickets and tettigoniid katydids can occur when the female consumes the externally attached spermatophore (see Fig. 13.22). If the spermatophore is small, a female may mate again more quickly than otherwise, replacing the sperm of her previous partner with that of another spermatophore (resource) donor. Or, as suggested earlier, if a male fails to guard his mate after copulation, this could indicate that he was a subdominant, satellite male; a female might use this information to mate again to receive the genes of a truly dominant male with demonstrated skills in sexual competition and defense.

Our point is that it is not inconceivable that females can assess the quality of their mates even after committing themselves to copulate with a particular individual. In species in which sperm precedence occurs, a female could rectify a perceived "mistake" by mating again with a superior male. In general, however, female insects probably benefit by evaluating male quality *prior* to copulation. Mating takes time and carries a risk of predation. Sperm precedence may be less than totally complete, in which case a female that begins to copulate has committed at least some of her eggs to her current partner. Therefore there should be selective gains for a female that avoids making mistakes in the first place. This could be the case if there were some accurate way to assess mate quality prior to insemination. Information about a male's value can come from two general sources, his interactions with other males and his courtship interactions with the female.

"Passive Choice"

In the preceding chapter we made the distinction between passive choice by females and active selection of mates. In a great many species competition among males determines which individuals monopolize key mating areas and enjoy high reproductive success. Female choice in these cases appears at first to be nonexistent, because dominant males seem to prevent females from mating with other individuals. However, males are in effect sorting themselves out, with only healthy, physically strong, or combatively skillful males available to females for mating. It is hard to see how a female could lose by accepting gametes from such an individual if males offer only their genes and no parental investment.

Actually, a modest form of female choice can operate even in this situation if, for example, a female were to refuse to mate with isolated individuals and instead copulate only with a male who was in a location with other males (Alexander, 1975). In this way the female would improve the likelihood of mating with a male that had been tested in intrasexual competition. She could reject isolated individuals by moving to areas with a high density of males. Alternatively a female might time her eclosion to coincide with the peak abundance of males. Or she might delay emission of a sex attractant or onset of a nuptial flight until a number of males had gathered nearby. Behavior of this sort would encourage male-male competition, thereby aiding the female to mate with a dominant individual.

Female Incitation of Male Aggression

Clear examples of female incitation of male aggression (Cox and Le Boeuf, 1977) are rare in insects but, for example, the pattern of production of attractant signals by females may sometimes generate conflict among males (see Greenfield, 1981, on strategic pheromonal signaling by females.) Female butterflies appear to incite pursuit by approaching a male and then flying off with the chance of attracting several pursuers (Rutowski, 1980; Rutowski et al., 1981). Female *Trimerotropis* grasshoppers engage in noisy display flights that may advertise their availability and thereby attract competing males (Wil-

13.3 An adult female of the New Zealand glowworm, *Arachnocampa luminosa,* showing the luminescent light organ at the tip of her abdomen, which she may use at times to incite male-male sexual competition. (Photograph by S. A. Rumsey; from A. M. Richards, 1960.)

ley and Willey, 1970). This also seems to be the function of the behavior of female glowworm flies (Fig. 13.3). The cave dwelling adult female uses a bioluminscent signal to announce her position to searching males, only to extinguish the light and move a short distance, there to repeat the enticement lure (A. M. Richards, 1960). Fighting for access to females occurs in this species, perhaps when several males are drawn to an area by a coyly signaling female.

Another speculative example of female incitation comes from the observation that females of the meloid beetle, *Phodaga alticeps,* that have not mated for some time will mount other females or males (Pinto, 1972). Homosexual mounting also occurs in domestic cattle, and Parker and Pearson (1976) suggest that it serves to attract the attention of dominant males, which are likely to try to monopolize access to females by driving off subordinate males. By mimicking a male sneak copulator, a female can draw a dominant male to her and mate with him when he discovers that the supposed intruder is a receptive female.

One of the better documented cases of female incitation among insects comes from Borgia's (1979, 1981) work on dung flies, *Scatophaga stercoraria.* In this species females gain if they reach the pinnacle of a cowpat, which is the optimal location for larval development and also the site at which large males are present. These males can guard the female from harassment while she lays her eggs. But small males sometimes grasp a female as she slips across the surface of the dung to the high point of the pat. If so, the female may perform rocking motions, which attract other males. Because large males can supplant smaller ones, the female's behavior increases the probability of a takeover by a more desirable mate for her (one that can protect the female from other males and from the damaging consequences of fights for possession of her).

Courtship and Active Female Choice

Although acceptance of the winner of male-male competition may provide a female with a superior male, it is also possible that the female may take a more active role in selecting a partner. The hypothesis we explore here is that courtship behavior may, prior to copulation, provide the female with the means to separate unacceptable from acceptable mates.

The traditional view of courtship as a reproductive isolating mechanism is based on the widespread evidence that species, even closely related ones, often have elaborate and distinctive (species-specific) courtship behaviors that precede copulation. This suggests that the distinctive features of a species' courtship have evolved to prevent mating errors between individuals of different species (see for example Dobzhansky, 1937; Mayr, 1963; N. G. Smith, 1966; Shorey, 1976; W. J. Smith, 1977; Ewing, 1979). The logic of this argument is as follows. Imagine that individuals in one species are not fully isolated reproductively from members of another species and that hybrids have reduced fitness. If so, selection would favor those members of species A that happened to exhibit distinctive courtship messages that lowered the risk of hybridization, with its fitness reducing consequences. The same selective pressures would apply to species B, and the end result would be the evolution of divergent, unambiguous species identifying courtship signals in both species that prevent hybridization mistakes. According to this view, separate species that once exhibited similar courtship traits would, in areas of geographic overlap, increasingly diverge from their ancestral patterns.

The argument is both persuasive and testable. Let us examine three predictions derived from the reproductive isolation hypothesis.

(1) Closely related species that live together and whose members therefore might have occasion to experience the disadvantageous effects of hybridization should differ more in courtship displays than other related species that live in complete geographic isolation from one another.

(2) If two related species exhibit *partial* overlap in their ranges, the individuals in the area of overlap (where the risk of hybridizing occurs) should show greater divergence in courtship signaling than the animals in the allopatric populations.

(3) The complexity and variety of courtship activities of a species should be greater for members of large species groups than for species with few close relatives.

Despite the logic of the argument, the empirical evidence does not clearly support any of these three predictions. First, although there are some cases, for example among the *Drosophila* (Bennet-Clark and Ewing, 1970), in which allopatric species do have more similar courtship signals than related species living together, there are few well-documented cases of nearly identical songs or flash patterns among allopatric or allochronic (seasonally segregated) species of acoustical insects (Alexander, 1967, 1969; T. J. Walker, 1974b; Morris and Gwynne, 1978) and bioluminescent insects (J. E. Lloyd, 1966, 1979b). Moreover, Pinto's (1980) recent work on the genus *Epicauta* suggests that courtship differences among 11 species of western North American blister beetles have not evolved in the context of reproductive isolation. Pinto compared interactions between sympatric pairs of species that occurred on the same plants, sympatric pairs separated by habitat differences, and allopatric pairs, and found a considerable degree of species discrimination in all cases. Species pairs that have experienced little or no selection to avoid hybrid matings nevertheless do not make mating errors (Table 13.1; see also Bryant, 1980).

Table 13.1 Courtship in *Epicauta* blister beetles, comparing males isolated in containers with conspecific females to those held with females of another species. All mean differences between conspecific and heterospecific pairings are statistically significant except *normalis* × *normalis* vs *normalis* × *jeffersi*. (From Pinto, 1980.)

Pairing	Mean no. courtship bouts/30 min	Mean min spent in courtship	Natural occurrence of heterospecific pairs
ventralis × *ventralis*	19.3	9.4	Sympatric; sexually active adults on same plant
ventralis × *andersoni*	2.7	1.4	
andersoni × *andersoni*	23.8	14.0	
andersoni × *ventralis*	0	—	
jeffersi × *jeffersi*	7.4	5.6	
jeffersi × *apache*	0.1	0.01	
apache × *apache*	15.1	5.1	
apache × *jeffersi*	1.4	0.6	
apache × *apache*	9.3	2.8	Sympatric; separated by habitat differences
apache × *maculata*	0	—	
maculata × *maculata*	10.6	7.4	
maculata × *apache*	7.3	3.0	
jeffersi × *jeffersi*	5.2	4.3	
jeffersi × *maculata*	0	—	
apache × *apache*	10.2	5.3	
apache × *phoenix*	0.4	0.1	
phoenix × *phoenix*	12.5	5.7	
phoenix × *apache*	0.4	0.1	
normalis × *normalis*	5.6	1.4	Allopatric
normalis × *jeffersi*	8.1	5.3	
normalis × *normalis*	13.7	6.6	
normalis × *apache*	0.4	0.03	

The second prediction states that species differences in courtship behavior in areas of geographic overlap should be common if reproductive signaling is the product of selection for avoidance of hybridization. But reproductive character displacement (Alexander, 1969: Littlejohn, 1969, 1981; Grant, 1972; Loftus-Hills, 1975) is actually extremely rare. Alexander (1969) in his review of pair formation signals of a thousand species of acoustical insects found only two possible examples of character displacement. Otte (1974) and T. J. Walker (1974b) have expanded the search within this group, but with no more success than Alexander. Evidence for courtship divergence in

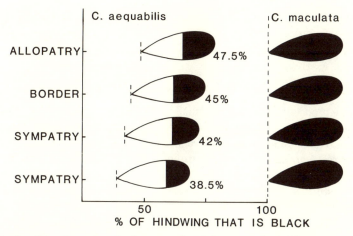

13.4   Possible character displacement in *Calopteryx* damselflies. In areas where *C. maculata* and *C. aequabilis* are fully sympatric, there is less black on the hind wing of *C. aequabilis*. This may make the males appear less like the entirely black-winged *C. maculata* male. (From Waage, 1980.)

zones of sympatry of two related species is also rare or nonexistent in the case of insect sex pheromones, flash patterns of fireflies (J. E. Lloyd, 1966, 1979b), and other visual signals—although Waage (1975, 1979c) has demonstrated a slight divergence in wing coloration in two species of *Calopteryx* in the region where the two damselflies are found together (Fig. 13.4).

Third, if reduction in mating errors across species were the driving force behind the evolution of courtship, we would expect that unusually complex mating antics would occur in taxa in which there were many closely related species. An abundance of potential hybridizers should favor individuals that avoided signal overlap. With many alternative patterns to avoid, the result should be increasing complexity of courtship. To the best of our knowledge, no one has produced a systematic and rigorous test of this prediction, but it is revealing that the insects with perhaps the most elaborate courtships now known, *Syrbula* grasshoppers (Otte, 1972) and the otitid fly *Physiphora demandata* (Alcock and Pyle, 1979), are members of genera with only two species members. Even more instructive is the observation that some females in both groups accept mates *without* preliminary courtship. What this suggests is that courtship may be largely irrelevant to species identification, thanks to the ability of females to detect subtle but reliable cues associated with conspecific males. If this is correct and applies widely among the insects, then elaborate and distinctive courtship procedures must serve some function other than reproductive isolation.

Sexual Selection, Species-Specific Courtship, and Speciation

Before proceeding further, we shall first consider an alternative explanation for the species-specific nature of courtship, for it is this feature that gave rise to the reproductive isolation hypothesis. The alternative view rests on the following argument. When a population becomes isolated genetically from its ancestral population, it usually experiences shifts in both natural and sexual selection. As we noted earlier, differences in the distribution of resources and predator pressure that affect the distibution of receptive females may have a profound influence on the evolution of the mating system of a species. If the conditions that affect success in competition among males change, then the superior manner of mate assessment by females will also be altered in the isolated population. Different vegetational properties of the environment could, for example, affect the transmission of acoustical signals, or variations in average humidity or wind speeds might alter the distribution of a pheromonal message, or heightened predation pressures could change the risk factor associated with certain visual displays. All of these factors and a host of others might lead to divergence in the signals used in competition for females and choice of superior mates.

In fact, there is no theoretical requirement that there be any environmental differences between the isolated population and the population from which it was derived for differences in reproductive behavior to begin to take place. All that is necessary is the chance occurrence in the isolated population of different mutations that influence the performance of individuals in sexual competition (M. J. West-Eberhard, manuscript). Sexual selection, acting on whatever genetic changes occur, can cause unusually rapid and extensive behavioral divergence between isolated and ancestral populations (Spieth, 1974a,b; Ringo, 1977; Carson, 1978). Two factors are involved. First, males that happen for any reason to be unusually successful in reproductive competition (and high variance in male reproductive success is commonplace) can have a great impact on the evolution of sexually related characters. Second, the spread of a successful mutant creates a novel social environment with new sexual selection pressures. This in turn can favor still another mutant with a counteradaptation, and so on in an endless chain of rapid changes (West-Eberhard, 1979).

The widespread occurrence of alternative mating tactics within species (see Chapter 9; Blum and Blum, 1979; Thornhill, 1981), especially geographic variation in mating behavior (Ehrman and Parsons, 1975; Bryant 1980), suggests that courtship behavior is an extremely labile attribute. This hypothesis is supported by laboratory studies that have employed artificial sexual selection to alter the premating isolating mechanisms of *Drosophila* in a relatively few generations (B. Wallace, 1954; Knight, Robertson, and Waddington, 1956). Ewing (1961) inadvertently altered the courtship behavior of *D. melanogaster* through artificial selection for small body size in males. Fruit flies employ wing vibration displays in courtship; Ewing found that the frequency with which males used this courtship component was a function of competition for females. When small males were paired individually with females, the average frequency of wing vibration remained the same as normal controls from generation to generation. But when he created a line of small males that were held in groups with females and so had to compete for access to mates, female choice favored individuals that provided extra wing vibration stimulation. Over the course of a few generations the average frequency of use of wing display by males in the competing line increased sharply (Fig. 13.5). Ewing points out that *quantitative* changes of this sort are often important in the reproductive isolation of species. Thus natural selection for small body size in one population of flies could have unanticipated, but correlated, effects on courtship behavior that might contribute to speciation.

One can conclude therefore that sexual selection has the power to promote rapid divergence of courtship signals in an isolated population. The changes may often be so great that when an isolated population meets the descendants of the ancestral population from which it was derived, they will not interbreed—showing, by definition, that speciation occurred after the populations were separated. According to this scenario, species-specific courtship and the reproductive isolation it produces may be *incidental* by-products of sexual selection operating in different socioecological environments. The distinctive features of the new species' mating behavior may have nothing whatever to do with the avoidance of hybridization. Instead they may have evolved in the context of competition among males for access to females and as a product of pressures created by females that discriminate among conspecific males. West-Eberhard (manuscript) has arrived at similar conclusions. J. E. Lloyd (1979b:330) speaks for those advocating a new perspective on courtship behavior that deemphasizes the reproductive isolation effect when he states, "My original interpretation of this elaborate pair-forming behavior of fireflies was that a number of

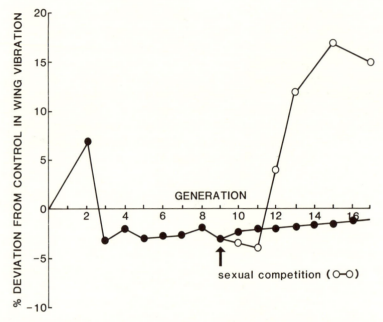

13.5   The effects of artificial selection for small body size on the courtship behavior of *Drosophila melanogaster*. See text for explanation. (From Ewing, 1961.)

sympatric, sibling species . . . make it necessary for reproductive isolation. It seems obvious now that sexual selection is the context that has produced it." We shall now examine how the sometimes elaborate properties of courtship may have evolved as a result of female choice operating on conspecific males.

The Potential Information Content of Courtship

Interactions among male and female insects prior to mating may be as brief as a few seconds or may last for many hours spread over a number of days (J. E. Lloyd, 1981). Even in the exceptionally short tussles between a male and a female house fly, the male may provide a battery of tactile, chemical, and auditory cues (Tobin and Stoffolano, 1973). Brevity of interaction is, therefore, not necessarily equivalent to simplicity of courtship; nonetheless, there are clearly degrees of complexity and variety among species. In the otitid fly, *Physiphora demandata,* individual males create unique chains of courtship signals by stringing together various combinations of seven different display movements, including rapid vibration of the wings while facing the female, drumming on the female's head with a foreleg, and the bizarre metathoracic leg-raise (Fig. 13.6). A courting male may bombard a female with one display after another, rarely repeating the same display twice in a row, for more than 30 minutes, by which time the female will have observed hundreds of discrete acts (Fig. 13.7). Otte (1972) reports that the grasshopper *Syrbula admirabilis* employs seven different body parts, including its wings, antennae, maxillary

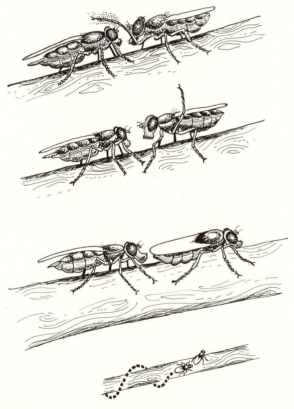

13.6   Three of the displays employed by courting males of the otitid fly *Physiphora demandata*. (*Top*) The male drums on the female's thorax with his foreleg. (*Middle*) The male stiffly raises his metathoracic leg on the side away from the female. (*Bottom*) The male turns to face away from the female. She may respond by placing her proboscis on the tip of the male's abdomen. She then appears to "pull" the male backward along his perch in a spiral course. (From Alcock and Pyle, 1979.)

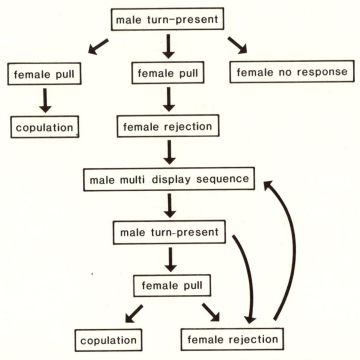

13.7   The pattern of use of courtship displays by *Physiphora demandata*. A male attempts to copulate with a female immediately upon encountering her. Only if she refuses does the male employ a series of displays before trying again to copulate. The pattern is repeated over and over until the male succeeds or the female departs. (From Alcock and Pyle, 1979.)

palpi, and hind legs to create 18 different courtship movements; on the other hand, another gomphocerine grasshopper, *Psoloessa delicatula*, uses only one body part and has a repertoire of just four courtship signal movements.

Some firefly females produce a simple continuous glow, which males use to locate mates; but in *Luciola obsoleta*, a New Guinean species, courtship is vastly more complex. It begins with males and females perched on vegetation, emitting a broad variety of flashes and flickers for prolonged periods. The next phase is pursuit of a glowing female by a flickering male, which may terminate when both land together on a leaf. This leads to still another interaction, replete with "flickers," "twinkles," "stepped flashes," and "beady glows," culminating sometimes with mounting and still more flash communication between the pair prior to copulation (J. E. Lloyd, 1972).

One way to deal with the puzzle created by the almost absurdly great differences between species in courtship behavior is to argue that many of the differences are essentially arbitrary. If some species have been subject to Fisherian runaway selection and others have not, much of modern courtship represents the evolutionary end product of the accidents of female preference for aesthetic features of males. According to the Fisherian argument, production of sexy sons becomes the sole or major benefit for females through arbitrary preferences for certain male characters.

As we have acknowledged elsewhere, runaway evolution of animal courtship is entirely feasible in theory and mathematically sound. As a working hypothesis, though, let us consider the possibility that females can select a mate with superior material resources or genes on the basis of male courtship. There is little convincing evidence in support of this hypothesis, and our comments are designed primarily to raise certain ideas for scrutiny. We begin by considering whether courtship might not contain indicators of the relative value of the material mating benefits a male can supply.

Courtship and Information about Nutritional Nuptial Gifts

Although we have seen that much of the assessment of food offered to female scorpionflies in the genera *Hylobittacus* and *Panorpa* occurs during copulation, females probably begin to judge the likelihood of receiving materials from a mate, and their quality, prior to copulation—during the premating courtship phase. For example, male *Hylobittacus* advertise their prey with an attractant pheromone and by holding the gift conspicuously with the hind legs (see Fig. 12.1). *Panorpa* males also advertise with chemical and visual displays. Therefore females have a chance to determine the size of the nuptial present prior to copulation and will avoid males that lack a gift. They may also sample the food briefly before permitting genital linkage to occur and can reject males whose gift is unpalatable or small, or whose contents have already been removed.

Similarly, in empidid flies males advertise their holdings by flying about at a landmark site with prey suspended beneath their body. Because in many species groups of males assemble in a swarm, females have the opportunity to compare the gifts available to them. Moreover, once having selected a male, the female receives the prey before mating so that she can assess its size and weight before making a decision to copulate.

The behavior of gift giving empidids and mecopterans has similarities to nuptial feeding during courtship in some birds (Schenkel, 1956; Stokes and Williams, 1971). Nisbet (1973) has shown that a male tern's performance in courtship feeding is a good predictor of his ability to feed his chicks when they hatch. Females that choose superior nuptial feeders enhance their reproductive success through the relatively high fledging success of their chicks.

In insects food can be transferred in ways other than through nuptial feeding, as we have noted in our discussions of the function of spermatophores, accessory gland products, and other secretions offered to females. In some cases the courtship of a male may help a female to assess the likelihood of receiving a relatively large nutritional benefit when copulating (Gwynne, 1982). Large males of the katydid *Conocephalus nigropleurum* produce significantly bigger spermatophores than small males (Fig. 13.8). Thus the potential for adaptive female choice exists in this species (assuming that materials in the spermatophore contribute to a female's nutritional welfare). *Conocephalus* females will move toward a tape recording of two males singing, in preference to a recording of a single male, even if the sound in-

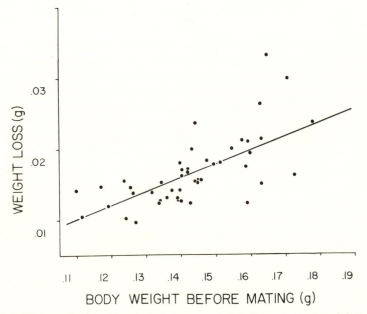

13.8   Large males provide relatively large spermatophores in the katydid *Conocephalus nigropleurum*. Heavier males lose more weight during mating than smaller individuals. (From Gwynne, 1982.)

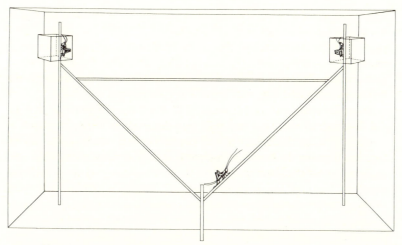

13.9 The design of a mate choice experiment with katydids (*Conocephalus nigropleurum*). The female was allowed to walk along a wooden Y-maze to a point where she had to choose between two calling males, 1 m apart, each in his own cage. (From Gwynne, 1982.)

tensities of the two tapes are equalized (Morris, Kerr, and Fullard, 1978).

This result suggested to Gwynne that in nature females may move into areas with high densities of singing males where they would be in a better position to have a choice between mates of different sizes or weights. Gwynne used virgin males and females for the most part in his laboratory choice experiments. He released 12 different virgin females singly in a cage onto a central stick between two singing males of different weights (Fig. 13.9). The males were placed in small screen cages, which had open fronts to allow the female access to them. In all 12 choice trials females mated with the larger male ($\chi^2 = 12$, $p < 0.01$) and so received a bigger spermatophore. In most cases the female moved directly to the larger male and mated with him. In a few cases, however, she moved toward and made physical contact with the small male at least once before moving toward and mating with the large male. Subsequent work on *Conocephalus* by Gwynne (personal communication) revealed that some females interacted with numerous calling males over several days, but did not mate; when mating occurred, the successful male was always larger than any the female had interacted with previously.

Courtship encounters apparently provide information that a female katydid needs to discriminate against small males in favor of large ones, perhaps because females can select males that produce relatively intense songs (a character associated with large size). The songs of *Conocephalus* katydids have both low- and high-frequency components; the high-frequency constituents become more attenuated (absorbed, diffused) over shorter distances than the low-frequency elements. By unconsciously comparing the proportion of high- and low-frequency sounds in a male's song, a female can judge how close a male is and so correct for distance effects on the intensity of songs of competing males. Moreover, in this and some other spermatophore donating orthopterans, a male contacted by a female often begins to communicate with her by shaking the leaf or stem on which he perches (G. K. Morris, 1980; P. D. Bell, 1980a,b; Gwynne, 1982). These tremulations set up plant-borne vibrations whose amplitude may be an indicator of male weight, which in turn is correlated with spermatophore size.

In *Orchelimum* katydids males offer a large spermatophore whose absolute weight is also proportional to that of the male that produced it. Females have been observed moving from one male to another for bouts of courtship before finally selecting one individual as a mate. Although males of this genus sing in groups, each maintains a territory, repelling other males with an acoustic display and occasionally fighting an opponent. Certain males control locations where encounters with females are higher. These dominant males weigh significantly more than their competitors. Because females almost always mate with a dominant male they receive a relatively large spermatophore (Feaver, 1977, forthcoming). And because females can inspect males visually, perhaps they can also accurately judge male body size through the cues provided by courting individuals.

Courtship and Information about Oviposition Sites

Advertisement by males of transferrable material benefits in courtship is not limited solely to nutritional resources. Just as is true for many birds, some male insects display in ways that direct the female's attention to a future possible egg laying site. Displays of this sort are characteristic of those territorial odonates that defend oviposition sites (Jacobs, 1955; Campanella and Wolf, 1974). For example, males of *Calopteryx aequabilis* perform a remarkable cross display upon arrival of a female in their territories; a male darts up from his perch and strikes out across the water, wings whirring in a distinctive manner, until he reaches the spot that is best for egg laying. There he hovers with his dark-tipped hind wings held out from his body, not moving, while his fore wings flutter rapidly (Fig. 13.10). A female often follows

13.10 The cross display of the damselfly *Calopteryx aequabilis,* which the male employs to show a prospective mate the location of a prime oviposition site. After inspecting the spot, the female may or may not fly up with the male to nearby vegetation to copulate. (Drawing by M. H. Stewart.)

a male, and after having seen the oviposition site, she may or may not enter into the next phase of courtship and eventually mate. It seems highly probable that females can judge the quality of an oviposition site from the air and that this information influences their mating decisions (Waage, 1973).

Females of the bark beetle *Ips confusus* can evidently judge the quality of a potential oviposition site indirectly through analysis of the sex pheromone of a calling male. Figure 13.11 shows that the attractiveness of a male's pheromone declines, the longer he has been living in a host tree and the more mates he has already drawn to his burrow (Borden, 1967). This result is entirely sensible from the female's perspective, since locations at which several other females have been nesting for some time are likely to provide much less food and space for her offspring than a nuptial burrow made recently by an unmated male colonizer. How the female discriminates among the pheromonal

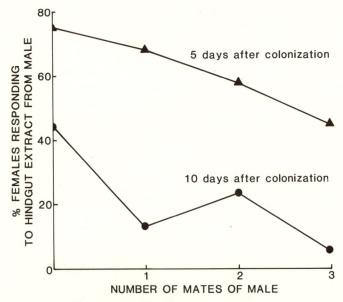

13.11   Females of the bark beetle *Ips confusus* can discriminate among males on the basis of (1) the number of mates he has and (2) the number of days he has resided in a host tree. Hindgut extracts from males that have recently colonized a tree and have no mates are far more likely to attract females than extracts from polygynous males late in the colonization phase. Each point on the graph is based on the percentage of a sample of caged females (N = 75 minimum) that approached an odor source baited with male hindgut extract. (From Borden, 1967.)

scents produced by males of different ages and mating histories is unknown.

Courtship Indicators of Male Parental Care

Courtship signals may also provide information about the probable quality of parental care or mate assistance that a male has to offer. The males of belostomatid water bugs, upon detecting an approaching female, usually perform a series of bobbing "push-ups," which create underwater waves that reach the female. These movements are the same as those used by the male to aerate the eggs after they have been glued to his back by a mate (R. L. Smith, 1979a,b; and see Fig. 13.21). Thus a female could potentially judge something about the egg brooding ability of a courting male by measuring the vigorousness or frequency of his push-ups. (There are many parallels with the courtship displays in fishes whose males provide exclusive parental care—for example, Tinbergen, 1951; D. Morris, 1954; Tavolga, 1969.)

In bark beetles of the genus *Dendroctonus* males may assist their mates in defense of oviposition galleries and in removing frass created in gallery excavations. Courtship in these beetles has interactive components—that is, pheromonal, acoustical, and tactile signals (Rudinsky et al., 1976; Ryker and Rudinsky, 1976: Ryker, in press). The acoustical elements include stridulatory chirps produced by scraping a pair of tiny prongs at the end of the abdomen across a series of small ridges on the underside of the wing covers of the insect. A distinctive and relatively loud chirp (but barely audible to a human listener at close range) is given when two males are competing for access to a gallery system (Fig. 13.12). Males produce the same "rivalry" chirp when they encounter a female, much to the surprise of Ryker and Rudinsky (1976), who anticipated that a different "courtship" stridulation would be employed in this context. Females apparently require the rivalry chirp signal if they are to mate with the male. If the male's stridulatory apparatus is experimentally removed, the silenced individual will approach the female but be repulsed and eventually ejected from her burrow. The chirp could be interpreted as a reproductive isolating mechanism, but given the elaborate pheromonal communication between members of bark beetle species coupled with species differences in host preferences, the likelihood of hybrid interactions *within* burrows must be extremely low.

An alternative hypothesis is that the female requires information about the aggressive abilities of a male before mating with him. The male's success in guarding the entrance will depend on his size, vigor,

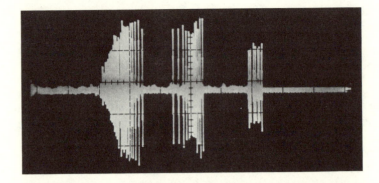

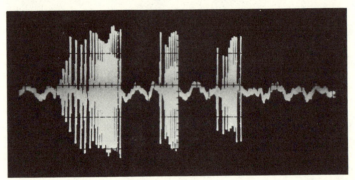

13.12   (*Above*) Premating courtship chirp of a male bark beetle in a gallery with a female. (*Below*) Rivalry (or aggressive) signal given by a male fighting with another male; no female is present. The two chirp signals are identical in the beetle *Dendroctonus pseudotsugae.* (Oscillograms courtesy of L. C. Ryker; from Rudinsky and Ryker, 1976. Reprinted with permission of Pergamon Press, Ltd.)

and forcefulness, which should also be related to his capacity for frass removal. This hypothesis is supported by direct observations made of courting males and females, which reveal that the first hour or so of courtship consists in highly aggressive pushing, shoving, and biting, which only gradually evolves into a less violent affair. Females may be attempting to assess the strength and defensive ability of a courting male both by analysis of his signals and through a direct physical test of his aggressivity.

Courtship and Information about Male Genetic Quality

Thus far we have focused on how females might evaluate attributes of potential mates in order to select a male that could provide superior

Table 13.2   The possible use by females of male courtship signals as indicators of the genetic value of a male's gametes.

Indicators of physiological soundness, good health, absence of deleterious mutations:
    Courtship persistence
    Intensity or complexity of display
    Use of more than one modality in display
    Ability to respond appropriately in signaling interaction with female

Indicators of social dominance:
    Possession of a territory
    Aggression and strength as demonstrated by large body size, ability to overcome sexual resistance of female, ability to avoid courtship interruptions
    Courtship persistence, intensity, or complexity

Indicators of resource gathering skill:
    Demonstration of body size or body weight
    Display of difficult-to-collect resource

Indicators of genetic complementarity with female:
    In outbred species, distinctiveness of display or other attributes relative to other males
    In inbred species, phenotypic similarity to female

material benefits. But male insects frequently offer only their gametes, in which case female choice, if it exists, can be based only on arbitrary characters *or* on differences among males related to their "genetic quality." Given a low but steady rate of mutations at all chromosomal loci, and gene flow among ecologically and genetically differentiated populations, it seems inevitable that there will be genetic variation among males that affects their genetic value to the female. Table 13.2 summarizes some attributes of males that might be reliable indicators of the male's relative genetic quality. We acknowledge again that there is little evidence that females actually do discriminate among males on the basis of these indicators—or that they benefit as a result. Nevertheless, we raise these possibilities because they strike us as a plausible way to interpret certain features of male courtship. Let us examine each of four categories of indicators of genetic quality.

Male Health. First, sexual reproduction could become advantageous for females if they were able to select males which had demonstrated that they were relatively immune to disease and parasites (Hamilton, Henderson, and Moran, 1981). If a female could transmit to her progeny heritable resistance to pathogens and parasites, her reproductive success would surely be enhanced. Looking at sexual reproduction somewhat differently, a female could also gain if she did *not* mate with a male that carried deleterious mutations. Numerous aspects of male courtship can be interpreted as a performance required by the female to test a male's vigor, endurance, and competence. Males that pass this test are likely to be relatively healthy and relatively free of damaging mutations.

Greenfield and Shaw (1982) note that males of many acoustical insects vary intraspecifically in the overall persistence with which they sing, the rate at which they produce the units of song, and the duration of each unit as well as its intensity. All of these factors *potentially* could be used by females to judge the physiological condition of calling males, because of the high energetic expense of acoustical signaling in crickets (Prestwich and Walker, 1981) and cicadas (Mac Nally and Young, 1981). (An alternative, and not mutually exclusive hypothesis, is that these characters are also correlated with dominance status—see below.) In Chapter 6 we discussed the apparent power preference of female mole crickets, which when flying through the air are far more likely to descend to the burrow of a loudly singing male than to that of a less noisy individual (Forrest, 1980, forthcoming). Any component of courtship that varies in the vigorousness with which it is performed could be the basis for female choice. Thus the wing vibrations that occur in many fly courtships, especially among the *Drosophila* (Bastock, 1956; Ewing, 1964), may convey information about a male's flight ability, size, or general strength. The same may be true for the flight displays that are so characteristic of some grasshopper courtships (Otte, 1970).

Pliske's (1975) study of mating in monarch butterflies also suggests that females may select males on the basis of their flight ability and strength. Monarch males typically pursue flying females, which they attempt to grasp on the dorsum of the thorax and ride to the ground, where copulation occurs. Although male monarchs occasionally display small pheromone dispensing hairpencils in courtship, these structures (Fig. 13.13) appear to play a less important role than in other species of Lepidoptera. Because monarchs often migrate great distances, flight ability may be of special significance to reproductive success, and therefore females may require a courtship activity that enables them to evaluate this ability in males.

The view of courtship as a physiological examination of a male also provides a way to interpret the complex and demanding courtship of *Physiphora demandata* (Alcock and Pyle, 1979), *Syrbula admirabilis* (Otte, 1972), and other insects (Lloyd, 1972, 1973). Burk (1981) has determined that among acalyptrate flies complexity of courtship is

13.13 Male hairpencils of (*left*) the monarch butterfly and (*right*) the closely related queen butterfly. Although the monarch is a larger animal, its hairpencils are smaller. (Photographs by T. Eisner.)

greater in species whose males offer genes only and no material benefits with copulation, a finding consistent with the notion that in the absence of material gains females attempt to choose males with few mutations or with superior genes. The capacity to execute elaborate behavioral displays should be correlated with neurophysiological soundness; a *Syrbula* grasshopper with an even slightly abnormal nervous system probably would have difficulty with one or more of the 18 different displays that a female may need before she becomes receptive. Likewise, it surely is physiologically challenging to produce an acoustical song based on four different wing stroking movements, each performed at high rapidity and all of them integrated in the development of a coherent message, as does the katydid *Amblycorpha uhleri* (Walker and Dew, 1972). If the female insists on prolonged courtship, the status of a male's energy reserves may be further tested by his ability to maintain a vigorous, yet varied, performance over time.

One can continue in this vein and argue that the frequent requirement for a multimodality courtship with combinations of acoustical, visual, olfactory, and tactile signals represents a means by which a female can assess the intrinsic condition of her partner. Again, a deleterious mutation is likely to have a pleiotropic effect on courtship behavior if the number of muscles, glands, or neural units involved in normal courtship is sufficiently large. P. D. Bell (1980a) has shown that males of the black-horned tree cricket (*Oecanthus nigricornis*) must generate acoustic, olfactory, and substrate vibrations as well as glandular secretions if they are to mate. Experimentally silenced males do not attract any females; those capable of calling, but with waxed-over metanotal glands upon whose secretions the females feed, are selected far less often than males whose courtship repertoire is complete.

Males of the parasitic wasp, *Nasonia vitripennis,* also engage in multimodal courtship with acoustical, chemical, and tactile cues. One can eliminate the acoustical component by placing a small amount of gum arabic on the top of a male's thorax. This treatment does not reduce the reproductive success of young males, but older, silenced wasps are less successful in mating than normal males of their age. Van den Assem and Putters (1980) speculate that older males are unable to produce as much pheromone; therefore females may require compensatory acoustical evidence of the male's physiological competence before mating.

Another way females may detect mutant, physiologically defective, or diseased males is to demand a complex interactive encounter. Thus duetting occurs in some drosophilid flies and cicadellid leafhoppers (Fig. 13.14; Ossiannilsson, 1949; Donegan and Ewing, 1980; Stirling, 1980). The ability of the male to keep up and provide the appropriate response at precisely the correct time could provide information useful as an indicator of his physiological state.

Male Dominance Status. A second category includes indicators that relate to male aggressive ability. In species in which males compete either for territories, for food resources, or directly for possession of females, cues linked to a male's social dominance can be used by females to choose genes that will enhance the success of their sons in intrasexual competition. (As noted in Chapter 6, female *Nauphoeta* prefer dominant males, which they distinguish by pheromonal cues—Breed, Smith, and Gall, 1980.) Dominance also must often be correlated with freedom from disease, general good health, and physiological integrity so that selection of a dominant male will have spillover benefits with respect to the transmission of these attributes as

13.14 Duetting courtship songs in the homopteran *Muellerianella brevipennis*. Males and females alternate acoustical signals in this species as shown, and in certain *Drosophila* fruit flies as well. (From Booij, 1982.)

well. Freeland (1981) has documented that dominant house mouse males are less likely to carry high parasite loads than subordinates. Females, by encouraging competition among males and selecting dominant mates, not only enjoy lowered risk of contracting parasites from a male but also may transfer pathogen resistance to their progeny. There is no reason why the same advantages should not apply to female insects that select dominant males. Territorial possession, for example, should be a major criterion of female choice in those lekking species in which males provide genes only. If a female can judge which site will be most strongly contested and if she can mate with its owner, she is nearly certain to acquire genes from a male who has been thoroughly challenged and therefore exhibits the various attributes that make social dominance possible for him.

A variety of other cues besides territorial possession may enable females to assess the status of available males. The aggressive nature of much of animal courtship may be selected for by females because it provides an indicator of a male's strength and size, just as female coyness offers a way in which to assess male courtship persistence. In addition, females may require males to provide signals that stimulate aggression in their rivals. Burk (forthcoming) has shown that females of the cricket *Teleogryllus oceanicus* discriminate against males that do not provide courtship stridulation. Subordinate males are reluctant to produce these signals even when confronted by a receptive female, because they are very likely to attract a dominant neighbor who may subject them to a damaging fight. As a result, the requirement that males stridulate means that females effectively choose dominant males capable of ignoring the risk of attracting rivals (Fig. 13.15).

Elaborate and prolonged courtship may have evolved in the context of dominance advertisement, because a male that can perform these conspicuous activities and yet remain undisturbed by other males is presumably a truly dominant individual that has thoroughly intimidated his opponents (Alcock and Pyle, 1979; Borgia, 1979). In the otitid *P. demandata* females usually leave if the male courting them must interrupt his signaling to deal with a territorial intruder (Alcock and Pyle, 1979). Similarly, some *Drosophila* females have a specific aggressive or repelling song that they employ when more than one male courts them simultaneously (Donegan and Ewing, 1980). Females of the cerambycid beetle *Monochamus scutellatus*, by ovipositing at an increasing rate as time passes (Fig. 13.16), "reward" those males that are able to maintain a prolonged pair bond with them. This is a species characterized by much aggression among males for possession of a female; the winner accompanies his mate from oviposition

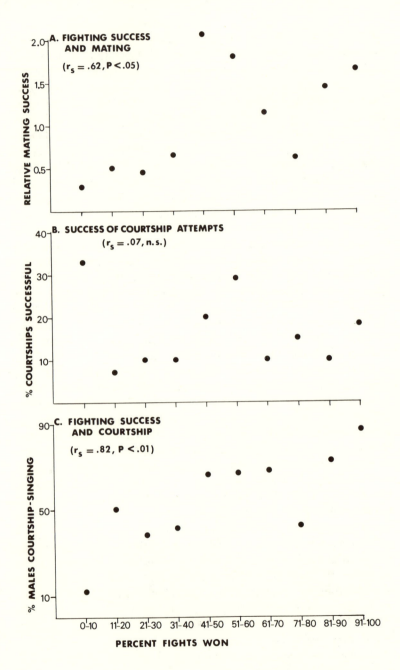

A. FIGHTING SUCCESS AND MATING

$(r_s = .62, P < .05)$

RELATIVE MATING SUCCESS

B. SUCCESS OF COURTSHIP ATTEMPTS

$(r_s = .07, n.s.)$

% COURTSHIPS SUCCESSFUL

C. FIGHTING SUCCESS AND COURTSHIP

$(r_s = .82, P < .01)$

% MALES COURTSHIP-SINGING

0-10 11-20 21-30 31-40 41-50 51-60 61-70 71-80 81-90 91-100

PERCENT FIGHTS WON

site to oviposition site, all the while attempting to prevent her being stolen by other males. The result of the female's egg laying pattern would be to select for unusually dominant males able to remain in contact with their mates for relatively long, uninterrupted periods.

Along these lines J. E. Lloyd (1981) suggests that females of some insects may exert selection favoring distinctive courtship signals. Males that are recognizable as individuals can be monitored by a female over a period of days to determine the male's staying power and capacity to retain control of a territory for a substantial time. Lloyd based his hypothesis on observations of territorial aggression among male bumble bees (*Bombus fervidus*). Males compete for control of perch sites near nest entrances from which receptive gynes sometimes emerge to copulate. Because perched males regularly fan their wings, Lloyd speculates that they may be transmitting unique pheromonal message to females within the nest. If a female is able to remember which odors she has perceived, she can determine how long various rivals are able to maintain control of the perch territory. This information could enable her to time her departure on a nuptial flight to coincide with the presence of a dominant individual that has been in control of the site for a relatively long period. Lloyd interprets in the same manner the distinctive visual signals of males of certain lekking fireflies and the individualistic acoustical messages of various territorial katydids. (As noted previously, in at least one katydid there is some evidence that females do inspect a number of males several days in succession before making a mating decision—D. T. Gwynne, personal communication.) The possibility that individual males have their own personalized signals as the result of sexual selection is an intriguing idea that deserves much more investigation.

Male Resource Gathering Skill. The third category of proximate cues that a female might use to assess the genetic quality of a male is interrelated with the first two. Courtship may provide an indication of a male's resource gathering ability. If a female is able to select a male with unusual success in securing food, her offspring could conceivably benefit. Good health, physiological soundness, and social dominance are all apt to be correlated with foraging skill, so that it would often be difficult (and probably unnecessary) to identify the relative

---

13.15 (A) Males of the cricket *Teleogryllus oceanicus* that regularly win fights with other males enjoy greater mating success than males that win less often. (B) This is not because successful fighters are unusually seductive courters, but (C) because males that win many fights sing their courtship songs regularly. (From Burk, forthcoming. Reprinted by permission of Westview Press, Denver, Colorado.)

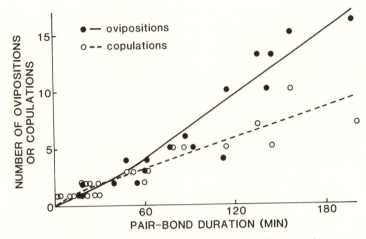

13.16   Females of the long-horned, wood-boring beetle *Monochamus scutellatus* oviposit more often, the longer males remain with them as guarding partners. (From Hughes, 1981.)

importance of these attributes to female choice. Again, demonstrations of body size and weight through substrate signaling, mounting, or grappling with the female, or through visual displays, may help females assess the past foraging competence of a male as measured by his current dimensions.

In addition, pheromonal analysis or sampling of glandular products could inform a female about the diet of a male, its quality or quantity. In oriental fruit moths, males that have been reared on an artificial diet generate a sex pheromone that is apparently chemically distinct from the scent released by individuals that have consumed apples during the larval stage (Baker, Nishida, and Roelofs, 1981). In two arctiid moths and the queen butterfly, the male sex pheromone has been definitely traced to pyrrolizine precursors in some of the food plants of these species (Pliske and Eisner, 1969; Conner et al., 1981; Schneider et al., 1982). Moreover, various male butterflies and moths feed as adults on withered plants that contain these toxic alkaloid precursors, presumably to gather additional material for use in chemical communication with females (Fig. 13.17; Boppré, 1981). Male queen butterflies whose hairpencils have been removed or that have been reared on pyrrolizine-deficient plants are less effective in inducing females to mate (Myers and Brower, 1969; Pliske and Eisner, 1969). The same is true for males of the arctiid *Utethesia ornatrix* that have been reared on pinto beans instead of the alkaloid-rich *Crotalaria,* their usual food plant (Conner et al., 1981). Even more dra-

13.17   Some butterfly males aggregate on dried weeds of heliotrope, perhaps to extract chemical precursors of the male sex pheromone from materials on the plants. (Photograph by L. P. Brower.)

matically, an alkaloid-laden diet is required for full development on the scent releasing structure of *Creatonotos* moths. Individuals reared on food plants without pyrrolizine substances have coremata as little as one-tenth the size of those fed alkaloid-rich food plants (Schneider et al., 1982).

There are at least five hypotheses for why odors derived from food plants may offer significant courtship information for females. The first two relate to potential material benefits offered by a male; the last three concern male genetic quality:

(1)   The odor may indicate something about the chemical composition of the male's nutrient investment in the spermatophores he has to offer. The hydrocarbon components of *Colias* male phero-

mones are identical to those found in the spermatophore (L. D. Marshall, 1980, 1982).

(2) The sex pheromone could inform a female of the defensive capacity of the potential mate. If copulation is lengthy, it may pay a female to mate with a well-protected individual unlikely to attract birds or other predators (Conner et al., 1981).

(3) A male's scent indicates that he (or his mother) has been able to find the food plant, an ability clearly relevant to the reproductive success of a female's daughters when they are searching for oviposition substrate.

(4) The scents constitute evidence that the individual has the capacity to consume particular food plants, a matter of special importance when the plants contain toxic defensive compounds.

(5) Finally, in some cases, males may demonstrate through their pheromones the ability as adults to collect distinctive scents from *rare* plants, as an indicator of their flight capacity and perceptual competence, both of which relate to the foraging chances of a female's progeny.

A hypothesis similar to (3) or (5) might account for the use, in the courtship of some empidid flies (Kessel, 1955), of empty silken balls of no nutritional benefit. These courtship artifacts are produced from salivary glands, which in turn derive their salivary contents from the nectar consumed by the males. The capacity to construct a silken ornament thus becomes an advertisement for the male's ability to accrue food resources (Fig. 13.18).

Male Genetic Complementarity. A fourth and final category of courtship information includes exhibition of the characteristics of the male that show that his genes would complement those of the female unusually well. We have already discussed the rare-male advantage phenomenon of *Drosophila,* which has been interpreted as the outcome of the ability of females to select mates that differ genetically from them. This could help the female produce highly heterozygous offspring, which would presumably enjoy certain benefits in an outbred species whose progeny faced unpredictable, variable ecological conditions. In species in which inbreeding was common but not necessarily universal, sisters might conceivably analyze the courtship of a suitor to determine his phenotypic resemblance to them. If a proximate cue such as a shared odor were an accurate correlate of shared genotype (Greenberg, 1979), a female could improve her chances of mating with a brother and could therefore generate homozygous offspring to take advantage of a predictably constant environmental situation (see Chapter 2).

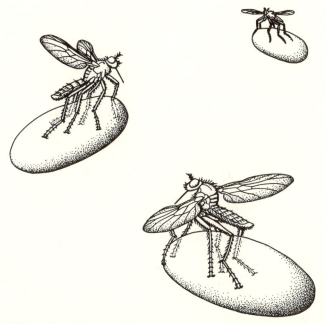

13.18 Balloon fly males with their nuptial gifts of silken threads, which are accepted by a female but not eaten or otherwise used, except perhaps as a means of assessing the physiological condition of males. (Drawing by M. H. Stewart.)

In summary, the pressure to make accurate identifications of the species and sex of an individual, Fisherian runaway processes, and random events may all be involved in the evolution of insect courtship. Given the substantial variety of information that can be conveyed to a female during courtship about the relative attributes of a male, we conclude that it is plausible that the behavior of a courting male provides the basis for an accurate analysis of the genetic value of his genes. We are aware that if females are attempting to judge this property, there will be selection favoring males that can deceive females into accepting them on the basis of indicators which signal that the male is more healthy, more socially dominant, or more genetically complementary than he actually is (Williams, 1966; Trivers, 1972; Otte, 1975; Dawkins, 1976; Dawkins and Krebs, 1978; Andersson, 1980). However, there is little likelihood that gross deceit is common because of strong counterselection favoring females that can accurately identify truly fit partners. Moreover, females are probably selected to use criteria in mate choice decisions that make it difficult for males to falisfy their status (Zahavi, 1975, 1977b; Trivers, 1976; Hailman, 1977; Dawkins and Krebs, 1978; Barnard and Burk, 1979). We surmise that courting males, as a general rule, are forced to provide information relevant to female reproductive success if they are ever to have a chance to reproduce.

Courtship and Mate Discrimination by Males

The arguments presented thus far have been based on the assumption that females should be the discriminating sex, using courtship information to select superior males from among the many suitors available to them. This expectation rests on the typically greater parental investment per offspring made by females; as a result, females (and their eggs) are usually a limited resource for males. But if the inequality in male and female parental investments declines, or if males transfer large amounts of valuable materials to a female during copulation, it is possible that males (and their nutritional gifts or parental services) may become a limited resource for females. The prediction follows that as male parental investment escalates, or as male nonpromiscuous mating effort (resource transfer) increases, males will become ever more selective, even to the point of sex-role reversals whereby they will be more discriminating than their females, which will compete among themselves for access to males (Trivers, 1972). In this chapter we shall explore whether the interactions between males and females sometimes help males to discriminate among potential mates of their species.

Before presenting evidence that males at times choose among females, we must caution that what may appear to be the reluctance of a male to mate with a female does not necessarily mean that the male is being selective. He may instead have entered into a general sexual refractory period following depletion of sperm or key resources needed to induce a female to use his sperm to fertilize her eggs. For example, a male house fly becomes less likely to try to mate as the number of previous matings accomplished in a short period increases. Leopold, Terranova, and Swilley (1971) have shown that females mated with males whose accessory glands have been depleted by a series of consecutive matings are far more likely to mate again promptly (with subsequent sperm precedence) than those that have received a large supply of accessory-gland fluids (Fig. 13.19). Even though a depleted male may still have sperm to transfer, the low probability that these will actually be used may not compensate him for the time and energetic costs of copulation. (A male that mates for a fifth

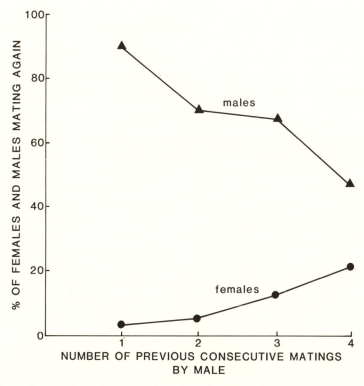

13.19 Repeated copulation reduces the willingness of male flies, *Musca domestica,* to copulate again. But if a female's partner has mated several times, the female is more likely to accept an additional copulation shortly after being inseminated. (From Leopold, Terranova, and Swilley, 1971.)

time in a row spends twice as long *in copula* as a male on his first copulation of the day.) The depleted male is therefore not being choosy, merely temporarily unreceptive. The same is true for certain species whose males undergo a period of sexual maturation following eclosion, during which time they will not mate even if presented with a receptive female. Thus in order to show that a male is truly selective in picking a female to court or mate, one must demonstrate that the male is fully receptive and able to reject some individuals while accepting others as his partners.

True discrimination by receptive males does occur, even in species in which males offer only their gametes to a female. The general principle is that courtship, copulation, and mate guarding all require an allocation of time and energy and risk taking. The problem facing an individual is how to allocate his lifetime supply of these commodities to fertilize more eggs than his rivals. Although a high degree of sexual motivation may promote contacts with females, completely indiscriminate courtship will rarely be advantageous, particularly if there are cues that enable males to identify relatively profitable females. Females can vary in their reproductive value to males for a host of reasons, summarized in Table 13.3. For example, there are very likely to be both receptive and totally nonreceptive females in a population. Males that direct a higher proportion of their courtship attempts to nonreceptive females will often secure a lower rate of fertilizations than those able to court receptive individuals more frequently. The same is true of males that court females which will accept their sperm only to replace those gametes with sperm from a rival male.

Even if copulation leads to some fertilization gain, there is still a question whether the male might gain more from an alternative mate. Given a choice between a virgin female and one that has mated 24 hours earlier, males of the desert fruit fly, *Drosophila mojavensis,* select the virgin 86 percent of the time (Markow, forthcoming). Sperm precedence studies with this species have shown that when the remating interval is 24 hours, females use the sperm of their more recent partner to fertilize about two-thirds of the eggs they lay from that time until the next mating. By choosing a virgin (which will fertilize all her eggs with his sperm), the male secures 50 percent more eggs

Table 13.3   The attributes of females that may affect their reproductive value to males.

	Attributes of females		
	Less preferred	More preferred	
No egg gain	Sperm rejecting (nonreceptive) Sperm displacing	Sperm accepting (receptive) Sperm using	Some egg gain
Some egg gain	Recent mate Less fecund • Small • Older • Nonvirgin	New mate More fecund • Large • Younger • Virgin	More egg gain
Inferior eggs	Small eggs Eggs of low genetic quality	Large eggs Eggs of high genetic quality	Superior eggs

than he would if he were to invest his time in a recently mated female.

Another way in which a male insect might gain more from his time investment with a single female is to avoid recent mates that he has already fertilized in favor of new partners. Earlier we reviewed this phenomenon in the bee *Lasioglossum rohweri* (Barrows, 1975a,b; Barrows, Bell and Michener, 1975) and the damselfly *Hetaerina vulnerata* (Alcock, 1982b). In these cases repeat matings with the same female did not appear to yield substantial genetic gains and wasted time that could be more profitably spent in other ways. Moreover, not all females are reproductively equal in the sense of having the same number of eggs to be laid after a mating. In insects ranging from aphids, wasps, scorpionflies, true flies, and lepidopterans, there are examples in which female body size is strongly positively correlated with fecundity (MacKerras, 1933; Marks, 1976; Thornhill, 1976b, 1981; Whitham, 1978, 1979; Suzuki, 1978; van den Assem, Gijswijt, and Nubel, 1980; Kempton, Lowe, and Bintcliffe, 1980). Direct evidence for male preference for larger females comes from Johnson's (1982) study of the brentid beetle, *Brenthus anchorago* (see Fig. 9.8).

Age is another factor that influences the reproductive value of a female. As a rule, the older the female, the less her expected future production of offspring. In *D. melanogaster* a 14-day-old female has half the reproductive value to a male of a 4-day-old individual (Boorman and Parker, 1976). Frequently, therefore, females that have already mated will be less desirable than virgin partners. In the eye gnat, *Hippelates collusor*, air that has been passed over 90-hour-old females is attractive to males (they will travel toward the odor source), whereas males avoid the odors of 120-hour-old females (even if these are virgins) (Adams and Mulla, 1968). We suspect that in nature the older females would almost certainly be mated and thus would not be receptive. Males that find older females repellent are able to spend more time searching for younger, unmated females that offer far greater genetic gain per unit of courtship time.

Possible Mate Discrimination by "Helpful" Males

In many of the examples just reviewed, males do not provide either parental services or material benefits to their mates. Therefore costly assistance of mates is not necessarily responsible for the evolution of mate discrimination by males. Nevertheless, high male parental investment and nonpromiscuous mating effort are expected to provide an added impetus for mate choice by males for two reasons. First, each copulation represents a commitment by the male to transfer some of his limited supply of goods or services to a female. If a male should happen to choose a female with relatively few eggs to fertilize, his helpful assistance will yield little return, while reducing the time and energy available for searching out other superior mates. Second, because males of this sort provide inducements to copulate, receptive females tend to make themselves available. This in turn has two consequences: (1) the cost of being choosy is reduced for a male because the time required to locate a replacement after rejecting a potential mate is low, but (2) there may be sperm accepting females that are receptive primarily in order to secure the material benefits that come with a copulation rather than to acquire sperm. For example, older females of some crickets are more likely to eat the male's spermatophore promptly after copulation than are younger ones (Khalifa, 1950). This would seem to favor cautiousness in making mating commitments and to promote the selection of females that are more likely to use a male's sperm to fertilize a large clutch of eggs rather than to eat the sperm packet and its contents.

Sexual cautiousness of males has evolved in katydids of the genus *Ephippiger*, which transfer enormous 1-gm spermatophores that constitute 25 to 30 percent of male body weight (Busnel and Dumortier, 1955). Males that have just copulated refuse to mate again for 3 to 5 days, during which they generate the materials for production of another spermatophore. Because of the costliness of a spermatophore, males that consistently select relatively fecund females should greatly outreproduce nonselective males. (An individual cannot compensate for an occasional unprofitable copulation by mating with large numbers of females.) It may be significant that female *Ephippiger* tremulate during courtship, shaking their bodies after having placed their legs on a male (Busnel, Pasquinelly, and Dumortier, 1955; Busnel, 1963; Dumortier, 1963). This courtship signal may provide information that a male can use to measure a female's weight, a factor often correlated with fecundity. Although males can use this information to reject lightweight mates, there is no evidence that they actually do so (but mate selectivity based on this factor is practiced by males of another tettigoniid katydid—as we shall see).

We can also speculate that male mantids choose among females, given the high costs of mating for males in some species. The well-known cautiousness of these males in approaching females (Roeder, 1967) has been interpreted as a stalking behavior designed to help the male get close enough to grasp a female and initiate copulation before he is killed. An alternative possibility is that the male may be assessing the size of the female, so that he can slip away uneaten if she proves to be a small individual of low fecundity. Unfortunately there has been little work on mate selection in male mantids.

In addition to visual and tactile assessment of female body weight and size, some resource donating males may be able to determine the size (and even physiological condition) of females by the pheromonal trails they produce. In one noctuid moth small females produce far less pheromone than large ones (Shorey and Gaston, 1965), a feature that males may exploit to analyze the reproductive profitability of a potential partner to whom they must transfer a bulky spermatophore.

As well as possibly preferring larger (more fecund) females, males that assist their mates sometimes prefer virgins to already mated individuals. Males of a spermatophore donating acridid grasshopper, *Melanoplus sanguinipes,* when presented with equal numbers of mated and virgin females, chose overwhelmingly to copulate with the virgin hoppers (Pickford and Gillott, 1972a,b). Females of this insect mate multiply but the relative receptivity of virgins versus mated females is not known. The preference of males for virgins could stem from signals provided by females that communicate something about their degree of sexual receptiveness. If so, males are evidently sensitive to these signals and couple with younger individuals of greater reproductive value, rather than with older females closer to the end of their reproductive efforts.

There are indications that male butterflies may prefer younger females, which have greater reproductive value than older individuals. Suzuki (1978) has shown that females of the pierid *Pieris rapae* lay about half their lifetime production of eggs in the first 5 days of their lives, although they may live 20 days. If this is a general pattern, the selection of younger females should be advantageous to males that offer their mates valuable materials in their spermatophores.

Male *Colias* butterflies invest about 6 percent of their body weight in the transfer of a spermatophore to a female (L. D. Marshall, 1980). This is a substantial donation and, as predicted, males court younger females for longer periods than they do older females (Rutowski, 1982). Males discriminate among potential partners on the basis of an age-related increase in ultraviolet reflectance from the wings of females.

Still other examples of choosiness by males that offer nuptial gifts or mate assistance include a tephritid fly, thynnine wasps, and certain bark beetles. In the fly *Rioxa pornia* the male attracts a female with a sex pheromone and then offers the attracted individual a salivary se-

13.20   The fly *Rioxa pornia* standing by a mound of saliva that he has produced after inspecting a female that has responded to his sex pheromone. (From Pritchard, 1967.)

cretion in the form of a small mound of foamy material (Fig. 13.20). Before depositing any foam, however, the male runs around to the rear of the attracted female and apparently assesses her suitability (Pritchard, 1967). In some cases after inspection the male refuses to produce a gift for the female but instead moves off and resumes pheromonal calling, later giving his store of saliva to another female.

Male thynnine wasps offer stored secretions or gathered nectar to their mates. In two species observed in Australia, Alcock (1981b,c) sometimes observed males approach pheromone releasing females only to abandon them without attempting to copulate. These apparent rejections may have occurred because males detected subtle signals from females that they would refuse to mate, but this seems unlikely because the females were releasing pheromones and did not make any overt movements as males came close. Perhaps males of these and other species can detect cues that indicate something about the potential fecundity of a female. If so, they may use the information to select females of relatively high fecundity.

A clear-cut case of male rejection of potential mates comes from studies of bark beetles. In a large number of these insects it is the male that constructs a small cavity (the nuptial chamber) in a trunk or limb of the host tree. He guards the entrance to the chamber and blocks it with his abdomen when a conspecific attempts to enter. In *Polygraphus rufipennis* the female tries to gain access to the chamber by rubbing the male's elytra with her mandibles. Even if he is not yet mated, a male does not automatically permit the female to enter, as the mean time between initial contact and entry of the first female is about 5 minutes (Rudinsky, Oester, and Ryker, 1978). Perhaps a male assesses some quality (such as size or strength) of a potential mate before permitting her to join him. Moreover, if a male has already acquired two or more females (which remain with him and excavate elaborate oviposition tunnels that run off from the nuptial chamber), he is still more reluctant to allow another female to join him. Females may signal for as much as 45 minutes, and some are forced to abandon the effort. In other bark beetles males will live with a limited number of mates; in *Ips paraconfusus* harems never are larger than three, and in *I. pini* the maximum is four (Birch, 1978).

For male bark beetles each mating may involve commitment of a certain amount of mate assistance in the form of tunnel cleaning. More important perhaps, a male's nuptial chamber is a jumping-off point for exploitation of the limited larval food resources in the vicinity. If a male allows too many females to gain access to these resources, the total number of surviving brood would fall because of intraspecific competition among the offspring of his mates (Raffa and Berryman, in press). Therefore there is a premium on the selection of a small number of superior females, and courtship in some way provides the basis for the male's selection process.

Sex-Role Reversals

The bark beetle example is suggestive of a sex-role reversal. It is the female who comes to the male and attempts through elaborate courtship to induce the male to permit her to copulate with him and so gain access to the valuable resources under his control. Sex-role reversals are expected when males expend more parental effort than females or when male mating effort limits female reproduction to a greater extent than female parental effort limits male reproduction. There are a number of apparent examples of such reversals in the insects; some are merely suggestive, others are better documented.

Among the speculative possibilities that deserve further study is *Drosophila americana,* whose males have on occasion been observed *not* to respond to the signal of sexual readiness by a female. Females have a courtship song that they produce before the male begins to sing. Confronted with an unreceptive male, a female may prolong her singing behavior in an apparent attempt to induce him to mate (Donegan and Ewing, 1980). One wonders whether males of this species transfer unusually large quantities of nutritional materials during copulation and therefore have become the object of sexual competition among females.

Studies of boll weevils offer another tantalizing suggestion of sex-role reversal (Cross and Mitchell, 1966). Whereas males do not respond to females more distant than a few centimeters, females track males over distances of many meters. Upon reaching a male, the female strokes him with her antennae prior to mating. Several females may converge upon a male, and a struggle for "possession" of him ensues, with intruders even attempting to wedge themselves between a mounted pair. Perhaps some males defend superior cotton bolls and permit their mates to oviposit there. If so, females may be competing for the limited resources that successful territorial males possess.

The two best examples of sex-role reversals in the insects occur in belostomatid waterbugs (R. L. Smith, 1976a,b, 1979a,b, 1980) and a tettigoniid katydid called the Mormon "cricket," *Anabrus simplex* (Gwynne, 1981, manuscript). Smith's work on waterbugs provides a clear case of the relation between sex-role reversal and male parental effort. The male belostomatid is a highly paternal insect that permits the female to glue her eggs onto his back (Fig. 13.21). He then broods the eggs until the young nymphs hatch, 2 to 4 weeks later. This requires that he expose the eggs to atmospheric air for much of the time and that he stroke the eggs with his legs and perform push-up movements to circulate water over his brood. These actions have costs for the paternal male. Eggs are large, adding weight and probably making the male less efficient at catching prey and more vulnerable to predators. Also, because a brooding male cannot spread his wings, he cannot abandon a deteriorating environment. Moreover, brooding reduces the ability of males to inseminate many females. Although the relative parental effort of the sexes has not been quantified for waterbugs, it is clear that males invest a great deal, perhaps more than fe-

13.21   Male waterbug, *Abedus herberti*, brooding a mass of eggs with a nymphal waterbug nearby. (Photograph by R. L. Smith.)

13.22   A female Mormon cricket, *An-abrus simplex*, copulating with a male (*A*) and then consuming the enormous spermatophylax provided by her mate (*B*). (Photographs by D. T. Gwynne.)

males do. Females may also have difficulty finding an unencumbered mate at times when the majority of the females are gravid.

The prediction from parental investment theory is that female waterbugs will court and compete for males. This prediction is met. Both sexes court, but unlike most animals it is the female waterbug that initiates interaction of the sexes and repeatedly approaches the coy male. Sometimes females are rejected by males, but usually the male courts in response to female courtship and eventually receives the female's eggs.

The other outstanding example of sex-role reversal involves a case in which male mating effort, rather than parental effort, may be limiting female reproductive success and thereby generating female competition for access to sexually cautious males. A male of the Mormon "cricket" provides an exeptionally large spermatophore to his mate, which she eventually consumes (Fig. 13.22). The donation deprives a male of about 25 percent of his total body weight, very much limiting the frequency with which an individual can expect to copulate, particularly if food resources are scarce. (Indeed it is possible that sometimes a male may be able to afford only one mating before his death.)

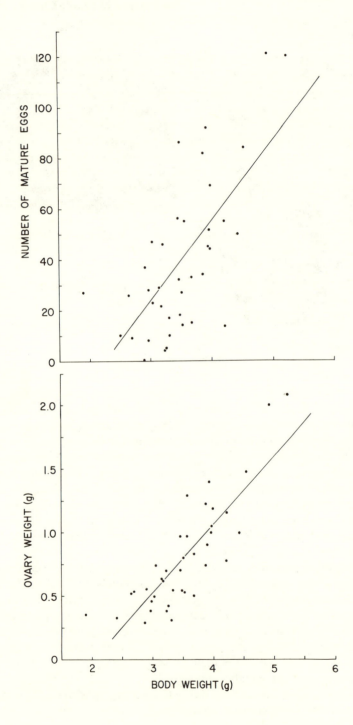

As a result, we expect mate choice by males and it occurs—but only under certain conditions.

The density and social organization of populations of Mormon crickets varies a great deal. In some areas "outbreaks" of these insects occur, and dense bands form with as many as several dozen of these flightless, voracious katydids per square meter marching across the countryside in search of food. As they move along, a male that is ready to mate ascends a plant and begins to sing. Females in the vicinity quickly respond and approach him. They often interact aggressively with one another by grappling as they converge upon the singer. The winner mounts the male in the position required to receive a spermatophore, but she will not necessarily receive one because the male can and does reject a large proportion of the females that reach him. Mate discrimination occurs, because a male can assess the weight of a female when she clambers onto his back and jerks up and down several times. The male has the option of uncoupling shortly after genital linkage, in which case he will not transfer his spermatophore. In his study of a dense band, Gwynne found that in a sample of 51 male-female encounters, the male exercised the rejection option 34 times (67 percent). Only once did a female terminate an encounter by rejecting the male. The accepted females were heavier on average than the rejected ones (some of which never obtained a partner), and female fecundity (ovary weight and number of mature eggs) was positively correlated with body weight (Fig. 13.23). Thus katydid males in the band behaved in ways that led them to secure a greater return on their investment in a spermatophore. They could afford to be selective because there were many more receptive females (of variable size) than calling males in their population.

The situation changes dramatically in populations of lower density. Males at a low-density site studied by Gwynne (manuscript) were evenly spaced due to acoustical signaling and fights between neighboring males were seen. There was no evidence of mate rejection by males, which eagerly descended from their calling perches to meet responsive females. Virtually all females secured mates in this population. Spermatophore sizes as a proportion of male weight at both locations were not different. Gwynne suggests that the predicted role reversal may only be expressed in Mormon crickets at high density, because only then do males become a limiting resource for females. At

---

13.23   The positive relation between female body weight and fecundity in the Mormon cricket. (From Gwynne, 1981. Copyright 1981 by the American Association for the Advancement of Science.)

the high-density site band members fought for desirable food plants, seeds, and dead insects, which were quickly eaten. Females at the low-density site had an abundant supply of food (sweet clover seeds) and males at this location were probably able to accrue more quickly the resources necessary to produce spermatophores. Nutritious spermatophores, therefore, may be more costly to construct and more critical to female reproduction when population density is high, leading to more intense sexual selection in females.

The extreme flexibility in mating behavior of this insect is further demonstrated by the ability of males and females to switch sex roles *temporarily*. When the members of a high-density nomadic band of Mormon crickets encounter a rich food source (such as a large patch of sweet clover), the females become more selective and the males become less choosy—as long as the food lasts.

The Mormon cricket story shows that the absolute size of the spermatophore is not the only factor that determines whether females will compete for chances to mate. But generally among the katydids, the greater the ratio of spermatophore weight to total weight of a clutch of eggs (an approximation of the relative contribution of males and females to reproduction), the more likely the males are to be selective, at least under some conditions (Gwynne, forthcoming b). These findings support the generalization that whichever sex provides the greater contribution to reproductive success will tend to become a limited resource for the other sex. Courtship then becomes a means for the limiting sex to determine the value of potential partners.

Although the speculation quotient of this chapter may alarm some readers, we feel that this risk is worth taking to promote a multifunctional view of courtship in insects. The behavior of courting males (and females) is potentially information rich. If there is variation in the reproductive value of individuals in a population, use of this information to discriminate among potential mates has important theoretical advantages for a courted insect. Enough evidence exists to indicate that insect nervous systems are sophisticated enough to enable females (and males) to make subtle discriminations among would-be partners and so advance their reproductive success. The challenge for the future is to test the hypothesis of "courtship as mate evaluation," to see whether it is generally applicable or is exercised only by the exceptional insect.

14

Female Mating Systems

In chapter 8 we categorized male mating systems in insects on the basis of how ecological factors influence the number of mates individuals can secure in a breeding season. In this final chapter we adopt the same approach to analyze the individual mating systems of female insects. We begin with a description of the different patterns of receptivity of females, ranging from those that mate only a single time to those that are always sexually accessible and have a steady succession of mates. We next consider the reproductive events, especially those related to interactions with males, that appear to be responsible for causing some female insects to become unwilling to copulate, permanently or temporarily. The mechanisms regulating female receptivity determine whether individuals mate with one or many males and thus control female mating systems. One can then ask what are the reproductive costs and benefits associated with the mechanisms that produce monogamy or the various forms of polyandry. Our argument is that if receipt of sperm were the only function of mating for a female, monogamy would be much more widespread that it actually is. The regular occurrence of polyandry in many insect groups is traced to a diversity of benefits females derive from multiple mating in certain environments.

Table 14.1   Examples of four different patterns of female receptivity in insects.

Mating pattern (order)	Reference
Single-mating females	
Hymenoptera: most solitary bees and wasps	Evans and West-Eberhard, 1970; Gordh, 1976; Alcock et al., 1978; Eickwort and Ginsberg, 1980
Diptera: most culicids	Giglioli and Mason, 1966; Craig, 1967
Muscid house fly	Murvosh, Fry, and LaBrecque, 1964; Riemann, Moen, and Thorson, 1967
Hippoboscid fly	Tarshis, 1958
Lepidoptera: some moths and butterflies	Burns, 1968; Pliske, 1973; L. E. Gilbert, 1976; Ehrlich and Ehrlich, 1978
Ephemeroptera: most species	Spieth, 1940
Multiple-mating females: a restricted single period of receptivity	
Hymenoptera: Some *Pogonomyrmex* ants	Hölldobler, 1976
The honey bee	Adams et al., 1977
Coleoptera: a cerambycid beetle	Hughes, 1981
Multiple-mating females: cyclic receptivity	
Diptera: *Drosophila* fruit flies	Boorman and Parker, 1976; Pyle and Gromko, 1978, 1981
Many tephritids	Christenson and Foote, 1960; Fletcher and Giannakakis, 1973
Some hippoboscids, neriids, and streblids	I. M. Cowan, 1943; Tarshis, 1958; Ross, 1961; Mangan, 1979
Lepidoptera: many moths and butterflies	Ouye et al., 1965; Burns, 1968; Pease, 1968; Marks, 1976; Byers, 1978; Sims, 1979
Orthoptera: many species	Haskell, 1958; Alexander, 1961; A. M. Richards, 1973; Hartmann and Loher, 1974; Parker and Smith, 1975; Loher and Rence, 1978; Gwynne and Morris, forthcoming
Odonata: most species	Jacobs, 1955; Waage, 1973, forthcoming; Corbet, 1980
Mecoptera: most species	Thornhill, 1977, 1981; Byers and Thornhill, 1983
Heteroptera: some pentatomids and delphacids	Tostowaryk, 1971; Oh, 1979; Nilakhe, 1976
Most belostomatids	R. L. Smith, 1976a, 1979a,b

Table 14.1   (*continued*)

Coleoptera: some meloids	Gerber and Church, 1976
Lampyrid fireflies	J. E. Lloyd, 1979b, 1980b
Scarab beetles	Landa, 1969; Milne, 1960
Coccinellid beetles	T. W. Fisher, 1959
Curculionid beetle	Mayer and Brazzel, 1963
Brentid beetle	Johnson, 1982

Multiple-mating females: continuous receptivity

Hymenoptera: some andrenid and megachilid bees	Rozen, 1958; Alcock, Eickwort and Eickwort, 1977; Severinghaus, Kurtak, and Eickwort, 1981
Some sphecid wasps	Cross, Stith, and Bauman, 1975; Brockmann, 1980; Colville and Colville, 1980
Heteroptera: some lygaeids	Loher and Gordon, 1980; Sillén-Tullberg, 1981
Pentatomid bug	Harris and Todd, 1980b
Naucorid water bug	Constantz, 1973
Orthoptera: phasmid walking-sticks	Sivinski, 1978, 1980b

Patterns of Receptivity in Female Insects

In order to understand why mating systems of female insects vary, it is helpful to examine the different patterns of female receptivity and the proximate mechanisms that regulate these patterns. The most obvious distinction is between females that mate just once during their lifetime and those that mate many times (Table 14.1). We have earlier cited females of the digger bee and the lovebug as examples of the single-mating (monogamous) pattern, whereas females damselflies, dung flies, and scorpionflies represent the multiple-mating (polyandrous) pattern. We can further divide the latter group into species whose females exhibit (1) a restricted bout of multiple copulations, usually early in their lives, followed by lifelong nonreceptivity, or (2) cyclic receptivity, or (3) more or less continuous receptivity.

The three multiple-mating categories blend into one another to some extent (Fig. 14.1), but at one end of the spectrum are species like those harvester ants (*Pogonomyrmex*) whose females often copulate with several males on the day of the nuptial flight but are not receptive thereafter (Hölldobler, 1976). Cyclically receptive females, in

A <u> | _____</u>
    SINGLE-MATING: Centris pallida

B1 _____<u>||| _____</u>
    MULTIPLE-MATING: Pogonomyrmex rugosus

B2 _<u>▌▐▐▐▐ _____</u>
    MULTIPLE-MATING: Apis mellifera

C1 __<u>|_____|_____|_____|_____</u>
    MULTIPLE-MATING: Drosophila melanogaster

C2 __<u>||_____|||_____|_____|||_____</u>
    MULTIPLE-MATING: Calopteryx maculata

D1 _____<u>||| | ||| || ||| ||| ||||| ||| || ||||| | |||</u>
    MULTIPLE-MATING: Anthidium maculosum

D2 _<u>▆▆▆▆▆▆▆ ▐ ▐▐ ▆▆▆▆▆ ▐▐ ▐ ▆▆▆▆ _</u>
    MULTIPLE-MATING: Nezara viridula

TIME ➡

14.1 The diversity of mating patterns by female insects. The horizontal line represents the life span of the female. The vertical bars represent copulations; their width indicates the duration of the mating. (A) Single-mating species; (B1, B2) species with restricted bouts of multiple mating; (C1, C2) species with cyclic bouts of receptivity; (D1, D2) continuously receptive females.

contrast, alternate between intervals of receptivity and sexually refractory periods for much or all of their adulthood. Thus the fruit fly *Drosophila melanogaster* (Pyle and Gromko, 1978) and many butterflies (for example, Sims, 1979) mate a total of two or three times at widely separated intervals. Shortly after each mating the female rejects males and will continue to do so for a period of days (or weeks), with receptivity only gradually regained. A variant on this theme is exhibited by some damselflies and the dung fly, which alternate periods of feeding with bouts of oviposition and mating. Just prior to and during the egg laying interval, a female will copulate with one or more males, depending on how many individuals can grasp her while she still has eggs to lay.

The period of receptivity is lengthened to such an extent in some species that the female is more or less continuously available for mating. During each day-long period in which she forages for pollen, a female of the bee *Anthidium maculosum* will mate with any male able to grasp her at the food plant (Alcock, Eickwort, and Eickwort, 1977). Likewise, females of the pentatomid *Nezara viridula* combine feeding with mating and have one partner after another with no obvious sexual refractory pause. A female will mate for several days with some males, or more briefly with others, but will almost always have a partner (Harris and Todd, 1980b).

Mechanisms That Control Female Receptivity

Females in categories other than the continuously receptive groups are sexually unresponsive to males at some point in the course of their lives. These females have internal mechanisms that regulate their receptivity. Table 14.2 lists some likely candidates for the various cues that may activate the mechanisms that reduce receptivity, at least temporarily.

First, there is evidence that some female insects must be mated within a limited period, usually early in their lives. If this does not happen, the female will automatically become nonreceptive after the interval passes. Females of the ichnuemonid wasp *Pimpla instigator* will mate during the first 5 to 7 days after eclosion, then become permanently nonreceptive whether or not they have copulated (Khalifa, 1949). An invariant schedule of receptivity also appears to be the rule for females of one chalcidoid wasp that are receptive immediately upon emergence, with a subsequent decline and eventual termination of receptivity, even if they never mate (Lessmann, 1974).

One explanation for the evolution of this kind of mechanism is that

Table 14.2   Events that may trigger mechanisms which reduce an insect female's receptivity.

Passage of time (the female has a preprogrammed schedule of declining receptivity)

Stimulation by a courting male

Mechanical stimulation by a copulating partner

Filling of the sperm storage organs

Acquisition of seminal fluid

Acquisition of a spermatophore

Acquisition of a nuptial gift

Depletion of the supply of mature eggs suitable for oviposition it occurs only in species in which the likelihood that a female will mate in the preprogrammed receptive period is essentially 100 percent. This may well apply to many parasitic wasps because (1) females lay clusters of eggs in a host, (2) the clusters always include some male progeny, and (3) brother-sister matings are standard practice. Therefore the chance that a female will not be mated shortly after reaching adulthood is probably infinitesimally small in species of this sort.

On the other hand, a preprogrammed steep decline in the willingness to mate over time (without reference to insemination) could mean that a female retains a certain amount of receptivity even after an initial mating. Thus a once-mated female might not immediately foreclose the possibility of a second copulation. If she were to encounter a second male with unusually desirable properties in the period before receptivity was automatically terminated, she could mate with him. With this system a female would be sure to mate at least once but would retain for a while the option of accepting a superior or additional mate, should one become available. Mertins (1980) reports that sometimes females of the typically monogamous bethyliid wasp *Laelis pedatus* will copulate again, after mating immediately upon eclosion.

In most insects, however, activation of sexual refractoriness in females is not time dependent, but instead is linked directly to sensory stimulation received by the female during courtship or copulation. In certain chalcidoid wasps (*Nasonia* and *Lariophagus*) it is possible to remove the courting male when the female adopts her distinctive mating posture. Even though she has not yet copulated, the female will shortly become permanently nonreceptive (van den Assem, 1970; Barrass, 1979). (Normally, of course, receipt of these cues would *always* be followed by insemination.)

The relation between acquisition of sperm (and allied materials) from a male and the onset of a sexual refractory period is still more obvious in those female insects that possess mechanisms which directly link aspects of copulation with a change in receptivity. For example, in some species the mechanical stimulation provided by a copulatory partner may activate a mechanism that inhibits further mating. Circumstantial evidence for such a system comes from observation of some female insects, including the screwworm fly *Cochliomyia hominivorax* and various *Drosophila*, which become nonreceptive instantly upon uncoupling following a very short copulation. Leopold (1976) argues that the rapidity of the effect makes it unlikely that a biochemical signal is transferred from the male to female, absorbed by the female, and translated into a change in receptivity, as these events would probably require more than a few minutes.

Not all female insects make use of the tactile information provided during copulation to trigger a decline in sexual readiness. Some male bittacids (*Bittacus strigosus* and *Hylobittacus apicalis*) were subjected to an experiment in which the penis was capped with warm beeswax. The wax, once hardened, prevented the transfer of sperm and accessory-gland fluids to the females but did not prevent intromission and the tactile stimulation associated with copulation. The males with sealed penes copulated for the normal length of time but their mates remained responsive to conspecific male pheromones, unlike a group of females who were mated with uncapped males (Thornhill and J. B. Johnson, unpublished).

Even insects that do modify their receptivity on the basis of tactile signals received during copulation probably do not usually rely *solely* on these cues. A female *Drosophila* experiences an immediate but short-term decline in receptivity following copulation, probably due to mechanical stimulation associated with insertion of the male's aedeagus. Her receptivity is quickly regained if she has not received both sperm and seminal fluid, derived from the male's accessory glands (A. Manning, 1962, 1967; Fowler, 1973; Gilbert, Richmond, and Sheehan, 1981; Gromko, Gilbert, and Richmond, forthcoming). The materials provided by the male from his accessory glands act to bridge the time gap between copulation and migration of the sperm into the spermathecae, preventing unnecessary copulation during this time. Once the sperm have fully entered their storage places, the female is in some way able to monitor the supply of sperm she has on

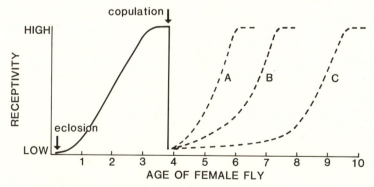

14.2   The pattern of receptivity of females of the fruit fly, *Drosophila melanogaster*. (*A*) The female will regain receptivity relatively quickly if she receives only mechanical stimulation from a copulatory partner but no sperm or associated fluids. (*B*) The receipt of accessory-gland fluids prolongs the period of female nonreceptivity. (*C*) The combination of mechanical stimulation, accessory-gland materials, and transferred sperm results in the slowest recovery of female receptivity.

hand (perhaps stretch receptors on the spermathecae inform the insect's nervous system of the degree of distension of these organs). As the spermathecae become depleted, the degree of receptivity rises; most females will mate again after about 7 days, if given the opportunity to lay eggs freely after the first copulation and if males are provided at the appropriate time (Fig. 14.2).

Sperm supply in the spermatheca probably also regulates female receptivity in the Mediterranean fruit fly (Nakagawa et al., 1971) and the moths *Atteva punctella* (O. R. Taylor, 1967) and *Acrolepiopsis assectella* (Thibout, 1975). However, in all these cases it is difficult to rule out an influence from accessory-gland materials.

The role of accessory-gland products has definitely been established for a number of species other than *Drosophila*. In the bittacids sexual receptivity is decreased by a factor produced by these glands. If a suffi-

14.3 The cockroach *Nauphoeta cinerea*. The female shown here has approached a pheromone dispensing male and has begun to feed upon secretions on the male's back before allowing copulation and spermatophore transfer to occur. (Photograph by L. M. Roth.)

cient amount is not passed to the female during copulation, she will remain willing to mate again. Females that terminate copulation prematurely, because they have not received a full nuptial meal, probably have not received a full complement of the "anti-receptivity factor" and therefore search out another male with whom to mate (Thornhill, 1976c; Thornhill and J. B. Johnson, unpublished).

If a spermatophore is passed during copulation, females probably often use its presence (or the presence of materials contained within it) as a cue for the regulation of receptivity. In *Nauphoeta* the male positions his spermatophore within the female's bursa copulatrix and cements it in place, after which the female refuses to mate (Fig. 14.3). If, however, one experimentally removes the spermatophore before it is sealed in the bursa, the female retains her readiness to cop-

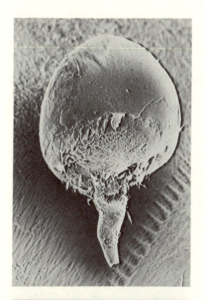

14.4 (*Above left*) A freshly transferred spermatophore of a *Colias* butterfly. (*Above right*) A partially deflated spermatophore that has been within a female's bursa copulatrix for some time. (*Left*) A collapsed spermatophore whose presumably nutritious contents have been absorbed by the female. (From Rutowski et al., 1981.)

Table 14.3   Comparison of spermatophore condition in females of the checkered white butterfly that solicited courtship versus a random sample of females. (From Rutowski, 1980.)

Females	Percent of females in spermatophore condition[a]—			Sample size
	I	II	III	
Random sample	17	17	66	47
Approached conspecific	4	20	76	78

[a] Condition I spermatophores were white and spherical; condition II were partly collapsed but still white; condition III were fully collapsed and often clear.
$\chi^2 = 6.35$; $p < 0.02$.

ulate (Roth, 1962). Hartmann and Loher (1974) have shown that in the grasshopper *Chorthippus curtipennis* the spermatophore is received and emptied of its contents in 1 to 5 hours, after which the female actively repels suitors. Males of the cricket *Teleogryllus commodus* transfer a prostaglandin synthesizing enzyme complex to their mates via the spermatophore. The material helps stimulate oviposition (Loher et al., 1981), but whether it also plays a role in suppressing locomotion and female response to male calling songs (two other consequences of mating) is not known (Loher, 1981). In the cecropia moth (and some other Lepidoptera) there is a seminal product within the spermatophore that appears in the bursa after copulation. It acts to trigger the secretion of a hormone that directs the corpora cardiaca to release yet another hormone, which in turn inhibits release of the sex pheromone by the female (Riddiford and Ashenhurst, 1973; see also O. R. Taylor, 1967; Obara, Tateda, and Kurabara, 1975).

The reciprocal of these observations is that there is an *increase* in receptivity related to depletion of a spermatophore in some insects (Table 14.3 and Fig. 14.4). For example, females of certain *Colias* butterflies and the checkered white, *Pieris protodice,* are more likely to approach conspecific males when the spermatophore within the bursa copulatrix is depleted than are females with a fresh spermatophore (Rutowski, 1980; Rutowski et al., 1981). Sugawara (1979) has shown that in *Pieris rapae* there are stretch receptors that monitor the distension of the bursa copulatrix and so provide information about the volume of materials contained within the organ. Once the nutritional resources contained within a spermatophore have been exhausted, females may become receptive to a new nutrient donation from a male.

It is probable that depletion of the nutritional substances passed by a male to a female results in a resumption of receptivity by female thynnine wasps. After securing a nuptial meal (Fig. 14.5), the female immediately descends underground for a period of days to search out scarab beetle larvae on which to lay her eggs (Ridsdill Smith, 1970). Eventually she resurfaces and solicits another copulation (by releasing a sex pheromone) and receives another feeding, which enables her to fuel a new bout of bettle hunting. If females are experimentally separated from their mates before completion of a nuptial feeding, they will generally resume calling at once and will readily accept another mating and another nuptial meal (Alcock, 1981b,c).

Finally, some of the Odonata provide a special case in which the onset of sexual receptivity appears to be triggered by the internal development of a complete clutch of mature eggs. Although during the time they are laying their eggs most female damselflies and dragonflies will accept any male that captures them as a partner, once the eggs have been laid females become unreceptive and will evade or struggle free from males that grasp them. Females of the calopterygid damselfly *Hetaerina vulnerata* that are approaching the water to oviposit or flying from one egg laying site to another, will stop flying if taken in midair by a male and will descend to the ground to copulate. But after completion of oviposition the female, now not receptive, will continue to fly vigorously if she is captured and will prevent the male from landing with her. Because cooperation by the female is essential

14.5 (*Left*) A species of micropezid fly in which the male appears to regurgitate food to a mate during copulation. (Photograph by E. S. Ross.) (*Right*) A species of thynnine wasp in which males regurgitate droplets of nectar to their mates while in copula. (Photograph by J. Alcock.)

for sperm transfer, males have little to gain by holding a recalcitrant female and they therefore release her after a brief interaction. She will then spend the next several days feeding in a sexual refractory phase before approaching the stream for another bout of egg laying and mating.

The Costs and Benefits of Single versus Multiple Mating

Differences in the internal mechanisms regulating female receptivity are the proximate basis for variation in the mating systems of females. They determine the frequency of mating by a female and thus the number of partners she is likely to have. Here we concern ourselves with the ultimate reasons why some insect females are monogamous while others are polyandrous. Reasonably complete discussions of this issue are found in Alcock et al. (1978), Parker (1979), and W. F. Walker (1980) and are abstracted here.

If we reexamine the events that are often related to frequency of mating by females (Table 14.2), it is apparent that the acquisition of sperm or allied materials are the primary factors that control receptivity in female insects. This is hardly surprising, for two of the major functions of copulation from the female's perspective are to acquire sperm to fertilize her eggs and (often) to secure the nutritional benefits present in seminal fluids and spermatophores. The ability of the female to detect (unconsciously) whether she has received these goods from a copulation is clearly adaptive because it enables her to regulate the frequency of her copulations in relation to her need for sperm or nutritional materials.

One would predict that in cases in which females received only gametes and little or nothing in the way of material benefits from copulation, monogamy would be the rule. This prediction is derived from the evidence that in most cases males will transfer more than enough sperm to fertilize a female's expected lifetime production of eggs. Sperm are small and economically manufactured; therefore a male can afford to transfer a large number to a female. Individuals that fail to provide a number equal to or greater than the number of eggs possessed by a female miss opportunities to fertilize some eggs. They generally experience a selective disadvantage relative to males that provide an abundance of sperm.

If single matings do satisfy the female's sperm requirements as expected, females that mate multiply and receive only unneeded sperm in return will usually gain no benefits to offset the several costs of copulation, which include:

(1) The time (and energy) devoted to courtship and copulation;
(2) The time required to repel unwanted suitors attracted to the female when she is in an intermediate state of receptivity;
(3) Possible increased vulnerability to predators (in cryptically colored species).

Of these costs the most universal is the time allocation required for each mating. Because courtship and copulation are often prolonged, repeated matings could consume a significant fraction of an adult female's life, particularly if she were short-lived and continuously receptive.

In addition, if a female employs the pattern of cyclical receptivity, at different times she will be totally nonreceptive, just slightly receptive, or highly receptive. Selection will therefore favor males that at least test the receptivity of conspecific females (especially if highly receptive females are rare, because slightly receptive females can sometimes be induced to copulate). Males may force truly nonreceptive individuals to expend considerable time and energy in order to repel them. By contrast, if a female is unambiguously and consistently nonreceptive after one mating, selection may favor males that avoid females which give any indication of a lack of readiness to mate. This would save the male time and reduce or eliminate the costs for the female of dealing with unnecessary suitors.

Several studies have shown a decline in the reproductive success of females that mate more often than the norm for their species, either because of the time costs of multiple mating or because of the sexual harassment of courting males. Gerber and Church (1976) found that female meloid beetles confined to a cage with sexually active males had shorter lives than those that could escape male attentions. In another laboratory study Nielson and Toles (1968) showed that female cicadellids that were housed with males and copulated often had only one-quarter the production of offspring enjoyed by females that mated just once. Finally, Pyle and Gromko (1981) conducted an artificial selection experiment with *Drosophila melanogaster,* which produced a line of females that remated at shorter intervals than the wild type. Quick rematers had a somewhat reduced number of progeny relative to controls.

The third possible cost of multiple mating, increased predation, is highly speculative. No one has demonstrated that copulating females are more vulnerable to attack than single females in camouflaged species—and in some cases, pairs may actually be safer than solitary individuals (Sivinski, 1980b).

An additional and probably very minor cost associated with cyclical and continuous receptivity is the physiological expense required for the development and maintenance of the mechanisms that permit the female to respond sexually to more than one courting male. The magnitude of this expense undoubtedly pales in comparison with the time costs of multiple mating.

A Classification of Insect Polyandry

Although female monogamy is widespread in insects (and vastly more common than male monogamy), multiple mating by females is not rare. Thus the disadvantageous aspects of polyandry (presumably the time demands of multiple courtships and copulations) must regularly be offset by certain advantages. In Table 14.4 we present a classifica-

Table 14.4   A classification of female mating systems in insects, with an identification of the hypothetical benefits associated with acceptance of more than one mating partner.

I. Monogamy: the female mates with a single male during one breeding season.
II. Polyandry: the female mates multiply, receiving sperm from more than one male during one breeding season.
   A. Sperm-replenishment polyandry
      1. The female adds to her depleted or inadequate sperm supplies.
      2. The female avoids the costs of storing and maintaining large quantities of sperm from a single donor.
   B. Material-benefit polyandry
      1. The female's partner provides nutritional benefits or other material resources in return for mating.
      2. The female's partner helps reduce the risk of predation or helps her reduce competition for a resource.
      3. The female's partner provides protection from other sexually active males.
   C. Genetic-benefit polyandry
      1. The female replaces sperm of a genetically inferior mate with the gametes of a genetically superior individual.
      2. The female adds sperm from a genetically different male to her sperm supplies to increase the genetic diversity of her offspring.
   D. Convenience polyandry
      1. The female avoids the costs of trying to prevent superfluous copulations.

tion of insect polyandry that emphasizes the variety of possible benefits that may be experienced by multiple-mating females.

Sperm-Replenishment Polyandry

Perhaps the most obvious advantage to mating more than once would
be replenishment of sperm supplies that have become depleted as a
result of oviposition. There are a number of probable cases of sperm-
replenishment polyandry. Oh (1979) notes that females of a delphacid
bug are not likely to regain their receptivity until many days have
passed since the last mating, and then only after a lengthy bout of oviposition. Much the same pattern has been found in a variety of tephritid flies (Cunningham et al., 1971; Fletcher and Giannakakis, 1973;
Prokopy and Hendrichs, 1979). Nielson and McAllan (1965) show
that once-mated females of a species of *Rhagoletis* have about half the
fertility of multiply-mated individuals.

Perhaps the best-documented example of sperm-replenishment
polyandry is provided by *Drosophila melanogaster*. The female fruit
fly's storage organs hold between five hundred and a thousand sperm
(Fowler, 1973; Gromko, Gilbert, and Richmond, forthcoming). This
figure corresponds well with the average number of progeny (528)
produced by single-mated females in a study conducted by Pyle and
Gromko (1978), especially when a certain amount of "inefficiency" in
the use of sperm to fertilize eggs is taken into account. Females given

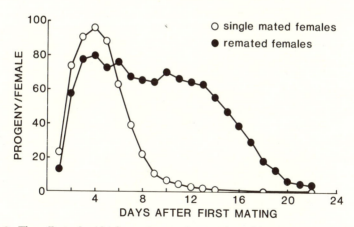

14.6 The effect of multiple mating on the number of fertilized eggs laid per
day by a female of *Drosophila melanogaster*. Eggs laid per day by a single-
mated female decline after day 5. Eggs laid per day by a female that has regular
access to males and copulates at about weekly intervals remain high for much
longer. (From Pyle and Gromko, 1978.)

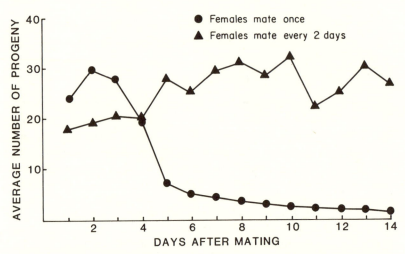

14.7   The effect of multiple mating on the number of eggs laid per day in the fruit fly *Drosophila mojavensis*. In this species females normally mate about once every 48 hours. That frequent mating is necessary to maintain high fertility is shown by the rapid decline in the number of eggs laid per day by females that have mated just once. (From Markow, forthcoming.)

the opportunity to mate more than once usually did so about 6 days after the first mating, at a time when stored sperm were still present but in rapidly diminishing volume. Those that had the chance to add to their sperm supplies averaged 1,035 offspring, almost exactly twice that of the single-mated, experimental group (Fig. 14.6).

The hypothesis that females of *D. melanogaster* mate multiply to maintain high levels of fertility is strengthened by a comparative study of mating frequency and fertility in another fruit fly, *D. mojavensis* (Markow, forthcoming). In this species females that are given free access to males generally mate at 48-hour intervals, not every fifth or sixth day. If a female copulates only once, her fertility begins to decline by the fourth day after mating and is close to nil by the fifth day (Fig. 14.7). A second mating 24 hours after the first does not raise female reproductive success, but if the second mating comes 48 hours after the initial copulation, high levels of fertility are maintained for 5 days instead of 3. (If a female is permitted to mate at regular 48-hour intervals, she can maintain a constant rate of production of offspring for 2 weeks or more.) Although the schedule of female receptivity is different in *D. mojavensis* and *D. melanogaster,* in both species females appear to accept more than one mate in order to replenish their sperm supplies and keep producing fertile eggs.

The discovery that female fruit flies, to fertilize all their eggs, need more sperm than they can store from a single copulation is puzzling when one realizes that males of *D. melanogaster* transfer two thousand to four thousand gametes per copulation. Females of this species "waste" 50 to 75 percent of the gametes they receive in a single mating. One can make the argument that the storage process is part of the female's internal selection mechanism for weeding out inferior gametes from the superior sperm that she gets from a male. But it is also possible that there are advantages to be gained from reducing the physiological expense involved in storing and maintaining large numbers of sperm. The storage organs require metabolic energy for their construction, and there is added evidence that the female must supply nutrients in some cases to transport and maintain the sperm in a viable state (Pesson, 1950; Davey, 1958; Davey and Webster, 1967). If these are significant expenses for the female, it could be that the need for sperm replenishment, which requires multiple mating, is a by-product of the evolution of small, energy conserving storage organs.

Material-Benefit Polyandry

The case of *Drosophila melanogaster,* and perhaps others in the sperm-replenishment category, might also be placed in the material-benefit group, illustrating that there may be multiple benefits to multiple mating by females of some insects. Although it is currently unclear in *D. melanogaster* to what extent accessory-gland donations by the male contribute to female fecundity, it is likely that these secretions are helpful to females. As discussed earlier, females become less receptive upon receipt of the accessory-gland products and will quickly regain their willingness to mate if these are not present in the male ejaculate. This suggests that the female gains more from mating than the sperm provided by a male. In other *Drosophila,* which mate as many as four times in a single morning, the possibility that males donate useful nutrients in seminal fluids is even greater (Markow, forthcoming).

Material-benefit polyandry is probably the most common form of multiple mating by female insects. The benefits received from copulation range from the materials in seminal fluids, to nutritious spermatophores in lepidopterans and orthopterans, to the nuptial gifts offered by empidid flies, mecopterans, and thynnine wasps, to access to superior oviposition or feeding sites in odonates, dung flies, and some megachilid bees, to the egg brooding services offered by male belostomatid water bugs. It is striking that females of all the insects on this list are known to mate regularly throughout their lives. In some species in which nutritional resources change hands during copulation, females may never have to secure food directly, but can rely solely on materials gathered by their many partners.

A special case in which acceptance of a mate provides a food gain for a female has been described by McLain (1981). He has shown that female soldier beetles, *Chauliognathus pennsylvanicus*, become less selective in the choice of a mate when competition for food from various wasps is high (Fig. 14.8). The wasps assault the beetles to drive them from the flowers on which they are feeding. They are more likely to attack singletons than pairs. As a result, unmated females are more likely to be knocked from their food plant and lose valuable foraging time. This favors rapid remating. The females not only accept a male sooner than they would otherwise (that is, before having mature ova) but also are more likely to permit a small male to mount them than in the absence of wasp competitors (Table 14.5).

In a similar vein, aposematic insects may also benefit through pairing if the larger, warningly colored stimulus offered by two individuals is more likely to deter a predator than the cues provided by a single female (W. F. Walker, 1980).

Finally, a male may help protect his mate not only against competitors and predators of other species but also against sexual harassment by conspecific males. We have already suggested that a major benefit of continuous receptivity by foraging females of the andrenid bee *Nomadopsis puellae* (Rutowski and Alcock, 1980) and by ovipositing fe-

14.8   Soldier beetle, *Chauliognathus pennsylvanicus*, copulating on goldenrod while the female feeds. (Photograph by D. K. McLain.)

Table 14.5   The average dry weight of copulating and nonmating males of the soldier beetle, *Chauliognathus pennsylvanicus,* in the presence and absence of wasps that can chase beetles from flowers on which they are foraging. (From McLain, 1981.)

	Times wasps chase beetles from plant (percent)	Average weight of males[a]
Wasps present (N = 139)		
Single beetle	68	11.06 (single males)
Pair	14	11.31 (mating males)
Wasps absent		
Single beetle		11.66 (single males)
Pair		13.48 (mating males)

[a] Based on samples of 230 to 460 individuals.

males of the dung fly *Scatophaga stercoraria* (Borgia, 1981) is to attract a male who will defend them against other males. This leads to a reduction in interruptions of foraging and oviposition, as well as a decrease in the chances that a sexually enthusiastic male will inadvertently damage a female in his efforts to couple with her.

Genetic-Benefit Polyandry

The third category of polyandry is considerably more speculative than the first two types. It is possible (but largely unproven) that females that mate with more than one male can gain purely genetic advantages by securing sperm from two or more individuals. The honey bee, *Apis mellifera,* provides perhaps the best-documented example (see review by Page, 1980). Although in the Hymenoptera haploid individuals develop into males and diploid genotypes typically become females, in the honey bee and some others diploid males are a possibility. These individuals arise because there is a sex-determining locus with multiple alleles, and diploid bees that are homozygous at this locus are males. The worker bees detect diploid male larvae and eat them.

A queen that mates with only a single male may acquire sperm with a sex allele identical to one of hers. If so, half of her diploid progeny will be worthless diploid males. This provides a strong selective force favoring queens that mate with more than one male in order to reduce the proportion of diploid males and thereby increase the number of valuable workers generated. Queens that make many workers are more likely to have food reserves to last through a winter and a large enough force to supply a daughter with her workers. (In large colonies a daughter queen is produced who remains in the hive with a proportion of the worker force, while the queen and the remainder of her worker daughters leave in a swarm to find a new homesite.)

The model developed by Page (1980) shows that if there are more than five sex alleles in most populations (as is thought to be the case), then females in order to maximize their fitness should mate with a fairly large number of males (Fig. 14.9). The estimated actual number is about 17 partners for each female over several nuptial flights.

Less strongly supported hypotheses of genetic benefit through multiple mating have been advanced in earlier chapters. For example, by remaining receptive during a preovipositional mating bout, a female cricket may be able to test the dominance status of her first partner, who is likely to be replaced (by a territory owner) if he happens to be a subordinate, satellite male.

A similar argument can be advanced to explain why some female odonates remain continuously receptive throughout the time they are laying their eggs. By retaining her readiness to mate, a female can test her original mate's ability to retain control of her for periods that range from a few minutes to an hour or more. If he is to prevent a takeover, a male often must be able to transport a female to an oviposition site and

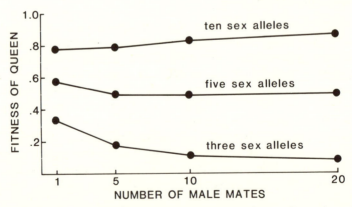

14.9 The effect of mating frequency on the fitness of queen honey bees that occur in populations with different numbers of sex alleles. Given certain assumptions, females that mate with a relatively large number of males may enjoy higher fitness than those that mate with few partners. (From Page, 1980.)

remain in tandem while she lays her eggs, or perch or hover by her while she is so engaged. Presumably only physiologically competent individuals can succeed in meeting these demands successfully.

As we have indicated, a common primary function of the mechanisms of receptivity in insects is to enable females to assess the quantity of sperm (and any associated materials) provided by a male. But in damselflies and dragonflies, the female's pattern of receptivity may permit her to gain access to prime oviposition sites controlled by males (a material benefit) and to provide the female with a high-quality mate who will fertilize all or most of her eggs during an oviposition bout (a genetic benefit). To this end, female receptivity has come to be linked with the period of egg laying rather than with the quantity of sperm stored within the spermathecae (Waage, 1979b, forthcoming).

Under some circumstances retention of receptivity may lead to a genetic benefit distinct from the replacement of genetically inferior sperm with superior gametes. By mating with more than one male, a female may acquire equally fit but *genetically different* complements of sperm from several individuals and thereby generate more genetic diversity in her young than if she were to mate with a single partner only. To the extent that it is advantageous to produce offspring that vary genetically, multiple mating by females should be adaptive, although there is some question about how much additional zygotic diversity multiple mating confers on a female (Williams, 1975).

Convenience Polyandry

A final category of polyandry is one that is based on the recognition that males can make it costly for females to resist their sexual attempts. Conflict between the sexes is widespread and manifests itself particularly dramatically in such things as the forced copulations of *Panorpa* scorpionflies. Females that have been grasped by a would-be "rapist" can probably refuse to permit the male ever to copulate with them by preventing entry of his aedeagus. But in some cases this response requires an extremely high degree of energy expenditure and, perhaps more important still, a considerable loss of time. The rapist male has little to gain by releasing an unwilling female, because his only option is to search for and capture yet another rejecting partner. It may be advantageous for him to struggle for a very long time with any female he has been fortunate enough to capture. At some point his persistence may make it adaptive for the female to cut her losses and accept copulation in order to terminate the interaction with no further time and energy wasted. Such a mating would be a copulation of convenience, because the female gains no sperm-replenishment, material, or genetic benefits from the encounter.

Another less dramatic case involves females of the eumenid wasp *Abispa ephippium*, whose male mating system was described in Chapter 8. Females of this species require water for the construction of large and elaborate mud nests, and males therefore patrol for mates at streams and ponds. If a female can avoid detection by a male, as often happens, she does not seek one out but instead fills her crop with water as quickly as possible and departs at once. But if she is spotted and grasped by a male, she makes no attempt to repel him or avoid mating; instead, she continues to drink, then flies up with the male to a tree or bush, copulates, and leaves. The female gains no material benefit, such as access to water, by permitting copulation to occur, for males are not territorial at the stream. Instead, they patrol the area with other males in broadly overlapping home ranges. The only apparent benefit to the female from mating appears to be avoidance of the cost of trying to escape from a male. A wasp that tried to evade males on her way to water and then fled at the approach of a male, or struggled with the male that captured her, would in all likelihood expend more energy and lose more time than a female that quietly accepted the 2 to 3 minutes required for complete copulation (Smith and Alcock, 1980). Thus it may simply be cheaper in some cases to mate than to try to prevent it from happening. There may also be a considerable convenience component in the frequent matings of the continuously receptive *Anthidium* bees (Alcock, Eickwort, and Eickwort, 1977) and pentatomid bugs (Harris and Todd, 1980b).

Conclusion

We have attempted in this and earlier chapters to analyze the diversity of insect mating systems from the perspective of individual selection theory. The effort to determine how an individual can gain superior genetic representation in the next generation by, for example, being continuously receptive, or by guarding a mate after copulating, or by offering a mate food strikes us as a highly productive venture. We recognize that the hypotheses presented here are, for the most part, working hypotheses that are not thoroughly tested. But the ideas we have described have the virtue of being consistent with modern evolutionary theory and can potentially be subjected to more rigorous tests. Given the past successes of the application of Darwinian thinking to problems in biology, we are confident that many further advances in the understanding of insect biology will be developed within this framework.

The other side of the coin is that advances in the study of insect behavior are very likely to contribute to further development of evolutionary theory itself. The wonderful diversity in the behavior and ecology of insects provides an unparalleled opportunity to examine the impact of evolution on behavior. It is probably not coincidental that admirers of insects have been among the most notable of evolutionary biologists. Charles Darwin began his career as an amateur beetle collector and was an enthusiastic bumble bee observer. In the past 20 years the modern Darwinians—R. D. Alexander, R. Dawkins, W. D. Hamilton, G. A. Parker, and E. O. Wilson—have each developed important new evolutionary ideas and syntheses based on their knowledge of and fondness for insect behavioral biology. There appears to be positive feedback between research on insect behavior and studies of evolutionary processes. The potential for continuing development of this relationship is high. We will be gratified indeed if our book encourages others to explore the evolutionary basis of insect reproductive behavior.

References

Adams, J. B., and J. W. McAllan. 1958. Pectinase in certain insects. *Canadian Journal of Zoology* 36:305–308.

Adams, J. E., D. Rothman, W. E. Kerr, and Z. L. Paulino. 1977. Estimation of sex alleles and queen matings from diploid male frequencies in a population of *Apis mellifera*. *Genetics* 86:583–596.

Adams, T. S., and M. S. Mulla. 1968. Ovarian development, pheromone production and mating in the eye gnat, *Hippelates collusor*. *Journal of Insect Physiology* 14:627–635.

Addicott, J. F. 1979. On the population biology of aphids. *American Naturalist* 114:760–762.

Afzelius, B. A., B. Baccetti, and R. Dallai. 1976. The giant spermatozoan of *Notonecta*. *Journal of Submicroscopic Cytology* 8:149–161.

Aiken, R. B. 1981. The relationship between body weight and homosexual mounting in *Palmacorixa nana* Walley (Heteroptera: Corixidae). *Florida Entomologist* 64:267–271.

Alcock, J. 1972. The behavior of a stinkbug, *Euschistus conspersus* Uhler (Hemiptera: Pentatomidae). *Psyche* 78:215–228.

———— 1973. The mating behavior of *Empis barbatoides* Melander and *Empis poplitea* Loew (Diptera: Empididae). *Journal of Natural History* 7:411–420.

———— 1975. Territorial behavior by males of *Philanthus multimaculatus* (Hymenoptera: Sphecidae) with a review of territoriality in male sphecids. *Animal Behaviour* 23:889–895.

———— 1976. Courtship and mating in *Hippomelas planicosta* (Coleoptera: Buprestidae). *Coleopterist's Bulletin* 30:343–348.

———— 1978. Notes on male mate-locating behavior in some bees and wasps of Arizona (Hymenoptera: Anthophoridae, Pompilidae, Sphecidae, Vespidae). *Pan-Pacific Entomologist* 54:215–225.

———— 1979a. *Animal behavior: an evolutionary approach.* 2nd ed. Sunderland, Massachusetts: Sinauer.

———— 1979b. Selective mate choice by females of *Harpobittacus australis* (Mecoptera: Bittacidae). *Psyche* 86:213–217.

———— 1979c. The evolution of intraspecific diversity in male reproductive strategies in some bees and wasps. In *Sexual selection and reproductive competition in insects,* ed. M. S. Blum and N. A. Blum. New York: Academic Press.

———— 1979d. The relation between female body size and provisioning behavior in the bee *Centris pallida. Journal of the Kansas Entomological Society* 52:623–632.

———— 1979e. Multiple mating in *Calopteryx maculata* (Odonata: Calopterygidae) and the advantage of non-contact guarding by males. *Journal of Natural History* 13:439–446.

———— 1979f. The behavioural consequences of size variation among males of the territorial wasp *Hemipepsis ustulata* (Hymenoptera: Pompilidae). *Behaviour* 71: 322–335.

———— 1981a. Lek territoriality in a tarantula hawk wasp *Hemipepsis ustulata* (*Hymenoptera: Pompilidae*). *Behavioral Ecology and Sociobiology* 8: 309–317.

———— 1981b. Notes on the reproductive behavior of some Australian thynnine wasps (Hymenoptera: Tiphiidae). *Journal of the Kansas Entomological Society* 54:681–693.

———— 1981c. Seduction on the wing. *Natural History* 90:36–41.

———— 1982a. Natural selection and communication in bark beetles. *Florida Entomologist* 65:17–32.

———— 1982b. Post-copulatory mate guarding by males of the damselfly *Hetaerina vulnerata* Selys (Odonata: Calopterygidae). *Animal Behaviour* 30:99–107.

Alcock, J., and S. L. Buchmann. Manuscript. The significance of post-insemination display by males of *Centris pallida* (Hymenoptera: Anthophoridae).

Alcock, J., and D. W. Pyle. 1979. The complex courtship behavior of *Physiphora demandata* (F.) (Diptera: Otitidae). *Zeitschrift für Tierpsychologie* 49:352–362.

Alcock, J., and J. E. Schaefer. 1983. Hill top territoriality in a Sonoran Desert bot fly (Diptera: Cuterebridae). *Animal Behaviour.*

Alcock, J., G. C. Eickwort, and K. R. Eickwort. 1977. The reproductive behavior of *Anthidium maculosum* (Hymenoptera: Megachilidae) and the evolutionary significance of multiple copulations by females. *Behavioral Ecology and Sociobiology* 2:385–396.

Alcock, J., C. E. Jones, and S. L. Buchmann. 1976. Location before emergence of the female bee, *Centris pallida,* by its male (Hymenoptera; Anthophoridae). *Journal of Zoology* 179:189–199.

———— 1977. Male mating strategies in the bee *Centris pallida* Fox (Hymenoptera: Anthophoridae). *American Naturalist* 111:145–155.

Alcock, J., E. M. Barrows, G. Gordh, L. J. Hubbard, L. L. Kirkendall, D. Pyle, T.L. Ponder, and F. G. Zalom. 1978. The ecology and evolution of male reproductive behaviour in the bees and wasps. *Zoological Journal of the Linnean Society of London* 64:293–326.

Alexander, R. D. 1961. Aggressiveness, territoriality and sexual behavior in field crickets (Orthoptera: Gryllidae). *Behaviour* 17:130–223.

———— 1962. Evolutionary change in cricket acoustical communication. *Evolution* 16:443–467.

———— 1964. The evolution of mating behavior in arthropods. *Royal Entomological Society of London Symposium* 2:78–94.

———— 1967. Acoustical communication in arthropods. *Annual Review of Entomology* 12:495–526.

———— 1969. Comparative animal behavior and systematics. In *Systematic biology: proceedings of an international conference*. Washington, D. C.: National Academy of Sciences.

———— 1975. Natural selection and specialized chorusing behavior in acoustical insects. In *Insects, science and society*, ed. D. Pimental. New York: Academic Press.

———— 1977. Evolution, human behavior, and determinism. *Proceedings Biennial Meeting of the Philosophy of Science Association* (1976) 2:3–21.

———— 1979. *Darwinism and human affairs*. Seattle: University of Washington Press.

Alexander, R. D., and G. Borgia, 1978. Group selection, altruism and the levels of organization of life. *Annual Review of Ecology and Systematics* 9:449–474.

———— 1979. On the origin and basis of the male-female phenomenon. In *Sexual selection and reproductive competition in insects*, ed. M. S. Blum and N. A. Blum. New York: Academic Press.

Alexander, R. D., and W. L. Brown. 1963. Mating behavior and the origin of insect wings. *Occasional Papers Museum of Zoology, University of Michigan* 628:1–19.

Alexander, R. D., and T. E. Moore, 1962. The evolutionary relationships of 17-year and 13-year cicadas and three new species (Homoptera, Cicadidae, *Magicicada*). *Miscellaneous Publications of the Museum of Zoology, University of Michigan*, 121:1–59.

Alexander, R. D., and D. Otte. 1967. The evolution of genitalia and mating behavior in crickets (Gryllidae) and other Orthoptera. *Miscellaneous Publications of the Museum of Zoology, University of Michigan* 133:1–62.

Alexander, R. D., and P. W. Sherman. 1977. Local mate competition and parental investment in social insects. *Science* 196:494–500.

Alexander, R. D., J. L. Hoogland, R. D. Howard, K. M. Noonan, and P. W. Sherman. 1979. Sexual dimorphism and breeding systems in pinnipeds, ungulates, primates and humans. In *Evolutionary biology and human social behavior: an anthropological perspective*, ed. N. A. Chagnon and W. Irons. North Scituate, Massachusetts: Duxbury Press.

Alford, D. V. 1975. *Bumblebees*. London: Avis-Poynter.

Allen, N. 1955. Mating habits of the tobacco hornworm. *Journal of Economic Entomology* 48:526–528.

Anderson, J. R. 1974. Symposium on reproduction of arthropods of medical and veterinary importance. II. Meeting of the sexes. *Journal of Medical Entomology* 11:7–19.

Andersson, M. 1980. Why are there so many threat displays? *Journal of Theoretical Biology* 86:773–781.

Antonovics, J. 1976. The nature of limits to natural selection. *Annals of the Missouri Botanical Garden* 63:224–247.

Anzenberger, G. 1977. Ethological study of African carpenter bees of the genus *Xylocopa* (Hymenoptera: Anthophoridae). *Zeitschrift für Tierpsychologie* 44:337–374.

Arrow, G. J. 1951. *Horned beetles*. The Hague: Dr. W. Junk.

Ashraf, M., and A. A. Berryman. 1969. Biology of *Scolytus ventralis* (Coleoptera: Scolytidae) attacking *Abies grandis* (Pinaceae) in northern Idaho. *Melanderia* 2:1–22.

van den Assem, J. 1970. Courtship and mating in *Lariophagus distinguendus* (Först) Kurdj. (Hymenoptera: Pteromalidae). *Netherlands Journal of Zoology* 20:329–352.

van den Assem, J., and G. D. E. Povel. 1973. Courtship behaviour of some *Muscidifurax* species (Hym., Pteromalidae), a possible example of a recently evolved ethological isolating mechanism. *Netherlands Journal of Zoology* 23:329–352.

van den Assem, J., and F. A. Putters. 1980. Patterns of sound produced by courting chalcidoid males and its biological significance. *Entomologia Experimentalis et Applicata* 27:293–302.

van den Assem, J., M. J. Gijswijt, and B. K. Nubel. 1980. Observation on courtship and mating strategies in a few species of parasitic wasps (Chalcidoidea). *Netherlands Journal of Zoology* 30:208–227.

van den Assem, J., F. Jachmann, and P. Simbolotti. 1980. Courtship behaviour of *Nasonia vitripennis* (Hymenoptera: Pteromalidae): some qualitative experimental evidence for the role of pheromones. *Behaviour* 75:301–307.

Austad, S. N., W. T., Jones, and P. M. Waser. 1979. Territorial defence in speckled wood butterflies: why does the resident always win? *Animal Behaviour* 27:960–961.

Axelrod, R., and W. D. Hamilton. 1981. The evolution of cooperation. *Science* 211:1390–96.

Baccetti, B., and B. A. Afzelius. 1976. *The biology of the sperm cell*. Basel: S. Karger.

Baker, R. R. 1972. Territorial behaviour of the nymphalid butterflies, *Aglais urticae* and *Inachis io*. *Journal of Animal Ecology* 41:453–469.

Baker, T. C., and R. T. Cardé. 1979. Endogenous and exogenous factors affecting prereproductive periodicity of female calling and male sex pheromone response in *Grapholita modesta* (Busck). *Journal of Insect Physiology* 25:943–950.

Baker, T. C., R. Nishida, and W. L. Roelofs. 1981. Close-range attraction of female oriental fruit moths to herbal scent of male hairpencils. *Science* 214:1359–61.

Baldwin, F. T., and E. H. Bryant. 1981. Effect of size upon mating performance in geographic strains of the housefly, *Musca domestica* L. *Evolution* 35:1134–41.

Banegas, A. D., and H. Mourier. 1967. Laboratory observation on the life history and habits of *Dermatobia hominis* (Diptera: Cuterebridae). I. Mating behavior. *Annals of the Entomological Society of America* 60:878–881.

Barfield, C. S., and W. W. Gibson. 1975. Observations of the life history of *Hypothyce mixta* Howden (Coleoptera: Scarabaeidae). *Coleopterist's Bulletin* 29:251–256.

Barnard, C. J., and T. Burk. 1979. Dominance hierarchies and the evolution of "individual recognition." *Journal of Theoretical Biology* 81:65–73.

Barrass, R. 1979. The survival value of courtship in insects. In *Sexual selection and reproductive competition in insects,* ed. M. S. Blum and N. A. Blum. New York: Academic Press.

Barrows, E. M. 1975a. Individually distinctive odors in an invertebrate. *Behavioral Biology* 15:57–64.

———— 1975b. Mating behavior in halictine bees (Hymenoptera: Halictidae). III. Copulatory behavior and olfactory communication. *Insectes Sociaux* 22:307–332.

Barrows, E. M., W. J. Bell, and C. D. Michener. 1975. Individual odor differences and their social functions in insects. *Proceedings of the National Academy of Sciences U.S.A.* 72:2824–28.

Barrows, E. M., M. R. Chabot, C. D. Michener, and T. P. Snyder. 1976. Foraging and mating behavior in *Perdita texana* (Hymenoptera: Andrenidae). *Journal of the Kansas Entomological Society* 49:275–279.

Barth, R. H., Jr., 1964. Mating behavior of *Byrsotria fumigata* (Guerin) (Blattidae, Blaberinea). *Behaviour* 23:1–30.

———— 1968. The mating behavior of *Gromphadorhina portentosa* (Schaum) (Blattaria, Blaberoidae, Blaberidae, Oxyhaloinae): an anomalous pattern for a cockroach. *Psyche* 75:124–131.

Barton-Browne, L. 1958a. The frequency of mating of the Australian sheep blow fly, *Lucilia cuprina. Australian Journal of Science* 20:185.

———— 1958b. The relation between ovarian development and mating in *Lucilia cuprina. Australian Journal of Science* 20:239–240.

Bastock, M. 1956. A gene mutation which changes a behavior pattern. *Evolution* 10:421–439.

Bateman, A. J. 1948. Intra-sexual selection in *Drosophila. Heredity* 2:349–368.

Beatty, R. A. 1975. Genetics of animal spermatozoa. In *Gamete competition in plants and animals,* ed. D. L. Mulcahy. Amsterdam: North Holland.

Beatty, R. A., and P. S. Burgoyne. 1971. Size classes of the head and flagellum of *Drosophila* spermatozoa. *Cytogenology* 10:177–189.

Beer, F. M. 1970. Notes and observations on the Buprestidae (Coleoptera). *Coleopterist's Bulletin* 24:39–41.

Beeson, C. F. C., and B. M. Bhatia. 1939. On the biology of the Cerambycidae (Coleopt.). *Indian Forest Records, Entomology* 5:1–235.

Bell, G. 1978. The handicap principle in sexual selection. *Evolution* 32:872–885.

Bell, P. D. 1979a. Acoustic attraction of herons by crickets. *Journal of the New York Entomological Society* 87:126–127.

———— 1979b. Mate choice and mating behavior in the black-horned tree cricket, *Oecanthus nigricornis* (Walker). Master's thesis, University of Toronto.

———— 1980a. Multimodal communication by the black-horned tree cricket, *Oecanthus nigricornis* (Walker) (Orthoptera: Gryllidae). *Canadian Journal of Zoology* 58:1861–68.

———— 1980b. Transmission of vibrations along plant stems: implications for insect communication. *Journal of the New York Entomological Society* 88:210–216.

Bellinger, P. F. 1954. Attraction of zebra males by female pupae. *Journal of the Lepidopterist's Society* 8:102.

Bennet-Clark, H. C. 1970. The mechanism and efficiency of sound production in mole crickets. *Journal of Experimental Biology* 52:619–652.

Bennet-Clark, H. C., and A. W. Ewing. 1970. The lovesong of the fruit fly. *Scientific American* 223:84–93.

Bennet-Clark, H. C., Y. Leroy, and L. Tsacas. 1980. Species and sex-specific songs and courtship behaviour in the genus *Zapronius* (Diptera: Drosophilidae). *Animal Behaviour* 28:230–255.

Bennett, G. F. 1961. On three species of Hippoboscidae (Diptera) on birds in Ontario. *Canadian Journal of Zoology* 39:379–406.

Bentur, J. F., and S. B. Mathad. 1975. The dual role of mating in egg production and survival in the cricket, *Plebiogryllus guttiventris* Walker. *Experimentia* 31:539–540.

Bequaert, J. C. 1953. The Hippoboscidae or louse-flies (Diptera) of mammals and birds. Part I. Structure, physiology and natural history. *Entomologica Americana* 32:1–209.

Bergström, G., and J. Tengö. 1978. Linalool in mandibular gland secretion of *Colletes* bees (Hymenoptera: Apoidea). *Journal of Chemical Ecology* 4:437–449.

Bernstein, H., G. S. Byers, and R. E. Michod. 1981. Evolution of sexual reproduction: importance of DNA repairs, complementation and variation. *American Naturalist* 117:537–549.

Bick, G. H., and J. C. Bick. 1963. Behavior and population structure of the damselfly *Enallagma civile* (Hagen) (Odonata: Coenagrionidae). *Southwestern Naturalist* 8:57–84.

Birch, M. C. 1978. Chemical communication in pine bark beetles. *American Scientist* 66:409–419.

Birch, M. C., and D. L. Wood. 1975. Mutual inhibition of the attractant pheromone response by two species of *Ips* (Coleoptera: Scolytidae). *Journal of Chemical Ecology* 1:101–113.

Blickle, R. L. 1959. Observations on the hovering and mating of *Tabanus bishoppi* Stone. *Annals of the Entomological Society of America* 52: 183–190.

Blum, M. S., and N. A. Blum, eds. 1979. *Sexual selection and reproductive competition in insects.* New York: Academic Press.

Blum, M. S., S. Z. Glowska, and S. Taber III. 1962. Chemistry of the drone system. II. Carbohydrates in the reproductive organs and semen. *Annals of the Entomological Society of America* 55:135–139.

Boggs, C. L. 1981a. Selection pressures affecting male nutrient investment at mating in heliconiine butterflies. *Evolution* 35:931–940.

——— 1981b. Nutritional and life history determinants of resource allocation in holometabolous insects. *American Naturalist* 117:692–709.

Boggs, C. L., and L. E. Gilbert. 1979. Male contribution to egg production in butterflies: evidence for transfer of nutrients at mating. *Science* 206:83–84.

Boggs, C. L., and W. Watt. 1981. Population structure of pierid butterflies. IV. Genetic and physiological investment in offspring by male *Colias*. *Oecologia* 50:320–324.

Bohart, G. E., and E. A. Cross. 1955. Time relations in the nest construction and life cycle of the alkali bee. *Annals of the Entomological Society of America* 48:403–406.

Bohart, G. E., W. P. Stephen, and R. K. Eppley. 1960. The biology of *Heterostylum robustum* (Diptera: Bombyliidae), a parasite of the alkali bee. *Annals of the Entomological Society of America* 53:425–436.

Bohart, G. E., P. F. Torchio, Y. Maeta, and R. W. Rust. 1972. Notes on the biology of *Emphoropsis pallida* Timberlake. *Journal of the Kansas Entomological Society* 45:381–392.

Boldyrev, B. T. 1915. Contributions à l'étude de la structure des spermatophores et des particularités de la copulation chez Locustodea et Gryllodes. *Horae Societatis Entomologicae Rossicae* 6:1–245.

Bonhag, P. F., and J. R. Wick. 1953. The functional anatomy of the male and female reproductive systems of the milkweed bug, *Oncopeltus fasciatus* (Pallas) (Heteroptera: Lygaeidae). *Journal of Morphology* 93:177–283.

Booij, C. J. H. 1982. Biosystematics of the *Muellerianella* complex (Homoptera, Delphacidae), interspecific and geographic variation in acoustic behaviour. *Zeitschrift für Tierpsychologie* 58:31–52.

Boorman, E., and G. A. Parker. 1976. Sperm (ejaculate) competition in *Drosophila melanogaster,* and the reproductive value of females to males in relation to female age and mating status. *Ecological Entomology* 1: 145–155.

Boppré, M. 1981. Adult Lepidoptera "feeding" at withered *Heliotropium* plants (Boraginaceae) in East Africa. *Ecological Entomology* 6: 449–452.

Borden, J. H. 1967. Factors influencing the response of *Ips confusus* (Coleoptera: Scolytidae) to male attractant. *Canadian Entomologist* 99: 1164–93.

Borgia, G. 1979. Sexual selection and the evolution of mating systems. In *Sexual selection and reproductive competition in insects,* ed. M. S. Blum and N. A. Blum. New York: Adademic Press.

———— 1980. Sexual competition in *Scatophaga stercoraria:* Size- and density-related changes in male ability to capture females. *Behaviour* 75:185–206.

———— 1981. Mate selection in the fly *Scatophaga stercoraria:* female choice in a male-controlled system. *Animal Behaviour* 29:71–80.

———— 1982. Experimental changes in resource structure and male density: size-related differences in mating success among male *Scatophaga stercoraria. Evolution* 36:307–315.

Bornemissza, G. F. 1966a. Specificity of male sex attractants in some Australian scorpionflies. *Nature* 209:723–733.

———— 1966b. Observations on the hunting and mating behaviour of two species of scorpionflies (Bittacidae: Mecoptera). *Australian Journal of Zoology* 14:371–382.

Borror, D. J., and R. E. White. 1970. *A field guide to the insects of America north of Mexico.* Boston: Houghton Mifflin.

Boyce, A. M. 1934. Bionomics of the walnut husk fly, *Rhagoletis completa. Hilgardia* 8:363–579.

Bradbury, J. W. 1977. Lek mating behavior in the hammerheaded bat. *Zeitschrift für Tierpsychologie* 45:225–255.

———— 1981. The evolution of leks. In *Natural selection and social behavior:*

recent research and theory, ed. R. D. Alexander and D. W. Tinkle. New York: Chiron Press.

Bradbury, J. W., and Vehrencamp, S. L. 1977. Social organization and foraging in emballonurid bats. III. Mating systems. *Behavioral Ecology and Sociobiology* 2:1–17.

Breed, M. D., and C. D. Rasmussen. 1980. Behavioural strategies during intermale agonistic interactions in a cockroach. *Animal Behaviour* 28: 1063–1069.

Breed, M. D., S. K. Smith, and B. G. Gall. 1980. Systems of mate selection in a cockroach species with male dominance hierarchies. *Animal Behaviour* 28:130–134.

Breed, M. D., C. Meaney, D. Deuth, and W. J. Bell. 1981. Agonistic interactions of two cockroach species, *Gomphadorhina portentosa* and *Supella longipalpa* (Orthoptera [Dictyoptera]: Blaberidae, Blatellidae). *Journal of the Kansas Entomological Society* 54:197–208.

Britton, E. B. 1970. Coleoptera. In *Insects of Australia*, ed. D. F. Waterhouse. Melbourne: Melbourne University Press.

Brockmann, H. J. 1980. Diversity in the nesting behavior of mud-daubers (*Trypoxylon politum* Say: Sphecidae). *Florida Entomologist* 63:53–64.

———— Manuscript. Mate conflict over offspring sex in a mud-daubing wasp.

Brockmann, H. J., A. Grafen, and R. Dawkins. 1979. Evolutionarily stable nesting strategy in a digger wasp. *Journal of Theoretical Biology* 77:473–496.

Brome, L. B., R. J. Bartell, A. C. M. van Gerwen, and L. A. Lawrence. 1976. Relationship between protein ingestion and sexual receptivity in females of the Australian sheep blowfly *Lucilia cuprina*. *Physiological Entomology* 1:235–240.

Brossut, R., P. Dubois, J. Rigaud, and L. Sreng. 1975. Étude biochemique de la secretion des glandes tergales des Blatteria. *Insect Biochemistry* 5:719–732.

Brower, L. P. 1963. The evolution of sex-limited mimicry in butterflies. *Proceedings of the 16th International Congress of Zoology* 4:173–179.

Brower, L. P., J. V. Z. Brower, and F. P. Cranston. 1965. Courtship behavior of the queen butterfly, *Danaus gilippus berenice* (Cramer). *Zoologica* 50:1–39.

Brown, J. L., and G. H. Orians. 1970. Spacing patterns in mobile animals. *Annual Review of Ecology and Systematics* 1:239–262.

Brown, L. 1980. Aggression and mating success in males of the forked fungus beetle, *Bolitotherus cornutus* (Panzer) (Coleoptera: Tenebrionidae). *Proceedings of the Entomological Society of Washington* 82:430–434.

Bryant, E. H. 1980. Geographic variation in components of mating success of the housefly *Musca domestica* L., in the United States. *American Naturalist* 116:655–669.

Buck, J. B., and E. Buck. 1966. Biology of synchronous flashing fireflies. *Nature* 211:562–564.

———— 1968. Mechanism of rhythmic synchronous flashing of fireflies. *Science* 159:1319–27.

Buckell, E. R. 1928. Notes on the life history and habits of *Melittobia chalybii* Ashmead (Chalcidoidea: Elachertidae). *Pan-Pacific Entomologist* 5: 14–22.

Burk, T. E. 1981. Signaling and sex in acalyptrate flies. *Florida Entomologist* 64:30–43.

———— 1982. Evolutionary significance of predation on sexually signalling males. *Florida Entomologist* 65:90–104.

———— Forthcoming. Female choice and male aggression in a field cricket (*Teleogryllus oceanicus*): the importance of courtship song. In *Orthopteran mating systems: sexual competition in a diverse group of insects*, ed. D. T. Gwynne and G. K. Morris. Boulder, Colorado: Westview Press.

———— In press. Behavioral ecology of mating in the Caribbean fruit fly, *Anastrepha suspensa* (Loew) (Diptera: Tephritidae). *Florida Entomologist*.

Burnet, B., K. Connolly, M. Kearney, and R. Cook. 1973. Effects of male paragonial gland secretion on sexual receptivity and courtship behavior of female *Drosophila melanogaster*. *Journal of Insect Physiology* 19: 2421–31.

Burns, J. M. 1968. Mating frequency in natural populations of skippers and butterflies as determined by spermatophore counts. *Proceedings of the National Academy of Sciences U.S.A.* 61:852–859.

Bush, G. L. 1966. The taxonomy, cytology, and evolution of the genus *Rhagoletis* in North America (Diptera, Tephritidae). *Bulletin of the Museum of Comparative Zoology, Harvard University* 134:431–562.

Busnel, R-G. 1963. On certain aspects of animal acoustic signals. In *Acoustic behaviour of animals*, ed. R-G. Busnel. New York: Elsevier.

Busnel, R-G., and B. Dumortier. 1955. Étude du cycle génital du mâle d'*Ephippiger* (Orthoptères) a des signaux acoustiques synthétiques. *Comptes Rendus de la Société Biologie* 149:11.

Busnel, R-G., F. Pasquinelly, and B. Dumortier. 1955. La trémulation du corps et la transmission aux supports des vibrations en résultant comme moyen d'information acourte portée des Ephippigères mâles et femelles. *Bulletin du Société Zoologique Française* 80:18–22.

Byers, G. W., and R. Thornhill. 1983. Biology of the Mecoptera. *Annual Review of Entomology* 28:203–228.

Byers, J. R. 1978. Biosystematics of the genus *Eoxoa* (Lepidoptera: Noctuidae). X. Incidence and level of multiple mating in natural and laboratory populations. *Canadian Entomologist* 110:193–200.

Cade, W. 1975. Acoustically orienting parasitoids: fly phonotaxis to cricket song. *Science* 190:1312–13.

———— 1979a. Effect of male-deprivation on female phonotaxis in field crickets (Orthoptera: Gryllidae). *Canadian Entomologist* 111:741–744.

———— 1979b. The evolution of alternative male reproductive strategies in field crickets. In *Sexual selection and reproductive competition in insects*, ed. M. S. Blum and N. A. Blum. New York: Academic Press.

———— 1980. Alternative male reproductive behaviors. *Florida Entomologist* 63:30–44.

———— 1981a. Field cricket spacing, and the phonotaxis of crickets and parasitoid flies to clumped and isolated cricket songs. *Zeitschrift für Tierpsychologie* 55:365–375.

———— 1981b. Alternative male strategies: genetic differences in crickets. *Science* 212:563–564.

Camenzind, R. 1962. Untersuchungen über die bisexuelle Fortpflanzung einer paedeogenetischen Gallmucke. *Revue Suisse de Zoologie* 69: 377–384.

Campanella, P. J. 1975. The evolution of mating systems in temperate zone dragonflies (Odonata: Anisoptera). II. *Libellula luctuosa* (Burmeister). *Behaviour* 54:278–309.

Campanella, P. J., and L. L. Wolf. 1974. Temporal leks as a mating system in a temperate zone dragonfly (Odonata: Anisoptera). I. *Plathemis lydia* (Drury). *Behaviour* 51:49–87.

Campbell, D. J., and E. Shipp. 1979. Regulation of spatial pattern in populations of the field cricket *Teleogryllus commodus* (Walker). *Zeitschrift für Tierpsychologie* 51:260–268.

Capinera, J. L. 1979. Qualitative variation in plants and insects: effect of propagule size on ecological plasticity. *American Naturalist* 114:350–361.

Carayon, J. 1964. Un cas d'offrande nuptiale chez les Heteroptères. *Comptes Rendus Hebedomadiares des Séances de l'académie des Sciences* 259:4815–18.

Carpenter, F. L., and R. E. MacMillen. 1976. Threshold model of feeding territoriality and test with a Hawaiian honeycreeper. *Science* 194:639–641.

Carson, H. L. 1978. Speciation and sexual selection in Hawaiian *Drosophila*. In *Ecological genetics: the interface,* ed. P. F. Bussard. New York: Springer-Verlag.

Catts, E. P. 1964. Field behavior of adult *Cephenemyia* (Diptera: Oestridae). *Canadian Entomologist* 96:579–585.

———— 1967. Biology of a California rodent bot fly *Cuterebra latifrons* Coquillett (Diptera: Cuterebridae). *Journal of Medical Entomology* 4:87–101.

———— 1979. Hilltop aggregation and mating behavior in *Gasterophilus* (Diptera: Gasterophilidae). *Journal of Medical Entomology* 16:461–464.

———— 1982. Biology of New World bot flies: Cuterebridae. *Annual Review of Entomology* 27:313–338.

Catts, E. P., and W. Olkowski. 1972. Biology of Tabanidae (Diptera): mating and feeding behavior of *Chrysops fuliginosus*. *Environmental Entomology* 1:448–453.

Cazier, M. A ., and E. G. Linsley. 1963. Territorial behavior among males of *Protoxaea gloriosa* (Fox) (Hymenoptera: Andrenidae). *Canadian Entomologist* 95:547–556.

Cazier, M. A., and M. A. Mortenson. 1964. Bionomical observations on tarantula hawks and their prey (Hymenoptera: Pompilidae: *Pepsis*). *Annals of the Entomological Society of America* 57:533–541.

Chapman, J. A. 1954. Studies on summit frequenting insects in western Montana. *Ecology* 35:41–49.

Chapman, R. F. 1971. *The insects: structure and function.* Elsevier: New York.

Charnov, E. L. 1978. Sex-ratio selection in eusocial Hymenoptera. *American Naturalist* 112:317–326.

Chopard, L. 1950. Sur l'anatomie et le dévelopement d'une blatte vivipare. *Proceedings of the 8th International Congress of Entomology, Stockholm* 1948:218–222.

Christenson, L. D., and R. H. Foote. 1960. Biology of fruit flies. *Annual Review of Entomology* 5:171–192.

Christiansen, K. 1964. Bionomics of Collembola. *Annual Review of Entomology* 9:147–178.

Chvála, M. 1978. The modified mating behaviour of *Empis albicans* Meig., with notes on the presumed origin of prey presentation. *Dipterologica Bohemoslovaca (Bratislava)* 1:55–67.

———— 1980. Swarming rituals in two *Empis* and one *Bicellaria* species. *Acta Entomologica Bohemoslovaca* 77:1–15.

Clausen, C. P. 1923. The biology of *Schizaspidia tenuicornis* Ashm., a eucharid parasite of *Camponotus*. *Annals of the Entomological Society of America* 16:195–217.

Clement, S. L., and R. P. Meyer. 1980. Adult biology and behavior of the dragonfly *Tanypteryx hageni* (Odonata: Petaluridae). *Journal of the Kansas Entomological Society* 53:711–719.

Clutton-Brock, T. H ., and P. H. Harvey. 1977. Primate ecology and social organization. *Journal of Zoology* 183:1–39.

Cognetti, G. 1961. Endomeiosis in parthenogenetic lines of aphids. *Experimentia* 17:168–169.

Cohen, J. 1971. Comparative physiology of gamete populations. *Advances in Comparative Physiology and Biochemistry* 4:268–380.

——— 1975. Gamete redundancy-wastage or selection? In *Gamete competition in plants and animals,* ed. D. L. Mulcahy. Amsterdam: North Holland.

——— 1977. *Reproduction.* London: Butterworth.

Cole, F. R. 1970. *The flies of western North America.* Berkeley: University of California Press.

Colville, R. E., and P. L. Colville. 1980. Nesting biology and male behavior of *Trypoxylon* (*Trypargilum*) *tenoctitlan* in Costa Rica (Hymenoptera: Sphecidae). *Annals of the Entomological Society of America* 73: 110–119.

Colwell, R. K. 1981. Group selection is implicated in the evolution of female-biased sex ratios. *Nature* 290:401–404.

Conner, W. E., T. Fisher, R. K. Vander Meer, A. Guerrero, D. Ghiringe, and J. Meinwald. 1980. Sex attractant of an arctiid moth (*Utetheisa ornatrix*): a pulsed chemical signal. *Behavioral Ecology and Sociobiology* 7:55–63.

——— 1981. Precopulatory sexual interaction in an arctiid moth (*Utetheisa ornatrix*): role of a pheromone derived from dietary alkaloids. *Behavioral Ecology and Sociobiology* 9:227–235.

Constantz, G. 1973. The mating behavior of a creeping water bug, *Ambrysus occidentalis* (Hemiptera: Naucoridae). *American Midland Naturalist* 92:230–239.

Corbet, P. S. 1957. The life history of the emperor dragonfly *Anax imperator* Leach (Odonata: Aeshnidae). *Journal of Animal Ecology* 26:1–69.

——— 1962. *A biology of dragonflies.* London: M. F. and G. Witherby.

——— 1980. Biology of Odonata. *Annual Review of Entomology* 25:189–217.

Corbet, P. S., and A. J. Haddow. 1962. Diptera swarming high above the forest canopy in Uganda, with special reference to Tabanidae. *Transactions of the Royal Entomological Society of London* 114:267–284.

Coulson, R.N. 1979. Population dynamics of bark beetles. *Annual Review of Entomology* 24:417–447.

Cowan, D. P. 1978. Behavior, inbreeding, and parental investment in solitary eumenid wasps (Hymenoptera: Vespidae). Ph.D. dissertation, University of Michigan.

——— 1979. Sibling matings in a hunting wasp: adaptive inbreeding? *Science* 205:1403–5.

——— 1981. Parental investment in two solitary wasps *Ancistrocerus adiabatus* and *Euodynerus foraminatus*. *Behavioral Ecology and Sociobiology* 9:95–102.

Cowan, I. M. 1943. Notes on the life history and morphology of *Cephenemyia jellisoni* Townsend and *Lipoptena depressa* Say, two dipterous parasites of the Columbian black-tailed deer (*Odocoileus hemionus columbianus* Richardson). *Canadian Journal of Research Section D* 21:171–187.

Cox, C. R., and B. J. LeBoeuf. 1977. Female incitation of male competition: a mechanism in sexual selection. *American Naturalist* 111:317–335.

Craig, G. B., Jr. 1967. Mosquitoes: female monogamy induced by male accessory gland substance. *Science* 156:1499–1501.

Crampton, G. C. 1931. The genitalia and terminal structures of the male of the archaic mecopteron, *Notiothauma reedi*, compared with related holometabola from the standpoint of phylogeny. *Psyche* 38:1–21.

Crankshaw, O. W., and R. W. Matthews. 1981. Sexual behavior among parasitic *Megarhyssa* wasps (Hymenoptera: Ichneumonidae). *Behavioral Ecology and Sociobiology* 9:1–8.

Crawford, C. S., and M. M. Dadone. 1980. Onset of evening chorus in *Tibicen marginalis* (Homoptera, Cicadidae). *Environmental Entomology* 8:1157–60.

Cross, E. A., M. G. Stith, and T. R. Bauman. 1975. Bionomics of the organ-pipe mud-dauber, *Trypoxylon politum* (Hymenoptera: Sphecidae). *Annals of the Entomological Society of America* 68:901–916.

Cross, W. H. 1973. Biology, control and eradication of the boll weevil. *Annual Review of Entomology* 18:17–46.

Cross, W. H., and H. C. Mitchell. 1966. Mating behavior of the female boll weevil. *Journal of Economic Entomology* 59:1503–7.

Crow, J. F. 1979. Genes that violate Mendel's rules. *Scientific American* 240:134–146.

Crozier, R. H. 1977. Evolutionary genetics of the Hymenoptera. *Annual Review of Entomology* 22:263–288.

———— 1979. Genetics of sociality. In *Social insects,* ed. H. R. Hermann. New York: Academic Press.

Cullen, E. 1957. Adaptations in the kittiwake to cliff nesting. *Ibis* 99:275–302.

Cullen, M. J. 1969. The biology of giant waterbugs (Hemiptera: Belostomatidae) in Trinidad. *Proceedings of the Royal Entomological Society of London* (A) 44:123–137.

Cunningham, R. T., G. J. Farias, S. Nakagawa, and D. L. Chambers. 1971. Reproduction in the Mediterranean fruitfly: depletion of stored sperm in females. *Annals of the Entomological Society of America* 64:312–313.

Daly, M., and M. Wilson. 1978. *Sex, evolution and behavior.* North Scituate, Massachusetts: Duxbury Press.

Darwin, C. 1859. *On the origin of species.* Facsimile of the first edition, 1964. Cambridge, Massachusetts: Harvard University Press.

———— 1872. *The expression of the emotions in man and animals.* London: Murray.

———— 1874. *The descent of man, and selection in relation to sex.* 2nd ed. New York: A. L. Burt.

Davey, K. G. 1958. The migration of spermatozoa in the female of *Rhodnius prolixus* Stål. *Journal of Experimental Biology* 35:694–701.

———— 1965. *Reproduction in the insects.* San Francisco: W. H. Freeman.

Davey, K. G., and G. F. Webster. 1967. The structure and secretion of the spermatheca of *Rhodnius prolixus* Stål: a histological study. *Canadian Journal of Zoology* 45:653–657.

Davies, N. B. 1978. Territorial defence in the speckled wood butterfly (*Pararge aegeria*): the resident always wins. *Animal Behavior* 26:138–147.

——— 1979. Game theory and territorial behaviour in speckled wood butterflies. *Animal Behaviour* 27:961–962.

Davis, J. W. F., and P. O'Donald. 1976. Sexual selection for a handicap: a critical analysis of Zahavi's model. *Journal of Theoretical Biology* 57:345–354.

Dawkins, R. 1976. *The selfish gene*. London: Oxford University Press.

——— 1978. Replicator selection and the extended phenotype. *Zeitschrift für Tierpsychologie* 47:61–76.

——— 1979. Twelve misunderstandings of kin selection. *Zeitschrift für Tierpsychologie* 51:184–200.

——— 1980. Good strategy or evolutionarily stable strategy? In *Sociobiology: beyond nature/nurture?* ed. G. W. Barlow and J. Silverberg. Boulder, Colorado: Westview Press.

Dawkins, R., and J. R. Krebs. 1978. Animal signals: information or manipulation? In *Behavioural ecology: an evolutionary approach,* ed. J. R. Krebs and N. B. Davies. Sunderland, Massachusetts: Sinauer.

DeLeon, D. 1935. The biology of *Coeloides dendroctoni* Cushman (Hymenoptera-Braconidae), an important parasite of the mountain pine beetle (*Dendroctonus monticolae* Hopk.). *Annals of the Entomological Society of America* 28:411–424.

DeLong, D. M. 1938. Biological studies of the leafhopper, *Empoasca fabae* as a bean pest. *U.S. Department of Agriculture Technical Bulletin* 618:1–60.

Diakonov, D. M. 1925. Experimental and biometrical investigations on dimorphic variability of *Forficula*. *Journal of Genetics* 15:201–232.

Dick, J. 1937. Oviposition in certain Coleoptera. *Annals of Applied Biology* 24:762–796.

Dobzhansky, Th. 1937. *Genetics and the origin of species*. New York: Columbia University Press.

Dodson, C. H. 1966. Ethology of some bees of the tribe Euglossini. *Journal of the Kansas Entomological Society* 39:607–629.

——— 1975. Coevolution of orchids and bees. In *Coevolution of animals and plants,* ed. L. E. Gilbert and P. H. Raven. Austin: University of Texas Press.

Dodson, G. 1978a. Behavioral, anatomical, and physiological aspects of reproduction in the Caribbean fruit fly, *Anastrepha suspensa* (Loew). Master's thesis, University of Florida.

——— 1978b. Morphology of the reproductive system in *Anastrepha suspensa* (Lowe) and notes on related species. *Florida Entomologist* 61:231–239.

——— 1982. Mating and territoriality in wild *Anastrepha suspensa* (Diptera: Tephritidae) in field cages. *Journal of the Georgia Entomological Society* 17:189–200.

Dodson, G., and L. D. Marshall. Manuscript. Mating patterns in an ambush bug (*Phymata fasciata*).

Dodson, G., G. K. Morris, and D. T. Gwynne. Forthcoming. Mating behavior in the primitive orthopteran genus *Cyphoderris* (Haglidae). In *Orthopteran mating systems: sexual competition in a diverse group of insects,* ed. D. T. Gwynne and G. K. Morris. Boulder, Colorado: Westview Press.

Dominey, W. J. 1980. Female mimicry in male bluegill sunfish—a genetic polymorphism? *Nature* 284:546–548.

Donegan, J., and A. W. Ewing. 1980. Duetting in *Drosophila* and *Zapronius* species. *Animal Behaviour* 28:1289.

Doolan, J. M. 1981. Male spacing and the influence of female courtship behavior in the bladder cicada, *Cystosoma saundersii* Westwood. *Behavioral Ecology and Sociobiology* 9:269–276.

Doolan, J. M., and Mac Nally, R. C. 1981. Population dynamics and breeding ecology in the cicada *Cystosoma saundersii:* the interaction between distributions of resources and intraspecific behaviour. *Journal of Animal Ecology* 50:925–940.

Doolittle, W. F., and C. Sapienza. 1980. Selfish genes, the phenotype paradigm, and genome evolution. *Nature* 284:601–603.

Downes, J. A. 1955. Observations on the swarming flight and mating of *Culicoides* (Diptera: Ceratopogonidae). *Transactions of the Royal Entomological Society of London* 106:213–236.

――― 1966. Observations on the mating behavior of the crab hole mosquito *Deinocerites cancer* (Diptera: Culicidae). *Canadian Entomologist* 98: 1169–77.

――― 1969. The swarming and mating flight of Diptera. *Annual Review of Entomology* 14:271–298.

――― 1970. The feeding and mating behaviour of the specialized Empidinae (Diptera); observations on four species of *Rhamphomyia* in the high arctic and a general discussion. *Canadian Entomologist* 102:769–791.

――― 1978. Feeding and mating in the insectivorous Ceratopogoninae (Diptera). *Memoirs of the Entomological Society of Canada* no. 104.

Downhower, J. F., and L. Brown. 1981. The timing of reproduction and its behavioral consequences for mottled sculpins *(Cottus bairdi)*. In *Natural selection and social behavior: recent research and new theory,* ed. R. D. Alexander and D. W. Tinkle. New York: Chiron Press.

Duffield, A. M. 1981. Biology of *Microdon fuscipennis* (Diptera: Syrphidae) with interpretations of the reproductive strategies of *Microdon* species found north of Mexico. *Proceedings of the Entomological Society of Washington* 83:716–724.

Dumortier, B. 1963. Ethological and physiological study of sound emissions in Arthropoda. In *Acoustic behaviour of animals,* ed. R-G. Busnel. New York: Elsevier.

Dunlap-Pianka, H., C. L. Boggs, and L. E. Gilbert. 1977. Ovarian dynamics in Heliconiine butterflies: programmed senescence vs. eternal youth. *Science* 197:487–490.

Dunn, J. A. 1959. The biology of the lettuce root aphid. *Annals of Applied Biology* 47:475–491.

Dybas, L. K., and H. S. Dybas. 1981. Coadaptation and taxonomic differentiation of sperm and spermathecae in featherwing beetles. *Evolution* 35:168–174.

Eberhard, W. G. 1974. The natural history and behaviour of the wasp *Trigonopsis cameronii* Kohl (Sphecidae). *Transactions of the Royal Entomological Society of London* 125:295–328.

――― 1975. The ecology and behavior of a subsocial pentatomid bug and two scelionid wasps: strategy and counterstrategy in a host and its parasites. *Smithsonian Contributions to Zoology* 205:1–39.

———— 1978. Mating swarms of a South American *Acropygia* (Hymenoptera: Formicidae). *Entomological News* 89:14–16.

———— 1979. The function of horns in *Podischnus agenor* (Dynastinae) and other beetles. In *Sexual selection and reproductive competition in insects,* ed. M. S. Blum and N. A. Blum. New York: Academic Press.

———— 1980. Horned beetles. *Scientific American* 242:166–182.

Edmunds, G. F., Jr., and Alstad, D. N. 1978. Coevolution in insect herbivores and conifers. *Science* 199:941–945.

Edmunds, G. F., Jr., and C. H. Edmunds. 1979. Predation, climate and mating of mayflies. In *Advances in Ephemeroptera biology,* ed. J. F. Flannagen and K. E. Marshall. New York: Plenum Press.

Edmunds, M. 1975. Courtship, mating, and possible sex pheromones in three species of Mantodea. *Entomologist's Monthly Magazine* 111:53–57.

Edwards, F. W. 1920. Some records of predacious Ceratopogoninae (Diptera). *Entomologist's Monthly Magazine* 56:203.

Eff, D. 1962. A little about the little-known *Papilio indra minori. Journal of the Lepidopterists' Society* 16:137–142.

Ehrlich, A. H., and P. R. Ehrlich. 1978. Reproductive strategies in butterflies. I. Mating frequency, plugging and egg number. *Journal of the Kansas Entomological Society* 51:666–697.

Ehrman, L., and P. A. Parsons. 1976. *The genetics of behavior.* Sunderland, Massachusetts: Sinauer.

Ehrman, L., and J. Probber. 1978. Rare *Drosophila* males: the mysterious matter of choice. *American Scientist* 66:216–222.

Eickwort, G. C. 1977. Male territorial behaviour in the mason bee *Hoplitis anthocopoides* (Hymenoptera: Megachilidae). *Animal Behaviour* 25: 542–554.

Eickwort, G. C., and K. R. Eickwort. 1973. Aspects of the biology of Costa Rican halictine bees. V. *Augochlorella endentata* (Hymenoptera: Halictidae). *Journal of the Kansas Entomological Society* 46:3–16.

Eickwort, G. C., and H. S. Ginsberg. 1980. Foraging and mating behavior in Apoidea. *Annual Review of Entomology* 25:421–446.

Eldredge, N., and S. J. Gould. 1972. Punctuated equilibria: an alternative to phyletic gradualism. In *Models in paleobiology,* ed. T. J. M. Schopf. San Francisco: Freeman, Cooper.

Ellertson, F. E. 1956. *Pleocoma oregonensis* Leach as a pest in sweet cherry orchards. *Journal of Economic Entomology* 49:431.

Eltringham, H. 1925. On the abdominal glands in *Heliconius erato* (Lepidoptera). *Transactions of the Entomological Society of London* 1925: 269–275.

Emlen, S. T., and L. W. Oring. 1977. Ecology, sexual selection, and the evolution of mating systems. *Science* 197:215–223.

Engelmann, F. 1970. *The physiology of insect reproduction.* New York: Pergamon Press.

Eriksen, C. H. 1960. The oviposition of *Enallagma exulans* (Odonata: Agrionidae). *Annals of the Entomological Society of America* 53:439.

Evans, A. R. Manuscript. The study of the behaviour of the Australian field cricket *Teleogryllus commodus* (Walker) (Orthoptera: Gryllidae) in the field and in habitat simulations.

Evans, H. E. 1966. *The comparative ethology and evolution of the sand wasps.* Cambridge, Massachusetts: Harvard University Press.

———— 1969. Phoretic copulation in Hymenoptera. *Entomological News* 80:113–124.

Evans, H. E., and K. M. O'Neill. 1978. Alternative mating strategies in the digger wasp *Philanthus zebratus* Cresson. *Proceedings of the National Academy of Science U.S.A.* 75:1901–3.

Evans, H. E., and M. J. West-Eberhard. 1970. *The wasps*. Ann Arbor: University of Michigan Press.

Everson, P. R., and J. F. Addicott. 1982. Mate selection strategies by male mites in the absence of intersexual selection by females: a test of six hypotheses. *Canadian Journal of Zoology* 60:2729–36.

Ewing, A. 1961. Body size and courtship behaviour in *Drosophila melanogaster*. *Animal Behaviour* 9:93–99.

———— 1964. The influence of wing area on the courtship behaviour of *Drosophila melanogaster*. *Animal Behaviour* 12:316–320.

———— 1979. Complex courtship songs in the *Drosophila funebris* species group: escape from an evolutionary bottleneck. *Animal Behaviour* 27:343–349.

Fabre, J. H. 1910. *Souvenirs entomologiques*. Paris: C. Delagrave.

Falconer, D. 1960. *Introduction to quantitative genetics*. New York: Ronald.

Farrow, R. A. 1963. The spermatophore of *Tetrix* Latrielle (Orthoptera: Tetrigidae). *Entomologist's Monthly Magazine* 99:217–223.

Feaver, M. 1977. Some aspects of the behavioral ecology of three species of *Orchelimum* (Orthoptera: Tettigoniidae). Ph.D. dissertation, University of Michigan.

———— Forthcoming. Pair formation in *Orchelimum nigripes* (Orthoptera: Tettigoniidae). In *Orthopteran mating systems: sexual competition in a diverse group of insects*, ed. D. T. Gwynne and G. K. Morris. Boulder, Colorado: Westview Press.

Fincke, O. M. 1982. Lifetime mating success in a natural population of the damselfly, *Enallagma hageni* (Walsh) (Odonata: Coenagrionidae). *Behavioral Ecology and Sociobiology* 10:293–302.

Fisher, R. A. 1958. *The genetical theory of natural selection*, 2nd ed. New York: Dover.

Fisher, T. W. 1959. Occurrence of spermatophores in certain species of *Chilocorus* (Coleoptera: Coccinellidae). *Pan-Pacific Entomologist* 35:205–208.

Fletcher, B. S. 1968. Storage and release of a sex pheromone by the Queensland fruit fly, *Dacus tryoni* (Diptera: Trypetidae). *Nature* 219:631–632.

Fletcher, B. S., and A. Giannakakis. 1973. Factors limiting the response of females of the Queensland fruit fly, *Dacus tryoni*, to the sex pheromone of the male. *Journal of Insect Physiology* 19:1147–55.

Forbes, R. S., and L. Daviault. 1965. The biology of the mountain-ash sawfly, *Pristiphora geniculata* (Htg.) (Hymenoptera: Tenthredinidae) in eastern Canada. *Canadian Entomologist* 96:1117–33.

Forrest, T. G. 1980. Phonotaxis in mole crickets: its reproductive significance. *Florida Entomologist* 63:45–53.

———— 1982. Acoustic communication and baffling behaviors of crickets. *Florida Entomologist,* 65:33–44.

———— Forthcoming. Calling songs and mate choice in mole crickets. In *Orthopteran mating systems: sexual competition in a diverse group of in-*

sects, ed. D. T. Gwynne and G. K. Morris. Boulder, Colorado: Westview Press.

Fowler, G. L. 1973. Some aspects of the reproductive biology of *Drosophila:* sperm transfer, sperm storage and sperm utilization. *Advances in Genetics* 17:293–352.

Frankie, G. W., S. B. Vinson, and R. E. Colville. 1980. Territorial behavior of *Centris adani* and its reproductive function in the Costa Rican dry forest (Hymenoptera: Anthophoridae). *Journal of the Kansas Entomological Society* 53:837–857.

Free, J. B. 1971. Stimuli eliciting mating behaviour of bumblebee *(Bombus pratorum* L.) males. *Behaviour* 40:55–61.

Freeland, W. J. 1981. Parasitism and behavioral dominance among male mice. *Science* 213:461–462.

Freeman, R. B. 1968. Charles Darwin on the routes of male bumble bees. *Bulletin of the British Museum of Natural History (History Series)* 3:177–189.

Fremling, C. R. 1970. Mayfly distribution as a water quality index. *United States Environmental Protection Agency, Water Pollution Control Research Series,* report no. 16030 DGH 11/70.

Friedel, T., and C. Gillott. 1977. Contribution of male-produced proteins to vitellogenesis in *Melanoplus sanguinipes. Journal of Insect Physiology* 23:145–151.

Friedlander, M., and G. Benz. 1981. The eupyrene-apyrene dichotomous spermatogenesis of Lepidoptera: organ culture study on the time of apyrene commitment in the codling moth. *International Journal of Invertebrate Reproduction* 3:113–120.

Friedlander, M., and H. Gitay. 1972. The fate of the normal-anucleated spermatozoa in inseminated females of the silkworm *Bombyx mori. Journal of Morphology* 138:121–130.

Fujisaki, K. 1981. Studies on the mating system of the winter cherry bug, *Acanthocoris sordidus* Thunberg (Heteroptera: Coreidae) II. Harem defense polygyny. *Researches on Population Ecology* 23:262–279.

Gabbutt, P. D. 1954. Notes on the mating behavior of *Nemobius sylvestris* (Bosc) (Orth., Gryllidae). *British Journal of Animal Behaviour* 2:854–888.

Gadgil, M. 1972. Male dimorphism as a consequence of sexual selection. *American Naturalist* 106:576–580.

Gerber, G. H. 1970. Evolution of the methods of spermatophore formation in pterygote insects. *Canadian Entomologist* 102:358–362.

Gerber, G. H., and N. S. Church. 1976. The reproductive cycles of male and female *Lytta nuttalli* (Coleoptera: Meloidae). *Canadian Entomologist* 108:1125–36.

Gerber, G. H., N. S. Church, and J. G. Rempel. 1971. The structure, formation, biochemistry, fate, and functions of the spermatophore of *Lytta nuttalli* Say (Coleoptera: Meloidae). *Canadian Journal of Zoology* 49:1595–1610.

——— 1972. The anatomy, histology, and physiology of the reproductive organ systems of *Lytta nuttalli* Say (Coleoptera: Meloidae). II. The abdomen and external genitalia. *Canadian Journal of Zoology* 50:645–660.

Gerber, H. S., and E. C. Klostermeyer. 1970. Sex control by bees: a voluntary act of egg fertilization during oviposition. *Science* 167:82–84.

Gerling, D., and H. R. Hermann. 1978. Biology and mating behavior of *Xylocopa virginica* L. (Hymenoptera, Anthophoridae). *Behavioral Ecology and Sociobiology* 3:99–111.

Ghiselin, M. T. 1974. *The economy of nature and the evolution of sex.* Berkeley: University of California Press.

Giglioli, M. E. C., and G. F. Mason. 1966. The mating plug in anopheline mosquitoes. *Proceedings of the Royal Entomological Society of London, Series A* 41:123–129.

Gilbert, D. G. 1981. Ejaculate esterase 6 and initial sperm use by female *Drosophila melanogaster*. *Journal of Insect Physiology* 27:641–650.

Gilbert, D. G., R. C. Richmond, and K. B. Sheehan. 1981. Studies of esterase 6 in *Drosophila melanogaster*. V. Progeny production and sperm use in females inseminated by males having active or null alleles. *Evolution* 35:21–37.

Gilbert, L. E. 1976. Postmating female odor in *Heliconius* butterflies: a male-contributed antiaphrodisiac? *Science* 193:419–420.

Gillett, J. D. 1971. *Mosquitoes.* London: Weidenfeld and Nicolson.

Gillies, M. T. 1956. A new character for the recognition of nulliparous females of *Anopheles gambiae*. *Bulletin of the World Health Organization* 15:451–459.

Given, B. 1954. Evolutionary trends in the Thynninae (Hymenoptera: Tiphiidae) with special reference to feeding habits of Australian species. *Transactions of the Royal Entomological Society of London* 105:1–10.

Glesener, R. R., and D. Tilman. 1978. Sexuality and the components of environmental uncertainty: clues from geographic parthenogenesis in terrestrial animals. *American Naturalist* 112:659–673.

Goeden, R. D., and D. M. Norris. 1964. Some biological and ecological aspects of the dispersal flight of *Scolytus quadrispinosus* (Coleoptera: Scolytidae). *Annals of the Entomological Society of America* 57:743–749.

——— 1965. Some biological and ecological aspects of ovipositional attack in *Carya* spp. by *Scolytus quadrispinosus* (Coleoptera: Scolytidae). *Annals of the Entomological Society of America* 58:771–777.

Goetghebuer, M. 1914. Notes à propos de l'accouplement de *Johannsenomyia* (*Ceratopogon*) *nitida* Mcq. *Annales de la Société Entomologique Belge* 58:202.

Gordh, G. 1976. *Goniozus gallicola* Fouts, a parasite of moth larvae, with notes on other bethylids (Hymenoptera: Bethylidae: Lepidoptera: Gelechiidae). *USDA, Agricultural Research Service, Technical Bulletin* no. 1524:1–27.

Gordon, B. R., and H. T. Gordon. 1971. Sperm storage and depletion in *Oncopeltus fasciatus*. *Entomologia Experimentalis et Applicata* 14:425–433.

Goss, G. J. 1977. The interaction between moths and pyrrolizidine alkaloid-containing plants including nutrient transfer via the spermatophore in *Lymire edwardsii* (Ctenuchidae). Ph.D. dissertation, University of Miami (Florida).

Gould, S. J. 1978. Sociobiology: the art of storytelling. *New Scientist* 80:530–533.

——— 1980. Is a new and general theory of evolution emerging? *Paleobiology* 6:119–130.

Gould, S. J., and N. Eldredge. 1977. Punctuated equilibria: the tempo and mode of evolution reconsidered. *Paleobiology* 3:115–151.

Gould, S. J., and R. C. Lewontin. 1979. The spandrels of San Marco and the Panglossian paradigm: a critique of the adaptationist program. *Proceedings of the Royal Society of London* 205:581–598.

Graham, N. P. H., and K. L. Taylor. 1941. Studies on some ectoparasites of sheep and their control. 1. Observations on the bionomics of the sheep ked (*Melophagus ovinus*). *Council of Science and Industrial Research Australia*, pamphlet 108:8–26.

Grant, P. R. 1972. Convergent and divergent character displacement. *Biological Journal of the Linnean Society* 4:39–68.

Greenberg, L. 1979. Genetic component of bee odor in kin recognition. *Science* 206:1095–97.

Greenfield, M. D. 1981. Moth sex pheromones: an evolutionary perspective. *Florida Entomologist* 64:4–17.

Greenfield, M. D., and K. C. Shaw. Forthcoming. Adaptive significance of chorusing in insects. In *Orthopteran mating systems: sexual competition in a diverse group of insects*, ed. D. T. Gwynne and G. K. Morris. Boulder, Colorado: Westview Press.

Gregory, G. E. 1965. The formation and fate of the spermatophore in the African migratory locust, *Locusta migratoria migratoides* Reiche and Fairmaire. *Transactions of the Royal Entomological Society of London* 117:33–66.

Grissell, E. E., and C. E. Goodpasture. 1981. A review of Nearctic Podagrionini, with description of sexual behavior of *Podagrion mantis* (Hymenoptera: Torymidae). *Annals of the Entomological Society of America* 74:226–241.

Gromko, M. H., D. G. Gilbert, and R. C. Richmond. Forthcoming. Sperm transfer and use in the repeat mating system of *Drosophila*. In *Sperm competition and the evolution of animal mating systems*, ed. R. L. Smith. New York: Academic Press.

Gross, M. R., and E. L. Charnov. 1980. Alternative life histories in bluegill sunfish. *Proceedings of the National Academy of Sciences U.S.A.* 77:6937–40.

Gross, M. R., and R. Shine, 1981. Parental care and mode of fertilization in ectothermic vertebrates. *Evolution* 35:775–793.

Guppy, J. C. 1961. Life history and behaviour of the armyworm, *Pseudaletia unipuncta* (Haw.) (Lepidoptera: Noctuidae), in eastern Ontario. *Canadian Entomologist* 93:1141–53.

Gupta, P. D. 1947. On the structure and formation of spermatophore in the cockroach *Periplaneta americana* (Linn.). *Indian Journal of Entomology* 8:79–84.

Gwynne, D. T. 1978. Male territoriality in the bumblebee wolf *Philanthus bicinctus* (Mickel) (Hymenoptera: Sphecidae): observations of the behaviour of individual males. *Zeitschrift für Tierpsychologie* 47:89–103.

——— 1980. Female defence polygyny in the bumblebee wolf, *Philanthus bicinctus* (Hymenoptera: Sphecidae). *Behavioral Ecology and Sociobiology* 7:213–225.

——— 1981. Sexual difference theory: mormon crickets show role reversal in mate choice. *Science* 213:779–780.

———— 1982. Mate selection by female katydids (Orthoptera: Tettigoniidae, *Conocephalus nigropleurum*). *Animal Behaviour* 30:734–738.

———— Forthcoming a. Male mating effort, confidence of paternity and insect sperm competition. In *Sperm competition and the evolution of animal mating systems*, ed. R. L. Smith. New York: Academic Press.

———— Forthcoming b. Male nutritional investment and the evolution of sexual differences in the Orthoptera. In *Orthopteran mating systems: sexual competition in a diverse group of insects*, ed. D. T. Gwynne and G. K. Morris. Boulder, Colorado: Westview Press.

———— Manuscript. Male nutrient investment, population density and sexual selection in Mormon crickets (*Anabrus simplex*, Orthoptera: Tettigoniidae).

Gwynne, D. T., and B. B. Hostetler. 1978. Mass emergence of *Prionus emarginatus* (Say) (Coleoptera: Cerambydidae). *Coleopterist's Bulletin* 32:347–348.

Gwynne, D. T., and G. K. Morris, eds. Forthcoming. *Orthopteran mating systems: sexual competition in a diverse group of insects*. Boulder, Colorado: Westview Press.

Gwynne, D. T., and K. M. O'Neill. 1980. Territoriality in digger wasps results in sex biased predation on males (Hymenoptera: Sphecidae, Philanthus). *Journal of the Kansas Entomological Society* 53:220–224.

Haas, A. 1960. Vergleichende Verhaltensstudien zum Paarungsschwarm solitärer Apiden. *Zeitschrift für Tierpsychologie* 17:402–416.

Haeger, J. S., and M. W. Provost. 1965. Colonization and biology of *Opifex fuscus*. *Transactions of the Royal Society of New Zealand* 6:21–31.

Hagen, K. S. 1962. Biology and ecology of predacious Coccinellidae. *Annual Review of Entomology* 7:289–326.

Hagley, E. A. C. 1965. On the life history and habits of the palm weevil *Rhynchophorus palmarium*. *Annals of the Entomological Society of America* 58:22–28.

Hailman, J. P. 1977. *Optical signals: animal communication and light*. Bloomington: Indiana University Press.

Halffter, G., and E. G. Matthews. 1966. The natural history of dung beetles of the subfamily Scarabaeinae (Coleoptera: Scarabaeidae). *Folia Entomologica Mexicana* 12–14:1–312.

Halliday, T. R. 1978. Sexual selection and mate choice. In *Behavioral ecology: an evolutionary approach*, ed. J. R. Krebs and N. B. Davies. Sunderland, Massachusetts: Sinauer.

Hamilton, W. D. 1967. Extraordinary sex ratios. *Science* 146:477–488.

———— 1971. Geometry for the selfish herd. *Journal of Theoretical Biology* 31:295–311.

———— 1978. Evolution and diversity under bark. In *Diversity of insect faunas*, ed. L. A. Mound and N. Waloff. London: Blackwell.

———— 1979. Wingless and fighting males in fig wasps and other insects. In *Sexual selection and reproductive competition in insects*, ed. M. S. Blum and N. A. Blum. New York: Academic Press.

———— 1980. Sex versus non-sex versus parasite. *Oikos* 35:282–290.

Hamilton, W. D., P. A. Henderson, and N. A. Moran. 1981. Fluctuation of environment and coevolved antagonist polymorphism as factors in the maintenance of sex. In *Natural selection and social behavior: recent re-

search and new theory, ed. R. D. Alexander and D. W. Tinkle. New York: Chiron Press.

Hamilton, W. J. III, R. E. Buskirk, and W. H. Buskirk. 1976. Social organization of the Namib Desert tenebrionid beetle *Onymacris rugatipennis*. *Canadian Entomologist* 108:305–316.

Hamlin, J. C. 1924. A review of the genus *Chelinidae* (Hemiptera-Heteroptera) with biological data. *Annals of the Entomological Society of America* 17:193–208.

Hanson, F. E., J. F. Case, E. Buck, and J. B. Buck. 1971. Synchrony and flash entrainment in a New Guinea firefly. *Science* 174:161–164.

Happ, G. M. 1969. Multiple sex pheromones of the mealworm beetle *Tenebrio molitor* L. *Nature* 222:180–181.

Hardee, D. D., W. H. Cross, and E. B. Mitchell. 1969. Male boll weevils are more attractive than cotton plants to boll weevils. *Journal of Economic Entomology* 62:165–169.

Hardee, D. D., E. B. Mitchell, and P. M. Huddleston. 1967. Laboratory studies of sex attraction in the boll weevil. *Journal of Economic Entomology* 60:1221–24.

Hardee, D. D., W. H. Cross, E. B. Mitchell, P. M. Huddleston, H. C. Mitchell, M. E. Merkl, and T. B. Davich. 1969. Biological factors influencing responses of the female boll weevil to the male sex pheromone in field and large-cage tests. *Journal of Economic Entomology* 62:161–165.

Hardeland, R. 1972. Species differences in the diurnal rhythmicity of courtship behavior within the *melanogaster* group of the genus *Drosophila*. *Animal Behaviour* 20:170–174.

Hardie, J. 1980. Juvenile hormone mimics the photo-periodic apterization of the alate gynopara of the aphid, *Aphis fabae*. *Nature* 286:602–604.

Harpending, H. C. 1979. The population genetics of interaction. *American Naturalist* 113:622–630.

Harris, R. L., E. D. Frazar, P. D. Grossman, and O. H. Graham. 1966. Mating habits of the stable fly. *Journal of Economic Entomology* 59:634–636.

Harris, V. E., and J. W. Todd. 1980a. Male-mediated aggregation of male, female and 5th-instar southern green stink bugs and concomitant attraction of a tachinid parasite, *Trichopoda pennipes*. *Entomologia Experimentalis et Applicata* 27:117–126.

———— 1980b. Temporal and numerical patterns of reproductive behavior in the southern green stinkbug, *Nezara viridula* (Hemiptera: Pentatomidae). *Entomologia Experimentalis et Applicata* 27:105–116.

Hartmann, R., and W. Loher. 1974. Control of sexual behaviour pattern "secondary defence" in the grasshopper, *Chorthippus curtipennis*. *Journal of Insect Physiology* 20:1713–28.

Hartung, J. 1981. Genome parliaments and sex with the Red Queen. In *Natural selection and social behavior: recent research and new theory*, ed. R. D. Alexander and D. W. Tinkle. New York: Chiron Press.

Harvey, P. A. 1934. Life history of *Kalotermes minor*. In *Termites and termite control*, ed. C. A. Kofoid, S. F. Light, A. C. Horner, M. Randall, W. B. Herms, and E. E. Bowe. 2nd ed. Berkeley: University of California Press.

Haskell, P. T. 1958. Stridulation and associated behaviour in certain Orthoptera. 2. Stridulation of females and their behaviour with males. *Animal Behaviour* 6:27–42.

Hassan, A. T. 1981. Coupling and oviposition behaviour in two macrodipla-cinid libellulids—*Aethriamantha rezia* (Kirby) and *Urothemis assignata* Selys (Libellulidae: Odonata). *Zoological Journal of the Linnean Society* 72:289–296.

Heatwole, H., D. M. Davis, and A. M. Wenner. 1962. The behaviour of *Megarhyssa*, a genus of parasitic hymenopterans (Ichneumonidae: Ephilatinae). *Zeitschrift für Tierpsychologie* 19:652–664.

Hedin, P. A., D. D. Hardee, A. C. Thompson, and R. C. Gueldner. 1974. An assessment of the lifetime biosynthesis potential of the male boll weevil. *Journal of Insect Physiology* 20:1707–12.

Heinrich, B., and Bartholomew, G. A. 1979. The ecology of the African dung beetle. *Scientific American* 241:118–127.

Henry, C. 1979. Acoustical communicatiion during courtship and mating in the green lacewing *Chrysopa carnea* (Neuroptera: Chrysopidae). *Annals of the Entomological Society of America* 72:68–79.

Hicks, C. H. 1929. *Pseudomasaris edwardsii* Cresson, another pollen-provisioning wasp, with further notes on *P. vespoides* (Cresson). *Canadian Entomologist* 61:121–125.

Hill, D. S. 1971. Wasps and figs. *New Scientist and Science Journal* 50:144–146.

Hinton, H. E. 1964. Sperm transfer in insects and the evolution of haemocoelic insemination. In *Insect reproduction,* ed. K. C. Highnam. Symposium of the Royal Entomological Society of London, no. 2.

Hirai, K., H. H. Shorey, and L. K. Gaston. 1978. Competition among courting male moths: male-to-male inhibitory pheromone. *Science* 202:644–645.

Hirshfield, M. F., and D. W. Tinkle. 1975. Natural selection and the evolution of reproductive effort. *Proceedings of the National Academy of Sciences U.S.A.* 72:2227–31.

Holland, G. P. 1955. Primary and secondary sexual characteristics of some *Ceratophyllinae,* with notes on the mechanism of copulation (Siphonaptera). *Transactions of the Royal Entomological Society of London* 107:233–248.

Hölldobler, B. 1976. The behavioral ecology of mating in harvester ants (Hymenoptera: Formicidae: *Pogonomyrmex*). *Behavioral Ecology and Sociology* 1:405–423.

Hölldobler, B., and C. P. Haskins. 1977. Sexual calling behavior in primitive ants. *Science* 195:793–794.

van Honk, C. G. J., H. H. W. Velthuis, and P. F. Roseler. 1978. A sex pheromone from the mandibular glands in bumblebee queens. *Experientia* 34:838–839.

Hoogland, J. L. 1981. The evolution of coloniality in white-tailed and black-tailed prairie dogs (Sciuridae: *Cynomys leucurus* and *C. ludovicianus*). *Ecology* 62:252–272.

Hook, A. W., and R. W. Matthews. 1980. Nesting biology of *Oxybelus sericeus* with a discussion of nest guarding by male sphecid wasps (Hymenoptera). *Psyche* 87:21–38.

Houston, T. F. 1970. Discovery of an apparent male soldier caste in a nest of a halictid bee. *Australian Journal of Zoology* 18:345–351.

Hrdy, S. B. 1979. Infanticide among animals: a review, classification, and examination of the implications for the reproductive strategies of females. *Ethology and Sociobiology* 1:13–40.

Hughes, A. L. 1981. Differential male mating success in the white spotted sawyer *Monochamus scutellatus* (Coleoptera: Cerambycidae). *Annals of the Entomological Society of America* 74:180–184.

Hughes, A. L., and M. K. Hughes. 1982. Male size, mating success and breeding habitat partitioning in the whitespotted sawyer *Monochamus scutellatus* (Say). *Oecologia* 55:258–263.

Hull, H. F., Jr., A. M. Wenner, and D. H. Wells. 1976. Reproductive behavior in an overwintering aggregation of monarch butterflies. *American Midland Naturalist* 95:10–19.

Humphries, D. A. 1967. The mating behaviour of the hen flea *Ceratophyllus gallinae* (Schrank) (Siphonaptera: Insecta). *Animal Behaviour* 15: 82–90.

Hurd, P. D., Jr., and E. G. Linsley. 1975. The principal *Larrea* bees of the southwestern United States (Hymenoptera: Apoidea). *Smithsonian Contributions to Zoology* 193:1–74.

———— 1976. The bee family Oxaeidae with a revision of the North American species (Hymenoptera: Apoidea). *Smithsonian Contributions to Zoology* 220:1–75.

Hurd, P. D., Jr., and J. A. Powell. 1958. Observations on the nesting habits of *Colletes stepheni* Timberlake (Hymenoptera: Apoidea). *Pan-Pacific Entomologist* 34:147–153.

Huxley, J. 1938. Darwin's theory of sexual selection and the data subsumed by it in light of recent research. *American Naturalist* 72:416–433.

Hynes, H. B. N. 1970. *The ecology of running waters.* Toronto: University of Toronto Press.

Jackson, R. R. 1980. The mating strategy of *Phidippus johnsoni* (Araneae: Salticidae). III. Intermale aggression and a cost-benefit analysis. *Journal of Arachnology* 8:241–249.

Jacobs, M. E. 1955. Studies on territorialism and sexual selection in dragonflies. *Ecology* 36:566–586.

Jacobson, M. 1972. *Insect sex pheromones.* New York: Academic Press.

Jaenike, J. 1978. An hypothesis to account for the maintenance of sex within populations. *Evolutionary Theory* 3:191–194.

Jaenson, T. G. T. 1979. Mating behaviour of males of *Glossina pallidipes* Austen (Diptera: Glossinidae). *Bulletin of Entomological Research* 69:573–588.

Janetos, A. C. 1980. Strategies of female mate choice: a theoretical analysis. *Behavioral Ecology and Sociobiology* 7:107–112.

Jansson, A. 1973. Stridulation and its significance in the genus *Cenocorixia* (Hemiptera, Corixidae). *Behaviour* 46:1–36.

Jaycox, E. R. 1967. Territorial behavior among males of *Anthidium banningense* (Hymenoptera: Megachilidae). *Journal of the Kansas Entomological Society* 40:565–570.

Jenni, D. A. 1974. Evolution of polyandry in birds. *American Zoologist* 14:129–144.

Johnson, L. K. 1982. Sexual selection in a tropical brentid weevil. *Evolution* 36:251–262.

Jones, M. D. R. 1964. Inhibition and excitation in the acoustic behaviour of *Pholidoptera. Nature* 203:322–323.

———— 1966. Alternation, synchronism, and rivalry between males. *Journal of Experimental Biology* 45:15–30.

Jordan, A. M. 1958. The mating behaviour of females of *Glossina palpalis* (R. D.) in captivity. *Bulletin of Entomological Research* 49:35–43.

Kamal, A. S. 1958. Comparative study of thirteen species of sarcosaphrophagous Calliphoridae and Sarcophagidae. *Annals of the Entomological Society of America* 51:261–271.

Kannowski, P. B., and R. L. Johnson. 1969. Male patrolling behaviour and sex attraction in ants of the genus *Formica*. *Animal Behaviour* 17:425–429.

Kay, R. E. 1969. Acoustic signalling and its possible relationship to assembling and navigating in the moth *Heliothis zea*. *Journal of Insect Physiology* 15:989–1001.

Kellen, W. R. 1959. Notes on the biology of *Halovelia mariannarum* Usinger in Samoa (Veliidae: Heteroptera). *Annals of the Entomological Society of America* 52:53–62.

Kempton, R. A., H. J. B. Lowe, and E. J. B. Bintcliffe. 1980. The relationship between fecundity and adult weight in *Myzus persicae*. *Journal of Animal Ecology* 49:917–926.

Kence, A. 1981. The rare-male advantage in *Drosophila:* a possible source of bias in experimental design. *American Naturalist* 117:1027–28.

Kennedy, J. S., and H. L. G. Stroyan. 1959. Biology of aphids. *Annual Review of Entomology* 4:139–160.

Kerr, W. E. 1974. Advances in cytology and genetics of bees. *Annual Review of Entomology* 19:253–268.

Kessel, E. L. 1955. The mating activities of balloon flies. *Systematic Zoology* 4:997–1004.

Khalifa, A. 1949. Spermatophore production in Trichoptera and some other insects. *Transactions of the Royal Entomological Society of London* 100:449–471.

———— 1950. Sexual behaviour in *Gryllus domesticus* L. *Behaviour* 2: 264–274.

Kiefer, B. I. 1969. Phenotypic effects of Y chromosome mutations in *Drosophila melanogaster*. I. Spermiogenesis and sterility in KL-1⁻ males. *Genetics* 61:157–166.

Kimsey, L. S. 1978. Nesting and male behavior in *Dynatus spinolae* (Lepeletier). *Pan-Pacific Entomologist* 54:65–68.

———— 1980. The behaviour of male orchid bees (Apidae, Hymenoptera, Insecta) and the question of leks. *Animal Behaviour* 28: 996–1004.

King, P. E., R. R. Askew, and C. Sanger. 1969. The detection of parasitized hosts by males of *Nasonia vitripennis* (Walker) (Hymenoptera: Pteromalidae) and some possible implications. *Proceedings of the Royal Entomological Society of London* 44A:85–90.

Kirk, V. M. 1971. Biological studies of the ground beetle, *Pterostichus lucublandus*. *Annals of the Entomological Society of America* 64:540–544.

Kirk, V. M., and B. J. Dupraz. 1972. Discharge by a female ground beetle, *Pterostichus lucublandus* (Coleoptera: Carabidae), used as defense against males. *Annals of the Entomological Society of America* 65:513.

Kirkpatrick, T. W. 1937. Studies on the ecology of coffee plantations in East Africa. II. The autecology of *Antestia* spp. (Pentatomidae) with a particular account of a strepsipterous parasite. *Transactions of the Royal Entomological Society of London* 86:247–343.

Knight, G., A. Robertson, and C. Waddington. 1956. Selection for sexual isolation within a species. *Evolution* 10:14–22.

Knowlton, N. 1979. Reproductive synchrony, parental investment, and evolutionary dynamics of sexual selection. *Animal Behaviour* 27:1022–33.

Krebs, J. R., and N. B. Davies, eds. 1978. *Behavioural ecology: an evolutionary approach.* Sunderland, Massachusetts: Sinauer.

———— 1981. *An introduction to behavioural ecology.* Sunderland, Massachusetts: Sinauer.

Krombein, K. V. 1967. *Trap-nesting wasps and bees: life histories, nests, and associations.* Washington, D. C.: Smithsonian Press.

Kullenberg, B., G. Bergström, B. Bringer, B. Carlberg, and B. Cederberg. 1973. Observations of scent marking by *Bombus* Latr. and *Psithyrus* Lep. males (Hymenoptera, Apidae) and localization of site of production of the secretion. *Zoon, Supplement* 1:23–44.

Labine, P. A. 1964. Population biology of the butterfly, *Euphydras editha.* I. Barriers to multiple inseminations. *Evolution* 18:335–336.

Lall, A. B., H. Seliger, W. H. Biggley, and J. E. Lloyd. 1981. Ecology of colors of bioluminescence in fireflies. *Science* 210:560–562.

Landa, V. 1960. Origin, development and function of the spermatophore in the cockchafer (*Melolontha melolontha* L.). *Acta Societatis Entomologicae Cechoslovenlae* 57:297–316.

Lande, R. 1976. The maintenance of genetic variation by mutation in a polygenic character with linked loci. *Genetical Research* 26:221–235.

———— 1981. Models of speciation by sexual selection on polygenic traits. *Proceedings of the National Academy of Sciences U.S.A.* 78:3721–25.

Las, A. 1980. Male courtship persistence in the greenhouse whitefly, *Trialeurodae vaporariorum* Westwood (Homoptera: Aleyrodidae). *Behaviour* 72:107–126.

Latimer, W. 1981. The acoustic behaviour of *Platycleis albopunctata* (Goeze) (Orthoptera, Tettigoniidae). *Behaviour* 76:182–206.

Lavigne, R. J., and F. R. Holland. 1969. Comparative behaviour of eleven species of Wyoming robber flies (Diptera: Asilidae). *Agricultural Experiment Station, University of Wyoming, Science Monograph* 18: 1–61.

Lea, A. O., and D. G. Evans. 1972. Sexual behavior of mosquitoes. 1. Age dependence of copulation and insemination in *Culex pipiens* complex and *Aedes taeniorhynchus* in the laboratory. *Annals of the Entomological Society of America* 65:285–289.

Lea, N. 1916. Notes on the Lord Howe Island phasma and an associated longicorn beetle. *Proceedings of the Royal Society of South Australia* 40: 145–147.

Lederhouse R. C., R. A. Morse, J. T. Ambrose, D. M. Burgett, W. E. Conner, L. Edwards, R. D. Fell, R. Rutowski, and M. Turrell. 1976. Crepuscular mating aggregations in certain *Ormia* and *Sitophaga. Annals of the Entomological Society of America* 69:656–658.

Lee, R. D. 1955. The biology of the Mexican chicken bug *Haematosiphon inodorus* (Duges) (Hemiptera: Cimicidae). *Pan-Pacific Entomologist* 31:47–61.

Lees, A. D. 1966. The control of polymorphism in aphids. *Advances in Insect Physiology* 3:207–277.

Lefevre, G., Jr., and U. B. Jonsson. 1962. Sperm transfer, storage, displacement, and utilization in *Drosophila melanogaster*. *Genetics* 47:1719–36.

Leonard, J. E., and L. Ehrman. 1976. Recognition and sexual selection in *Drosophila:* classification, quantification, and identification. *Science* 193:693–695.

Leopold, R, A. 1976. The role of male accessory glands in insect reproduction. *Annual Review of Entomology* 21:199–221.

Leopold, R. A., A. C. Terranova, and E. M. Swilley. 1971. Mating refusal in *Musca domestica:* effects of repeated mating and decerebration upon frequency and duration of copulation. *Journal of Experimental Zoology* 176:353–360.

Lessmann, D. 1974. Ein Beitrag zur Verbreitung und Lebensweise von *Megastigmus spermotrophus* Wachtl und *M. bipunctatus* Swederus (Hymenoptera, Chalcidoidea). *Zeitschrift für Angewandte Entomologie* 75:1–42.

Leston, D., and J. W. S. Pringle. 1963. Acoustical behaviour of Hemiptera. In *Acoustic behaviour of animals,* ed. R-G. Busnel. Amsterdam: Elsevier.

Levin, D. A. 1975. Pest pressure and recombination systems in plants. *American Naturalist* 109:437–451.

Lew, A. C., and H. J. Ball. 1980. Effect of copulation time on spermatozoa transfer of *Diabrotica virgifera* (Coleoptera: Chrysomelidae). *Annals of the Entomological Society of America* 73:360–361.

Lewis, T. 1973. *Thrips, their biology, ecology, and economic importance.* New York: Academic Press.

Lewontin, R. C. 1970. The units of selection. *Annual Review of Ecology and Systematics* 1:1–19.

——— 1978. Adaptation. *Scientific American* 239:212–230.

Leyrer, R. L., and R. E. Monroe. 1973. Isolation and identification of the scent of the moth, *Galleria mellonella,* and a reevaluation of its sex pheromone. *Journal of Insect Physiology* 19:2267–71.

Lin, N. 1963. Territorial behaviour in the cicada killer wasp *Sphecius speciosus* (Drury) (Hymenoptera: Sphecidae). *Behaviour* 20:115–133.

——— 1972. Territorial behaviour among males of the social wasp *Polistes exclamans* Viereck (Hymenoptera: Vespidae). *Proceedings of the Entomological Society of Washington* 74:148–154.

Linley, J. R., and G. M. Adams. 1972. A study of the mating behavior of *Culicoides melleus* (Coquillet) (Diptera: Ceratopogonidae). *Transactions of the Royal Entomological Society of London* 124:81–121.

——— 1974. Sexual receptivity in *Culicoides melleus* (Diptera: Ceratopogonidae). *Transactions of the Royal Entomological Society of London* 126:279–303.

Linsley, E. G. 1959. Ecology of Cerambycidae. *Annual Review of Entomology* 4:99–138.

Linsley, E. G., and J. W. MacSwain. 1942. Bionomics of the meloid genus *Hornia* (Coleoptera). *University of California Publications in Entomology* 7:189–206.

Litte, M. 1979. *Mischocyttarus flavitarsis* in Arizona: social and nesting biology of a polistine wasp. *Zeitschrift für Tierpsychologie* 50:282–312.

——— 1981. Social biology of the polistine wasp *Mischocyttarus labiatus:* survival in a Columbian rain forest. *Smithsonian Contributions to Zoology* 327:1–27.

Littlejohn, M. J. 1969. The systematic significance of isolating mechanisms. In *Systematic biology: proceedings of an international conference.* Washington, D.C.: National Academy of Sciences.

───── 1981. Reproductive isolation: a critical review. In *Evolution and speciation,* ed. W. R. Atchley and D. S. Woodruff. Cambridge: Cambridge University Press.

Lloyd, J. E. 1965. Aggressive mimicry in *Photuris:* firefly femmes fatales. *Science* 149:653–654.

───── 1966. Studies on the flash communication system in *Photinus* fireflies. *University of Michigan Museum of Zoology, Miscellaneous Publications* 130:1–95.

───── 1971. Bioluminescent communication in insects. *Annual Review of Entomology* 16:97–122.

───── 1972. Mating behavior of a New Guinea *Luciola* firefly: a new communicative protocol (Coleoptera: Lampyridae). *Coleopterist's Bulletin* 26:155–163.

───── 1973. Fireflies of Melanasia: bioluminescence, mating behavior, and synchronous flashing (Coleoptera: Lampyridae). *Environmental Entomology* 2:991–1008.

───── 1975. Aggressive mimicry in *Photuris:* signal repertoires by femmes fatales. *Science* 187:452–453.

───── 1979a. Mating behavior and natural selection. *Florida Entomologist* 62:17–34.

───── 1979b. Sexual selection in luminescent beetles. In *Sexual selection and reproductive competition in insects,* ed. M. S. Blum and N. A. Blum. New York: Academic Press.

───── 1980a. Insect behavioral ecology: coming of age in bionomics or compleat biologists have revolutions too. *Florida Entomologist* 63:1–4.

───── 1980b. Male *Photuris* fireflies mimic sexual signals of their females' prey. *Science* 210:669–671.

───── 1981. Sexual selection: individuality, identification, and recognition in a bumblebee and other insects. *Florida Entomologist* 64:89–118.

Lloyd, M., and H. S. Dybas. 1966. The periodical cicada problem: II, evolution. *Evolution* 20:466–505.

Loftus-Hills, J. J. 1975. The evidence for reproductive character displacement between the toads *Bufo americanus* and *B. woodhousei fowleri. Evolution* 28:368–369.

Loher, W. 1981. The effect of mating on female sexual behavior of *Teleogryllus commodus* Walker. *Behavioral Ecology and Sociobiology* 9:219–225.

Loher, W., and M. K. Chandrashekaran. 1970. Acoustical and sexual behavior in the grasshopper *Chimarocephala pacifica pacifica* (Oedipodinae). *Entomologia Experimentalis et Applicata* 13:71–84.

Loher, W., and H. T. Gordon. 1968. The maturation of sexual behavior in a new strain of the large milkweed bug *Oncopeltus fasciatus. Annals of the Entomological Society of America* 61:1566–72.

Loher, W., and F. Huber. 1966. Nervous and endocrine control of sexual behaviour in a grasshopper (*Gomphocerus rufus* L., Acridinae). *Symposium of the Society of Experimental Biology* 20:381–400.

Loher, W., and B. Rence. 1978. The mating behavior of *Teleogryllus commodus* (Walker) and its central and peripheral control. *Zeitschrift für Tierpsychologie* 49:225–259.

Loher, W., I. Ganjian, I. Kubo, D. Stanley-Samuelson, and S. S. Tobe. 1981. Prostaglandins: their role in egg-laying of the cricket *Teleogryllus commodus*. *Proceedings of the National Academy of Sciences U.S.A.* 78:7835–38.

Lorenz, K. Z. 1963. *On aggression*. New York: Bantam Books.

Low, B. S. 1978. Environmental uncertainties and the parental strategies of marsupials and placentals. *American Naturalist* 112:197–213.

Lum, P. T. 1961. The reproductive system of some Florida mosquitoes. II. The male accessory glands and their roles. *Annals of the Entomological Society of America* 54:430–433.

MacDonald, J. F., R. D. Akre, and W. B. Hill. 1974. Comparative biology and behavior of *Vespula atropilosa* and *V. pennsylvanica* (Hymenoptera: Vespidae). *Melanderia* 18:1–93.

Mackensen, O. 1951. Viability and sex determination in the honey bee (*Apis mellifera* L.). *Genetics* 36:500–509.

MacKerras, I. M. 1933. Observations on the life-histories, nutritional requirements and fecundity of blowflies. *Bulletin of Entomological Research* 24:353–362.

———— 1970. Skeletal anatomy. In *Insects of Australia*, ed. D. F. Waterhouse. Melbourne: Melbourne University Press.

Mac Nally, R., and D. Young. 1981. Song energetics of the bladder cicada, *Cystosoma saundersii*. *Journal of Experimental Biology* 90:185–196.

Magnus, D. 1958. Experimentelle Untersuchungen zur Bionomie und Ethologie des Kaisermantels *Argynnis paphia* L. (Lep. Nymph.). *Zeitschrift für Tierpsychologie* 15:397–426.

Maier, C. T., and G. P. Waldbauer. 1979. Dual mate-seeking strategies in male syrphid flies (Diptera: Syrphidae). *Annals of the Entomological Society of America* 72:54–61.

Mangan, R. L. 1979. Reproductive behavior of the cactus fly, *Odontoloxozus longicornis*, male territoriality and female guarding as adaptive strategies. *Behavioral Ecology and Sociobiology* 4:265–278.

Mangold, J. R. 1978. Attraction of *Euphasiopteryx ochracea*, *Corethrella* sp. and gryllids to broadcast songs of the southern mole cricket. *Florida Entomologist* 61:57–61.

Manning, A. 1962. The sperm factor affecting the receptivity in female *Drosophila melanogaster*. *Nature* 211:1321–22.

———— 1967. The control of sexual receptivity in female *Drosophila*. *Animal Behaviour* 15:239–250.

Manning, J. T. 1975. Male discrimination and investment in *Asellus aquaticus* (L.) and *A. meridianus* Racovitsza (Crustacea: Isopoda). *Behaviour* 55:1–14.

———— 1980. Sex ratio and optimal male time investment strategies in *Asellus aquaticus* (L.) and *A. meridianus* Racovitsza. *Behaviour* 74:265–273.

Marikovsky, P. I. 1961. Material on the sexual biology of the ant *Formica rufa* L. *Insectes Sociaux* 8:23–30.

Markl, H., B. Hölldobler, and T. Hölldobler. 1977. Mating behavior and sound production in harvester ants (*Pogonomyrmex*, Formicidae). *Insectes Sociaux* 24:191–212.

Markow, T. A. 1980. Rare male advantages among *Drosophila* of the same laboratory strain. *Behavior Genetics* 10:553–556.

————— Forthcoming. Mating systems of cactophilic *Drosophila*. In *Ecological genetics and evolution: the cactus-yeast-Drosophila model system*, ed. J. S. F. Barker and W. T. Starmer. New York: Academic Press.

Markow, T. A., M. Quaid, and S. Kerr. 1978. Male mating experience and competitive courtship success in *Drosophila melanogaster*. *Nature* 276: 821–822.

Marks, R. J. 1976. Mating behavior and fecundity of the red bollworm *Diparopsis castanes* Hmps. (Lepidoptera, Noctuidae). *Bulletin of Entomological Research* 66:145–158.

Marlatt, C. L. 1898. The periodical cicada. *United States Department of Agriculture Bulletin* 14:1–148.

Marshall, G. A. K. 1901. On the female pouch in *Acraea. Entomologist* 34: 73–75.

Marshall, L. D. 1980. Paternal investment in *Colias philodice-eurytheme* butterflies (Lepidoptera: Pieridae) Master's thesis, Arizona State University.

————— 1982. Male nutrient investment in the Lepidoptera: what nutrients should males invest? *American Naturalist* 120:27–35.

Marshall, L. D., and J. Alcock. 1981. The evolution of the mating system of the carpenter bee *Xylocopa varipuncta* (Hymenoptera: Anthophoridae). *Journal of Zoology* 193:315–324.

Mason, L. G. 1964. Stabilizing selection for mating fitness in natural populations of *Tetraopes. Evolution* 18:492–497.

Matthes, D. 1962. Excitatoren und Paarungsverhalten mitteleuropäischer Malachiiden (Coleopt., Malacodermata). *Zeitschrift für Morphologie und Ökologie der Tiere* 51:375–546.

Matthews, R. W. 1975. Courtship in parasitic wasps. In *Symposium on evolutionary strategies of parasitic insects and mites*, ed. P. Price. New York: Plenum Press.

Matthews, R. W., and J. R. Matthews. 1978. *Insect behavior*. New York: John Wiley.

Matthewson, J. A. 1968. Nest construction and life history of the eastern curcurbit bee, *Peponapis pruinosa* (Hymenoptera: Apoidea). *Journal of the Kansas Entomological Society* 41:255–261.

Mayer, M. S., and J. R. Brazzel. 1963. The mating behavior of the boll weevil, *Anthonomus grandis. Journal of Economic Entomology* 56:605–609.

Maynard Smith, J. 1956. Fertility, mating behavior and sexual selection in *Drosophila subobscura. Journal of Genetics* 54:261–279.

————— 1966. *The theory of evolution*. Baltimore: Penguin Books.

————— 1971. What use is sex? *Journal of Theoretical Biology* 30:319–335.

————— 1972. Game theory and the evolution of fighting. In *On evolution*, ed. J. Maynard Smith. Edinburgh: Edinburgh University Press.

————— 1974. The theory of games and evolution of animal conflict. *Journal of Theoretical Biology* 47:209–221.

————— 1976a. Evolution and the theory of games. *American Scientist* 64:41–45.

————— 1976b. Sexual selection and the handicap principle. *Journal of Theoretical Biology* 57:237–242.

————— 1978a. Optimization theory in evolution. *Annual Review of Ecology and Systematics* 9:31–56.

————— 1978b. *The evolution of sex*. Cambridge: Cambridge University Press.

Maynard Smith, J., and G. A. Parker. 1976. The logic of asymmetric contests. *Animal Behaviour* 24:159–175.

Maynard Smith, J., and G. R. Price. 1973. The logic of animal conflicts. *Nature* 246:15–18.

Mayr, E. 1963. *Animal species and evolution*. Cambridge, Massachusetts: Harvard University Press.

———— 1972. Sexual selection and natural selection. In *Sexual selection and the descent of man*, ed. B. Campbell. Chicago: Aldine.

Mays, D. 1971. Mating behavior of nemobiine crickets—*Hygronemobius, Nemobius* and *Pteronemobius* (Orthoptera: Gryllidae). *Florida Entomologist* 54:113–126.

McAlpine, D. K. 1970. Diptera. In *Insects of Australia,* ed. D. F. Waterhouse. Melbourne: Melbourne University Press.

———— 1975. Combat between males of *Pogonortalis doclea* (Diptera, Platystomatidae) and its relation to structural modification. *Australian Entomological Magazine* 2:104–107.

———— 1979. Agonistic behavior in *Achias australis* (Diptera, Platystomatidae) and the significance of eyestalks. In *Sexual selection and reproductive competition in insects,* ed. M. S. Blum and N. A. Blum. New York: Academic Press.

McAlpine, J. F., and D. D. Munroe. 1968. Swarming of lonchaeid flies and other insects, with descriptions of four new species of Lonchaeidae (Diptera). *Canadian Entomologist* 100:1154–78.

McCauley, D. E., and M. J. Wade. 1978. Female choice and the mating structure of a natural population of the soldier beetle, *Chauliognathus pennsylvanicus. Evolution* 32:771–775.

McLain, D. K. 1980. Female choice and the adaptive significance of prolonged copulation in *Nezara viridula* (Hemiptera: Pentatomaidae). *Psyche* 87:325–336.

———— 1981. Interspecific interference competition and mate choice in the soldier beetle *Chauliognathus pennsylvanicus. Behavioral Ecology and Sociobiology* 9:65–66.

Mead-Briggs, A. R., and J. A. Vaughan. 1969. Some requirements for mating in the rabbit flea, *Spilopsyllus cuniculi* (Dale) (Siphon., Pulicidae). *Journal of Experimental Biology* 51:495–511.

Meixner, A. J., and K. C. Shaw. 1979. Spacing and movement of singing *Neoconocephalus nebrascensis* males (Tettigoniidae: Copophorinae). *Annals of the Entomological Society of America* 72:602–606.

Mertins, J. W. 1980. Life history and behavior of *Laelis pedatus,* a gregarious bethylid ectoparasite of *Anthrenus verbasci. Annals of the Entomological Society of America* 73:686–693.

Meunchow, G. 1978. A note on the timing of sex in asexual/sexual organisms. *American Naturalist* 112:774–779.

Michelsen, A. 1963. Observations on the sexual behaviour of some longicorn beetles, subfamily Lepturinae (Coleoptera, Cerambycidae). *Behaviour* 22:152–166.

———— 1979. Insect ears as mechanical systems. *American Scientist* 67:696–706.

Michener, C. D. 1948. Observations of the mating behavior of harvester ants. *Journal of the New York Entomological Society* 56:239–242.

———— 1974. *The social behavior of bees.* Cambridge, Massachusetts: Harvard University Press.

Michener, C. D., and C. W. Rettenmeyer. 1956. The ethology of *Andrena erythronii* with comparative data on other species (Hymenoptera, Andrenidae). *University of Kansas Science Bulletin* 37:645–684.

Miller, D. R., and M. Kosztarab. 1979. Recent advances in the study of scale insects. *Annual Review of Entomology* 24:1–27.

Milne, A. 1960. Biology and ecology of the garden chafer, *Phyllopertha horticola* (L.). VII. The flight season: male and female behavior, and concluding discussion. *Bulletin of Entomological Research* 51:353–378.

Milne, L. J., and Milne, M. 1976. The social behavior of burying beetles. *Scientific American* 235:84–89.

Mitchell, P. L. 1980. Combat and territorial defense of *Acanthocephala femorata* (Hemiptera: Coreidae). *Annals of the Entomological Society of America* 73:404–408.

Mitchell, W. C., and R. F. L. Mau. 1971. Response of the female southern green stinkbug and its parasite *Trichopoda pennipes*, to male stink bug pheromones. *Annals of the Entomological Society of America* 64:856–859.

Mitter, C., D. J. Futuyuma, J. C. Schneider, and J. D. Hare. 1979. Genetic variation and host plant relations in a parthenogenetic moth. *Evolution* 33:777–790.

Mitzmain, M. B. 1910. Some new facts on the bionomics of the California rodent fleas. *Annals of the Entomological Society of America* 3:61–82.

Morris, D. 1954. The reproductive behaviour of the river-bullhead (*Cottus gobio* L.), with special reference to fanning activity. *Behaviour* 7:1–32.

Morris, G. K. 1971. Aggression in male conocephaline grasshoppers (Tettigoniidae). *Animal Behaviour* 19:132–137.

——— 1979. Mating systems, paternal investment and aggressive behavior of acoustic Orthoptera. *Florida Entomologist* 62:9–17.

——— 1980. Calling display and mating behaviour of *Copiphora rhinoceros* Pictet (Orthoptera: Tettigoniidae). *Animal Behaviour* 28:42–51.

Morris, G. K., and D. T. Gwynne. 1978. Geographical distribution and biological observations of *Cyphoderris* (Orthoptera: Haglidae) with a description of a new species. *Psyche* 85:147–167.

Morris, G. K., G. E. Kerr, and J. H. Fullard. 1978. Phonotactic preference of female meadow katydids (Orthoptera: Tettigoniidae: *Conocephalus nigropleurum*). *Canadian Journal of Zoology* 56:1479–87.

Morris, G. K., G. E. Kerr, and D. T. Gwynne. 1975. Ontogeny of phonotaxis in *Orchelimum gladiator*. (Orthoptera: Tettigoniidae: Conocephalinae). *Canadian Journal of Zoology* 53:1127–30.

Morton, E. S. 1975. Ecological sources of selection on avian sounds. *American Naturalist* 109:17–34.

Mühlenberg, M. 1973. Das Paarungsverhalten von *Ceratolaemus* Hesse sp. (Diptera, Bombyliidae, Cyrtosiinae) im Vergleich zu dem anderer Woolschweber. *Zeitschrift für Tierpsychologie* 33:437–460.

Mukai, T., H. Schaffer, and C. Cockeram. 1972. Genetic consequences of truncated selection at the phenotypic level in *Drosophila melanogaster*. *Genetics* 72:763–769.

Mullins, D. E., and C. B. Keil. 1980. Paternal investment of urates in cockroaches. *Nature* 283:567–569.

Murvosh, C. M., R. L. Fry, and G. C. LaBrecque. 1964. Studies on the mating behavior of the housefly, *Musca domestica* L. *Ohio Journal of Science* 64:264–271.

Myers, J., and L. P. Brower. 1969. A behavioural analysis of the courtship pheromone receptors of the queen butterfly, *Danaus gilippus berenice. Journal of Insect Physiology* 15:2117–30.

Nakagawa, S., G. J. Farias, D. Suda, R. T. Cunningham, and D. L. Chambers. 1971. Reproduction of the Mediterranean fruitfly; frequency of mating in the laboratory. *Annals of the Entomological Society of America* 69:949–950.

Nation, J. L. 1972. Courtship behavior and evidence of a sex attractant in the male Caribbean fruit fly, *Anastrepha suspensa. Annals of the Entomological Society of America* 65:1364–67.

Needham, J. G., J. R. Traver, and Y. Hsu. 1935. *The biology of mayflies.* Ithaca, New York: Comstock.

Neilson, W. T. A., and J. W. McAllan. 1965. Effects of mating on fecundity of the apple maggot, *Rhagoletis pomonella* (Walsh). *Canadian Entomologist* 97:276–279.

Nelson, S., A. D. Carlson, and J. Copeland. 1975. Mating-induced behavioural switch in female fireflies. *Nature* 255:628–629.

Nickerson, J. C., D. E. Snyder, and C. C. Oliver. 1979. Acoustical burrows constructed by mole crickets. *Annals of the Entomological Society of America* 72:438–440.

Nielsen, E. T. 1959. Copulation of *Glyptotendipes* (*Phytotendipes*) *paripes* Edwards. *Nature* 184:1252–63.

Nielsen, H. T. 1964. Swarming and some other habits of *Mansonia perturbans* and *Psorophora ferox* (Diptera: Culicidae). *Behaviour* 24:67–89.

Nielsen, H. T., and E. T. Nielsen. 1953. Field observations on the habits of *Aedes taeniorhynchus. Ecology* 34:141–156.

Nielson, M. W., and S. L. Toles. 1968. Observations on the biology of *Acinopterus angulatus* and *Aceratagallia curvata* in Arizona (Homoptera: Cicadellidae). *Annals of the Entomological Society of America* 61:54–56.

Nijholt, W. W. 1970. The effect of mating and the presence of the male ambrosia beetle, *Trypodendron lineatum*, on "secondary" attraction. *Canadian Entomologist* 102:894–897.

Nilakhe, S. S. 1976. Overwintering survival, fecundity, and mating behavior of the rice stink bug. *Annals of the Entomological Society of America* 69:717–720.

Nisbet, I. C. T. 1973. Courtship-feeding, egg-size and breeding success in common terns. *Nature* 241:141–142.

Norris, M. J. 1954. Sexual maturation in the desert locust with special reference to the effects of grouping. *Anti-Locust Bulletin* 18:1–44.

Nuorteva, P. 1954. Studies on the salivary enzymes of some bugs injuring wheat kernels. *Annales Entomologici Fennici* 20:102–124.

Nuttall, M. J. 1973a. Pre-emergence fertilization of *Megarhyssa nortoni nortoni* (Hymenoptera: Ichneumonidae). *New Zealand Entomologist* 5:112–117.

———— 1973b. Confirmation of pre-emergence fertilization of *Megarhyssa nortoni nortoni* (Hymenoptera: Ichneumonidae). *New Zealand Entomologist* 5:342–343.

Obara, Y., H. Tateda, and M. Kurabara. 1975. Mating behavior of the cabbage butterfly, *Pieris rapae crucivora* Boisduval. V. Copulatory stimuli inducing changes of female response patterns. *Dobut Zasshi* 84:71–76.

O'Donald, P. 1962. The theory of sexual selection. *Heredity* 17:541–552.

—————— 1980. *Genetic models of sexual selection*. Cambridge: Cambridge University Press.

Oh, R. J. 1979. Repeated copulation in the brown planthopper, *Nilaparvata lugens* Stål (Homoptera: Delphacidae). *Ecological Entomology* 4: 345–353.

Olander, R., and E. Palmén. 1968. Taxonomy, ecology and behaviour of the northern Baltic *Clunio marinus* Halid. (Dipt., Chironomidae). *Annales Zoologici Fennici* 5:97–110.

O'Neill, K. M. 1981. Male mating strategies and body size in three species of beewolves (Hymenoptera: Sphecidae; *Philanthus*). Ph.D. dissertation, Colorado State University.

O'Neill, K. M., and H. E. Evans. 1981. Predation on conspecific males by females of the beewolf *Philanthus basilaris* Cresson (Hymenoptera: Sphecidae). *Journal of the Kansas Entomological Society* 54:553–556.

Orgel, L. E., and F. H. C. Crick. 1980. Selfish DNA: the ultimate parasite. *Nature* 284:604–607.

Orians, G. H. 1969. On the evolution of mating systems in birds and mammals. *American Naturalist* 103:589–603.

Ossiannilsson, F. 1949. Insect drummers. *Opuscula Entomologica, Supplement* 10:1–145.

Oster, G. F., and E. O. Wilson. 1978. *Caste and ecology in the social insects*. Princeton: Princeton University Press.

Otte, D. 1970. A comparative study of communicative behavior in grasshoppers. *University of Michigan Museum of Zoology, Miscellaneous Publications* 141:1–168.

—————— 1972. Simple versus elaborate behavior in grasshoppers: an analysis of communication in the genus *Syrbula*. *Behaviour* 42:291–322.

—————— 1974. Effects and functions in the evolution of signaling systems. *Annual Review of Ecology and Systematics* 5:385–417.

—————— 1975. On the role of intraspecific deception. *American Naturalist* 109:239–242.

—————— 1977. Communication in Orthoptera. In *How animals communicate*, ed. T. A. Sebeok. Bloomington: Indiana University Press.

—————— 1979. Historical development of sexual selection theory. In *Sexual selection and reproductive competition in insects*, ed. M. S. Blum and N. A. Blum. New York: Academic Press.

—————— 1980. On theories of flash synchronization in fireflies. *American Naturalist* 116:587–590.

Otte, D., and A. Joern. 1975. Insect territoriality and its evolution: population studies of desert grasshoppers on creosote bushes. *Journal of Animal Ecology* 44:29–54.

Otte, D., and J. Loftus-Hills. 1979. Chorusing in *Syrbula* (Orthoptera: Acrididae): cooperation, interference, competition or concealment? *Entomological News* 90:159–165.

Otte, D., and J. Smiley. 1977. Synchrony in Texas fireflies with a consideration of male interaction models. *Biology of Behavior* 2:143–158.

Otte, D., and K. Stayman. 1979. Beetle horns: some patterns in functional morphology. In *Sexual selection and reproductive competition in insects*, ed. M. S. Blum and N. A. Blum. New York: Academic Press.

Ouye, M. T., R. S. Garcia, H. M. Graham, and D. F. Martin. 1965. Mating studies on the pink bollworm *Pectinophora gossypiella* (Lepidoptera: Gelechidae), based on the presence of spermatophores. *Annals of the Entomological Society of America* 58:880–882.

Pace, A. E. 1967. Life history and behavior of a fungus beetle, *Bolitotherus cornutus* (Tenebrionidae). *University of Michigan Museum of Zoology, Occasional Papers* 653:1–15.

Page, R. E., Jr. 1980. The evolution of multiple mating behavior by honey bee queens (*Apis mellifera* L.). *Genetics* 96:263–273.

Page, R. E., Jr., and R. A. Metcalf. 1982. Multiple mating, sperm utilization, and social evolution. *American Naturalist* 119:263–281.

Pamilo, P., S. Varvio-Aho, and A. Pekkarinen. 1978. Low enzyme gene variability in Hymenoptera as a consequence of haplodiploidy. *Hereditas* 88:93–99.

Parker, G. A. 1968. The sexual behaviour of the blowfly, *Protophormia terraenovae* R.—D. *Behaviour* 32:291–308.

———— 1970a. Sperm competition and its evolutionary effect on copula duration in the fly *Scatophaga stercoraria*. *Journal of Insect Physiology* 16:1301–28.

———— 1970b. The reproductive behaviour and the nature of sexual selection in *Scatophaga stercoraria* L. (Diptera: Scatophagidae). II. The fertilization rate and the spatial and temporal relationships of each sex around the site of mating and oviposition. *Journal of Animal Ecology* 39:205–228.

———— 1970c. Sperm competition and its evolutionary consequences in the insects. *Biological Reviews* 45:525–567.

———— 1970d. The reproductive behaviour and nature of sexual selection in *Scatophaga stercoraria* L. (Diptera: Scatophagidae). VII. The origin and evolution of the passive phase. *Evolution* 24:774–788.

———— 1972. Reproductive behaviour of *Sepsis cynipsea* (L.) (Diptera: Sepsidae). I. A preliminary analysis of the reproductive strategy and its associated behaviour patterns. *Behaviour* 41:172–206.

———— 1974a. Assessment strategy and the evolution of fighting behaviour. *Journal of Theoretical Biology* 47:223–243.

———— 1974b. Courtship persistence and female-guarding as male time-investment strategies. *Behaviour* 48:157–184.

———— 1974c. The reproductive behaviour and the nature of sexual selection in *Scatophaga stercoraria* L. (Diptera: Scatophagidae). IX. Spatial distribution of fertilization rates and evolution of male search strategy within the reproductive area. *Evolution* 28:93–108.

———— 1974d. The reproductive behaviour and the nature of sexual selection in *Scatophaga stercoraria* L. (Diptera: Scatophagidae). VIII. The behaviour of searching males. *Journal of Entomology* (A) 48:199–211.

———— 1978a. Searching for mates. In *Behavioural ecology, an evolutionary approach,* ed. J. R. Krebs and N. B. Davies. London: Blackwell.

———— 1978b. Evolution of competitive mate searching. *Annual Review of Entomology* 23:173–196.

———— 1978c. Selfish genes, evolutionary games and the adaptiveness of behaviour. *Nature* 274:849–855.

———— 1979. Sexual selection and sexual conflict. In *Sexual selection and reproductive competition in insects,* ed. M. S. Blum and N. A. Blum. New York: Academic Press.

Parker, G. A., and R. G. Pearson. 1976. Possible origin and adaptive significance of mounting behaviour shown by some female mammals in estrus. *Journal of Natural History* 10:211–245.

Parker, G. A., and J. L. Smith. 1975. Sperm competition and the evolution of the precopulatory passive phase behaviour in *Locusta migratoria migratorioides*. *Journal of Entomology* (A) 49:155–171.

Parker, G. A., and R. A. Stuart. 1976. Animal behavior as a strategy optimizer: evolution of resource assessment strategies and optimal emigration thresholds. *American Naturalist* 110:1055–76.

Parker, G. A., R. R. Baker, and V. G. F. Smith. 1972. The origin and evolution of gamete dimorphism and the male-female phenomenon. *Journal of Theoretical Biology* 36:529–553.

Parker, G. A., G. R. G. Hayhurst, and J. S. Bradley. 1974. Attack and defense strategies in reproductive interactions of *Locusta migratoria,* and their adaptive significance. *Zeitschrift für Tierpsychologie* 34:1–24.

Parsons, P. A. 1977. Lek behavior in *Drosophila* (*Hirtodrosophila*) *polypori* Malloch—an Australian rainforest species. *Evolution* 31:223–225.

Partridge, L. 1980. Mate choice increases a component of offspring fitness in fruit flies. *Nature* 283:290–291.

Payne, R. B., and K. Payne. 1977. Social organization and mating success in local song population of village indigo birds *Vidua chalybeata. Zeitschrift für Tierpsychologie* 45:113–173.

Pazella, V. M. 1979. Behavioral ecology of the dragonfly *Libellula pulchella* Drury (Odonata: Anisoptera). *American Midland Naturalist* 102:1–22.

Pease, R. W., Jr. 1968. The evolutionary and biological significance of multiple pairing in Lepidoptera. *Journal of the Lepidopterists' Society* 22:197–209.

Peckham, D. J. 1977. Reduction of miltogrammine cleptoparasitism by male *Oxybelus subulatus* (Hymenoptera: Sphecidae). *Annals of the Entomological Society of America* 70:823–828.

Peckham, D. J., and A. W. Hook. 1980. Behavioral observations on *Oxybelus* in southeastern North America. *Annals of the Entomological Society of America* 73:557–567.

Peckham, D. J., F. E. Kurczewski, and D. B. Peckham. 1973. Nesting behavior of nearctic species of *Oxybelus* (Hymenoptera: Sphecidae). *Annals of the Entomological Society of America* 66:647–661.

Periotti, M. E. 1973. The mitochondrial derivative of the spermatozoan of *Drosophila* before and after fertilization. *Journal of Ultrastructure Research* 44:181–198.

Pesson, P. 1950. Sur un phénomène de phoresie des spermatozoides par des cellules oviductaires, chez *Aspidiotus ostreaeformis* Curt. (Hemiptera-Homoptera-Coccoidae). *Proceedings of the 8th International Congress of Entomology, Stockholm* 1948:566–570.

Petit, C., and L. Ehrman. 1969. Sexual selection in *Drosophila. Evolutionary Biology* 3:177–223.

Pickford, R., and C. Gillott. 1971. Insemination in the migratory grasshopper, *Melanoplus sanguinipes* (Fabr.). *Canadian Journal of Zoology* 49:1583–88.

——— 1972a. Courtship behavior of the migratory grasshopper, *Melanoplus sanguinipes* (Orthoptera: Acrididae). *Canadian Entomologist* 104:715–722.

———— 1972b. Coupling behaviour of the migratory grasshopper, *Melanoplus sanguinipes* (Orthoptera: Acrididae). *Canadian Entomologist* 104: 873–879.

van der Pijl, L., and C. H. Dodson. 1966. *Orchid flowers: their pollination and evolution.* Coral Gables, Florida: University of Miami Press.

Pinto, J. D. 1972. A synopsis of the bionomics of *Phodaga alticeps* (Coleoptera: Meloidae) with special reference to sexual behavior. *Canadian Entomologist* 104:577–595.

———— 1973. Sexual behavior in the genus *Pleuropompha* LeConte: a new mating display in blister beetles (Coleoptera: Meloidae). *Canadian Entomologist* 105:957–969.

———— 1975. Intra- and interspecific courtship behavior in blister beetles of the genus *Tegrodera* (Meloidae). *Annals of the Entomological Society of America* 68:275–285.

———— 1980. Behavior and taxonomy of the *Epicauta maculata* group (Coleoptera: Meloidae). *University of California Publications in Entomology* 89:1–111.

Piper, G. L. 1976. Bionomics of *Euarestoides acutangulus* (Diptera: Tephritidae). *Annals of the Entomological Society of America* 69:381–386.

Pliske, T. E. 1973. Factors determining mating frequencies in some New World butterflies and skippers. *Annals of the Entomological Society of America* 66:164–169.

———— 1975. Courtship behavior of the monarch butterfly, *Danaus plexippus* L. *Annals of the Entomological Society of America* 68:143–151.

Pliske, T. E., and T. Eisner. 1969. Sex pheromone of the queen butterfly: biology. *Science* 164:1170–72.

Pollack, G. S., and R. R. Hoy. 1979. Temporal patterns as a cue for species-specific calling song recognition in crickets. *Science* 204:429–432.

Potter, D. A. 1981. Agonistic behavior in male spider mites: factors affecting frequency and intensity of fighting. *Annals of the Entomological Society of America* 74:138–143.

Potter, D. A., D. L. Wrensch, and D. E. Johnston. 1976. Guarding, aggressive behavior, and mating success in male two spotted spider mites. *Annals of the Entomological Society of America* 69:707–711.

Poulton, E. D. 1890. *The colors of animals: their meaning and use, especially considered in the case of insects.* London: Kegan Paul.

Power, H. W. 1976. On forces of selection and the evolution of mating types. *American Naturalist* 110:937–944.

Prestwich, K. N., and T. J. Walker. 1981. Energetics of singing in crickets: effect of temperature in three trilling species (Orthoptera: Gryllidae). *Journal of Comparative Physiology* (B) 143:199–212.

Pritchard, G. 1967. Laboratory observations on the mating behaviour of the island fruit fly *Rioxa pornia* (Diptera: Tephritidae). *Journal of the Australian Entomological Society* 6:127–132.

Prokopy, R. J., and J. Hendrichs. 1979. Mating behavior of *Ceratitis capitata* on a field-caged host tree. *Annals of the Entomological Society of America* 72:642–648.

Prouty, M. J., and G. R. Coatney. 1934. Further studies on the biology of the pigeon fly, *Pseudolynchia maura* Bigot (Diptera, Hippoboscidae). *Parasitology* 26:249–258.

Provost, M. W. 1958. Mating and male swarming in *Psorophora* mosquitoes.

Proceedings of 10th International Congress of Entomology, Montreal 2:553–561.

Provost, M. W., and J. S. Haeger. 1967. Mating and pupal attendance in *Deinocerites cancer* and comparisons with *Opifex fuscus* (Diptera: Culicidae). *Annals of the Entomological Society of America* 60:565–574.

Prozesky-Schulze, L., O. P. M. Prozesky, F. Anderson, and G. J. J. van der Merwe. 1975. Use of a self-made sound baffle by a tree cricket. *Nature* 255:142–143.

Pyle, D. W., and M. H. Gromko. 1978. Repeated mating by female *Drosophila melanogaster:* the adaptive importance. *Experientia* 34:449–450.

———— 1981. Genetic basis for repeated mating in *Drosophila melanogaster. American Naturalist* 117:133–146.

Raffa, K. F., and A. A. Berryman. In press. The role of host resistance in the colonization behavior and ecology of bark beetles (Coleoptera: Scolytidae). *Ecological Monographs.*

Reid, R. W. 1958. The behaviour of the mountain pine beetle, *Dendroctonus monticolae* Hopk., during mating, egg laying and gallery construction. *Canadian Entomologist* 90:505–509.

Renner, M. 1952. Analyze der Kopulationsbereitschaft des Weibchens der Feldheuschrecke *Euthystira brachyptera* Ocsk. in ihrer Abhängigkeit vom Zustand des Geschlechtsapparates. *Zeitschrift für Tierpsychologie* 9:122–154.

Rentz, D. C. 1975. Two new katydids from Costa Rica with comments on their life history strategies (Tettigoniidae: Pseudophyllinae). *Entomological News* 86:129–140.

Richards, A. M. 1960. Observations on the New Zealand glow-worm *Arachnocampa luminosa* (Skuse) 1890. *Transactions of the Royal Society of New Zealand* 88:559–574.

———— 1973. A comparative study of the giant wetas *Deinacrida heteracantha* and *D. fallai* (Orthoptera: Henicidae) from New Zealand. *Journal of Zoology* 169:195–236.

———— 1980. Sexual selection, guarding and sexual conflict in a species of Coccinellidae (Coleoptera). *Journal of the Australian Entomological Society* 19:26.

Richards, O. W. 1927. Sexual selection and allied problems in the insects. *Biological Reviews* 2:298–364.

Riddiford, L. M., and J. Ashenhurst. 1973. The switchover from virgin to mated behavior in female *Cecropia* moths: the role of the bursa copulatrix. *Biological Bulletin* 144:162–171.

Ridley, M. 1978. Paternal care. *Animal Behaviour* 26:904–932.

Ridsdill Smith, T. J. 1970. The behaviour of *Hemithynnus hyalinatus* (Hymenoptera: Tiphiidae), with notes on some other Thynninae. *Journal of the Australian Entomological Society* 9:196–208.

Riegert, P. W. 1965. Effects of grouping, pairing and mating on the bionomics of *Melanoplus bilituratus* (Walker) (Orthoptera: Acrididae). *Canadian Entomologist* 97:1046–51.

Riemann, J. G., D. O. Moen, and B. J. Thorson. 1967. Female monogamy and its control in the housefly, *Musca domestica* L. *Journal of Insect Physiology* 13:407–418.

Ringo, J. M. 1977. Why 300 species of Hawaiian *Drosophila?* The sexual selection hypothesis. *Evolution* 31:694–696.

Ritcher, P. O., and F. M. Beer. 1956. Notes on the biology of *Pleocoma dubitalis dubitalis* Davis. *Pan-Pacific Entomologist* 32:181–184.

Ritter, H., Jr. 1964. Defense of mate and mating chamber in a wood roach. *Science* 143:1459–60.

Roberts, R. B. 1971. Biology of the crepuscular bee *Ptiloglossa guinnae* n. sp. with notes on associated bees, mites and yeast. *Journal of the Kansas Entomological Society* 44:283–294.

Robertson, H. M. 1982. Mating behaviour and its relationship to territoriality in *Platycypha caligata* (Selys) (Odonata: Chlorocyphidae). *Behaviour* 79:11–27.

Roeder, K. D. 1967. *Nerve cells and insect behavior.* Cambridge, Massachusetts: Harvard University Press.

Ross, A. 1961. Biological studies on bat ectoparasites of the genus *Trichobius* (Diptera: Streblidae) in North America, north of Mexico. *Wasmann Journal of Biology* 19:229–246.

Roth, L. M. 1962. Hypersexual activity induced in females of the cockroach *Nauphoeta cinerea. Science* 138:1267–69.

——— 1967. Uricose glands in the accessory sex gland complex of male Blatteria. *Annals of the Entomological Society of America* 60:1203–11.

——— 1969. The evolution of tergal glands in the Blatteria. *Annals of the Entomological Society of America* 62:176–208.

Roth, L. M., and W. H. Stahl. 1956. Tergal and cercal secretions of *Blatta orientalis* L. *Science* 123:798–799.

Roth, L. M., and E. R. Willis. 1952. A study of cockroach behavior. *American Midland Naturalist* 47:66–129.

Rothschild, M. 1975. Recent advances in our knowledge of the order Siphonaptera. *Annual Review of Entomology* 20:241–259.

Rothstein, S. I. 1979. Gene frequencies and selection for inhibitory traits, with special emphasis on the adaptiveness of territoriality. *American Naturalist* 113:317–331.

Roubaud, E. 1910. Recherches sur la biologie des *Synagris* (Hymen.). Evolution de l'instinct chez les guêpes solitaires. *Annales de le Société Entomologique de France* 79:1–21. Translated in *Smithsonian Reports* 1910 (1911):507–525.

Rozen, J. G. 1958. Monographic study of the genus *Nomadopsis* Ashmead (Hymenoptera: Andrenidae). *University of California Publications in Entomology* 15:1–202.

——— 1965. The biology and immature stages of *Melitturga clavicornis* (Latrielle) and of *Sphecodes albilaris* (Kirby) and the recognition of the Oxaeidae at the family level (Hymenoptera, Apoidea). *American Museum Novitates* 2224:1–18.

———. 1969. Biological note on the bee *Tetralonia minuta* and its cleptoparasite *Morgania histriotransvaalensis* (Hymenoptera: Anthophoridae). *Proceedings of the Entomological Society of Washington* 71:102–107.

———. 1977. Biology and immature stages of the bee genus *Meganomia* (Hymenoptera, Melittidae). *American Museum Novitates* 2630:1–14.

Rudinsky, J. A. 1969. Masking of the aggregation pheromone in *Dendroctonus psuedotsugae* Hopk. *Science* 166:884–885.

Rudinsky, J. A., and L. C. Ryker. 1976. Sound production in Scolytidae: rivalry and pre-mating stridulation of the male Douglas fir beetle. *Journal of Insect Physiology* 22:997–1003.

Rudinsky, J. A., P. T. Oester, and L. C. Ryker. 1978. Gallery initiation and male stridulation of the polygamous spruce bark beetle *Polygraphus rufipennis*. *Annals of the Entomological Society of America* 71:317–321.

Rudinsky, J. A., L. C. Ryker, R. R. Michael, L. M. Libbey, and M. E. Morgan. 1976. Sound production in Scolytidae: female sonic stimulus of male pheromone release in two *Dendroctonus* beetles. *Journal of Insect Physiology* 22:1675–81.

Rutowski, R. L. 1978a. The form and function of ascending flights in *Colias* butterflies. *Behavioral Ecology and Sociobiology* 3:163–172.

———— 1978b. The courtship behaviour of the small sulphur butterfly, *Eurema lisa* (Lepidoptera: Pieridae). *Animal Behaviour* 26:892–903.

———— 1979. Courtship behavior of the checkered white, *Pieris protodice* (Pieridae). *Journal of the Lepidopterists' Society* 33:42–49.

———— 1980. Courtship solicitation by females of the checkered white butterfly, *Pieris protodice*. *Behavioral Ecology and Sociobiology* 7:113–117.

———— 1982. Mate choice and lepidopteran mating behavior. *Florida Entomologist* 65:72–82.

Rutowski, R. L., and J. Alcock. 1980. Temporal variation in male copulatory behavior in the solitary bee *Nomadopsis puellae* (Hymenoptera: Andrenidae). *Behaviour* 73:175–188.

Rutowski, R. L., C. E. Long, L. D. Marshall, and R. S. Vetter. 1981. Courtship solicitation by *Colias* females (Lepidoptera: Pieridae). *American Midland Naturalist* 105:334–340.

Ruttner, F. 1956. The mating of the honeybee. *Bee World* 37:3–15.

Ryan, R. B., R. W. Mortensen, and T. R. Torgersen. 1981. Reproductive biology of *Telenomus californicus* Ashmead, an egg parasite of the Douglas-fir tussock moth: laboratory studies. *Annals of the Entomological Society of America* 74:213–216.

Ryker, L. C. In press. The world of the Douglas-fir beetle. *Scientific American*.

Ryker, L. C., and J. A. Rudinsky. 1976. Sound production in Scolytidae: aggressive and mating behavior of the mountain pine beetle. *Annals of the Entomological Society of America* 69:677–680.

Sacca, G. 1964. Comparative bionomics of the genus *Musca*. *Annual Review of Entomology* 9:341–358.

Sakagami, S. F., H. Ubukata, M. Iga, and M. J. Toda. 1974. Observations of the behavior of some Odonata in the Bonin Islands, with considerations on the evolution of reproductive behavior in Libellulidae. *Journal of the Faculty of Science, Hokkaido University, Series VI, Zoology* 19:722–757.

Sakaluk, S. K., and Cade, W. H. 1980. Female mating frequency and progeny production in singly and doubly mated house and field crickets. *Canadian Journal of Zoology* 58:404–411.

Sanders, C. J. 1975. Factors affecting adult emergence and mating behavior of the eastern spruce budworm, *Choristoneura fumiferana* (Lepidoptera: Totricidae). *Canadian Entomologist* 107:967–977.

Satterthwait, A. F. 1931. *Anaphoidea calendra* Gahan, a mymarid parasite of eggs of weevils of the genus *Calendra*. *Journal of the New York Entomological Society* 39:171–190.

Saunders, E. 1909. *Bombi* and other aculeates collected in 1908 in the Berner

Oberland by the Rev. A. E. Eaton, M.A. *Entomologist's Monthly Magazine* 45:83–84.

Savolainen, E., and J. Syrajämäki. 1971. Swarming and mating of *Erioptera gemina* Tjeder (Dipt., Limoniidae). *Annales Entomologici Fennici* 37:79–85.

Scarbrough, A. G. 1978. Ethology of *Cerotainia albipilosa* Curran (Diptera: Asilidae) in Maryland: courtship, mating and oviposition. *Proceedings of the Entomological Society of Washington* 80:179–190.

Schal, C., and W. J. Bell. 1982. Ecological correlates of paternal investment of urates in a tropical cockroach. *Science* 218:170–172.

Schaller, F. 1971. Indirect sperm transfer by soil arthropods. *Annual Review of Entomology* 16:407–446.

Scheiring, J. E. 1977. Stabilizing selection for size as related to mating fitness in *Tetraopes*. *Evolution* 31:447–449.

Schenkel, R. 1956. Zur Deutung der Balzleistungen einiger Phasianiden und Tetraoniden. *Ornithologische Beobachter* 53:65–95.

Schmidt, E. 1975. Zur Klassifikation des Eiablageverhaltens der Odonaten. *Odonatologica* 4:177–183.

Schneider, D. 1964. Insect antennae. *Annual Review of Entomology* 9:103–122.

———— 1969. Insect olfaction: deciphering system for chemical messages. *Science* 163:1031–36.

Schneider, D., M. Boppré, J. Zweig, S. B. Horsley, T. W. Bell, J. Meinwald, K. Hansen, and E. W. Diehl. 1982. Scent organ development in *Creatonotos* moths: regulation by pyrrolizidine alkaloids. *Science* 215:1264–65.

Schöne, H., and J. Tengö. 1981. Competition of males, courtship behaviour and chemical communication in the digger wasp *Bembix rostrata* (Hymenoptera, Sphecidae). *Behaviour* 77:44–66.

Schowalter, T. D., and W. G. Whitford. 1979. Territorial behavior of *Bootettix argentatus* Bruner (Orthoptera: Acrididae). *American Midland Naturalist* 102:182–184.

Schuster, M. F. 1965. Studies on the biology of *Dusmetia sangwani* (Hymenoptera: Encyrtidae). *Annals of the Entomological Society of America* 58:272–275.

Schwagmeyer, P. L. 1979. The Bruce effect: an evaluation of male/female advantages. *American Naturalist* 113:932–939.

Scott, J. A. 1968. Hilltopping as a mechanism to aid the survival of low density species. *Journal of Research on the Lepidoptera* 7:191–204.

———— 1972. Mating of butterflies. *Journal of Research on the Lepidoptera* 11:99–127.

———— 1973a. Population biology and adult behavior of the circumpolar butterfly, *Parnassius phoebus* F. (Papilionidae). *Entomologia Scandica* 4:161–168.

———— 1973b. Adult behaviour and population biology of two skippers (Hesperiidae) mating in contrasting topographic sites. *Journal of Research on the Lepidoptera* 12:181–196.

———— 1974. Adult behavior and population biology of *Poladryas minuta,* and the relationship of the Texas and Colorado populations. *Pan-Pacific Entomologist* 50:9–22.

Searcy, W. A. 1979. Female choice of mates: a general model for birds and its application to red-winged blackbirds (*Agelaius phoeniceus*). *American Naturalist* 114:77–100.

Searcy, W. A., and Yasukawa, K. 1981. Does the "sexy son" hypothesis apply to mate choice in red-winged blackbirds? *American Naturalist* 117:343–348.

Selander, R. B. 1964. Sexual behavior in blister beetles (Coleoptera: Meloidae). I. The genus *Pyrota*. *Canadian Entomologist* 96:1037–82.

Selander, R. K. 1972. Sexual selection and dimorphism in birds. In *Sexual selection and the descent of man*, ed. B. Campbell. Chicago: Aldine.

Severinghaus, L., B. H. Kurtak, and G. C. Eickwort. 1981. The reproductive behavior of *Anthidium manicatum* (Hymenoptera: Megachilidae) and the significance of size for territorial males. *Behavioral Ecology and Sociobiology* 9:51–58.

Shapiro, A. M. 1970. The role of sexual behavior in density-related dispersal of pierid butterflies. *American Naturalist* 104:367–372.

Sharp, J. L., N. C. Leppla, D. R. Bennett, W. K. Turner, and E. W. Hamilton. 1974. Flight ability of *Plecia nearctica* in the laboratory. *Annals of the Entomological Society of America* 67:735–738.

Shaw, K. C. 1968. An analysis of the phonoresponse of males of the true katydid, *Pterophylla camellifolia* (Fabricius) (Orthoptera: Tettigoniidae). *Behaviour* 31:204–260.

Shields, O. 1967. Hilltopping. *Journal of Research on the Lepidoptera* 6:69–178.

Shields, O., and J. F. Emmel. 1973. A review of carrying pair behavior and mating times in butterflies. *Journal of Research on the Lepidoptera* 12:25–64.

Shields, W. M. 1982. *Philopatry, inbreeding and the evolution of sex*. Albany: State University of New York Press.

———— Forthcoming. Optimal inbreeding and the evolution of philopatry. In *The ecology of animal movement*, ed. I. R. Swingland and P. J. Greenwood. London: Oxford University Press.

Shinn, A. F. 1967. A revision of the bee genus *Calliopsis* and the biology and ecology of *C. andreniformis* (Hymenoptera, Andrenidae). *University of Kansas Science Bulletin* 46:753–936.

Shorey, H. H. 1973. Behavioral response to insect pheromones. *Annual Review of Entomology* 18:349–380.

———— 1976. *Animal communication by pheromones*. New York: Academic Press.

———— 1977. Pheromones. In *How animals communicate*, ed. T. A. Sebeok. Bloomington: Indiana University Press.

Shorey, H. H., and L. K. Gaston. 1964. Sex pheromones of noctuid moths. III. Inhibition of male response to the sex pheromone in *Trichoplusia ni* (Lepidoptera: Noctuidae). *Annals of the Entomological Society of America* 57:775–779.

———— 1965. Sex pheromones of noctuid moths. VII. Quantitative aspects of the production and release of pheromone by females of *Trichoplusia ni* (Lepidoptera: Noctuidae). *Annals of the Entomological Society of America* 58:604–608.

Shuster, S. M. 1981. Sexual selection in the Socorro isopod, *Thermosphaeroma thermophilum* (Cole) (Crustacea: Peracarida). *Animal Behaviour* 28:698–707.

Sigurjónsdóttir, H., and G. A. Parker. 1981. Dung fly struggles: evidence for assessment strategy. *Behavioral Ecology and Sociobiology* 8:219–230.

Silberglied, R. E. 1977. Communication in the Lepidoptera. In *How animals communicate,* ed. T. A. Sebeok. Bloomington: Indiana University Press.

Silberglied, R. E., and O. R. Taylor. 1978. Ultraviolet reflection and its behavioral role in courtship of sulfur butterflies *Colias eurytheme* and *Colias philodice* (Lepidoptera, Pieridae). *Behavioral Ecology and Sociobiology* 3:203–243.

Sillén-Tullberg, B. 1981. Prolonged copulation: a male "post-copulatory" strategy in a promiscuous species, *Lygaeus equestris* (Heteroptera: Lygaeidae). *Behavioral Ecology and Sociobiology* 9:283–289.

Silverstein, R. M. 1981. Pheromones: background and potential for use in insect pest control. *Science* 213:1326–32.

Silverstein, R. M., J. O. Rodin, and D. L. Wood. 1966. Sex attractants in frass produced by male *Ips confusus* in ponderosa pine. *Science* 154:509–510.

Simon-Thomas, R. T., and E. P. R. Poorter. 1972. Notes on the behaviour of males of *Philanthus triangulum* (F). (Hymenoptera, Sphecidae). *Tijdschrift voor Entomologie* 115:141–152.

Sims, S. R. 1979. Aspects of mating frequency and reproductive maturity in *Papilio zelicaon. American Midland Naturalist* 102:36–50.

Sivinski, J. 1978. Intrasexual aggression in the stick insects, *Diapheromera veliei* and *D. covilleae,* and sexual dimorphism in the Phasmatodea. *Psyche* 85:395–406.

——— 1980a. Sexual selection and insect sperm. *Florida Entomologist* 63:99–111.

——— 1980b. The effects of mating on predation in the stick insect *Diapheromera veliei* Walsh (Phasmatodea: Heteronemiidae). *Annals of the Entomological Society of America* 73:553–556.

——— 1981. "Love bites" in a lycid beetle. *Florida Entomologist* 64:541.

——— Forthcoming. Sperm in competition. In *Sperm competition and the evolution of animal mating systems,* ed. R. L. Smith. New York: Academic Press.

Smith, A. P., and J. Alcock. 1980. A comparative study of the mating systems of Australian eumenid wasps (Hymenoptera). *Zeitschrift für Tierpsychologie* 53:41–60.

Smith, D. C., and R. J. Prokopy. 1980. Mating behavior of *Rhagoletis pomonella* (Diptera, Tephritidae). VI. Site of early-season encounters. *Canadian Entomologist* 121:585–590.

Smith, N. G. 1966. Communication and relationships in the genus *Tyrannus. Nuttall Ornithological Club Publication* 6:1–250.

Smith, R. H., and M. R. Shaw. 1980. Haplodiploid sex ratios and the mutation rate. *Nature* 287:728–729.

Smith, R. L. 1976a. Brooding behavior of a male water bug *Belostoma flumineum* (Hemiptera: Belostomatidae). *Journal of the Kansas Entomological Society* 49:333–343.

——— 1976b. Male brooding behavior of the water bug *Abedus herberti* (Hemiptera: Belostomatidae). *Annals of the Entomological Society of America* 69:740–747.

——— 1979a. Paternity assurance and altered roles in the mating behaviour of a giant water bug, *Abedus herberti* (Heteroptera: Belostomatidae). *Animal Behaviour* 27:716–725.

———— 1979b. Repeated copulation and sperm precedence: paternity assurance for a male brooding water bug. *Science* 205:1029–31.

———— 1980. Evolution of exclusive postcopulatory paternal care in the insects. *Florida Entomologist* 63:65–78.

———— ed. Forthcoming. *Sperm competition and the evolution of animal mating systems.* New York: Academic Press.

Smith, R. L., and W. M. Langley. 1978. Cicada stress sound: an assay of its effectiveness as a predator defense mechanism. *Southwestern Naturalist* 23:187–196.

Smith, W. J. 1977. *The behavior of communicating: an ethological approach.* Cambridge, Massachusetts: Harvard University Press.

Snodgrass, R. E. 1946. The skeletal anatomy of fleas (Siphonaptera). *Smithsonian Miscellaneous Collections* 104:1–89.

Solomon, J. D., and W. W. Neel. 1973. Mating behavior in the carpenterworm moth, *Prionoxystus robiniae* (Lepidoptera: Cossidae). *Annals of the Entomological Society of America* 66:312–314.

Soper, R. S., G. E. Shewell, and D. Tyrell. 1976. *Colcondamyia auditrix* Nov. sp. (Diptera: Sarcophagidae), a parasite which is attracted by the mating song of its host, *Okanagana rimosa* (Homoptera: Cicadidae). *Canadian Entomologist* 108:61–68.

Southwood, T. R. E., and D. Leston. 1959. *Land and water bugs of the British Isles.* London: Frederick Warne.

Spiess, E. B., and H. L. Carson. 1981. Sexual selection in *Drosophila silvestris* of Hawaii. *Proceedings of the National Academy of Sciences U.S.A.* 78:3088–92.

Spieth, H. T. 1940. Studies on the biology of the Ephemeroptera. II. The nuptial flight. *Journal of the New York Entomological Society* 48:379–390.

———— 1952. Mating behavior within the genus *Drosophila. Bulletin of the American Museum of Natural History* 99:395–474.

———— 1966. Courtship behavior of endemic Hawaiian *Drosophila. University of Texas Publications* 6615:245–313.

———— 1968. Evolutionary implications of sexual behavior in *Drosophila. Evolutionary Biology* 2:157–193.

———— 1974a. Mating behavior and evolution of the Hawaiian *Drosophila.* In *Genetic mechanisms of speciation in insects,* ed. M. J. D. White. Sydney: Australian and New Zealand Book Co.

———— 1974b. Courtship behavior in *Drosophila. Annual Review of Entomology* 19:385–405.

———— 1981a. *Drosophila heteroneura* and *Drosophila silvestris:* head shapes, behavior and evolution. *Evolution* 35:921–930.

———— 1981b. Courtship behavior and evolutionary status of the Hawaiian *Drosophila primaeva* Hardy and Kaneshiro. *Evolution* 35:815–817.

Spieth, H. T., and W. B. Heed. 1975. The *Drosophila pinicola* species group (Diptera: Drosophilidae). *Pan-Pacific Entomologist* 51:287–295.

Stanton, D. S. In press. Muller's ratchet and the function of sexual recombination. *Evolution.*

Stay, B., and L. M. Roth. 1958. The reproductive behaviour of *Diploptera punctata* (Blattaria: Diplopteridae). *Proceedings of the 10th International Congress of Entomology, Montreal* 2:547–552.

Stearns, S. C. 1976. Life-history tactics: a review of the ideas. *Quarterly Review of Biology* 51:3–47.

Stebbins, G. L., and F. J. Ayala. 1981. Is a new evolutionary synthesis necessary? *Science* 213:967–971.

Steiner, A. L. 1978. Observations on spacing, aggressive and lekking behavior of digger wasp males of *Eucerceris flavocincta* (Hymenoptera: Sphecidae, Cercerini). *Journal of the Kansas Entomological Society* 51: 492–498.

Stephen, W. P., G. E. Bohart, and P. F. Torchio. 1969. *The biology and external morphology of bees.* Corvallis: Agricultural Experiment Station, Oregon State University.

Sternlicht, M. 1973. Parasitic wasps attracted by the sex pheromone of their coccid host. *Entomophaga* 18:339–342.

Stirling, P. D. 1980. Competition and coexistence among *Eupteryx* leafhoppers (Hemiptera: Cicadellidae) occurring on stinging nettles (*Urtica dioica*). *Journal of Animal Ecology* 49:793–805.

Stokes, A. W., and H. W. Williams. 1971. Courtship feeding in gallinaceous birds. *Auk* 88:543–559.

Stoltsfuz, W. B., and B. A. Foote. 1965. The use of froth masses in courtship in *Eutreta* (Diptera: Tephritidae). *Proceedings of the Entomological Society of Washington* 67:263–264.

Stoutamire, W. P. 1974. Australian terrestrial orchids, thynnid wasps and pseudocopulation. *American Orchid Society Bulletin* 43:13–18.

————— 1975. Pseudocopulation in Australian terrestrial orchids. *American Orchid Society Bulletin* 44:226–233.

Stresemann, E. 1914. Beiträge zur Kenntnis der Avifauna von Buru. *Novitates Zoologicae* 21:385–400.

Sugawara, T. 1979. Stretch reception in the bursa copulatrix of the butterfly, *Pieris rapae crucivora* and its role in behavior. *Journal of Comparative Physiology* 130:191–199.

Sullivan, R. T. 1981. Insect swarming and mating. *Florida Entomologist* 64:44–65.

Suomalainen, E., A. Saura, and J. Lokki. 1976. Evolution of parthenogenetic insects. *Evolutionary Biology* 9:209–257.

Suzuki, Y. 1978. Adult longevity and reproductive potential of the small cabbage white, *Pieris rapae crucivora* Boisduval (Lepidoptera: Pieridae). *Applied Entomology and Zoology* 13:312–313.

Svensson, B. G. 1980a. Species-isolating mechanisms in male bumble bees (Hymenoptera, Apidae). Ph.D. dissertation, Uppsala University, Sweden.

————— 1980b. Patrolling behaviour of bumble bee males in a subalpine/alpine area, Swedish Lapland. *Zoon* 7:67–94.

Sweeney, B. W., and R. L. Vannote. 1982. Population sychrony in mayflies: a predator satiation hypothesis. *Evolution* 36:810–821.

Sweet, M. H. 1964. The biology and ecology of the Rhyparochrominae of New England (Heteroptera: Lygaeidae). Part I. *Entomologica Americana* 43:1–124. Part II. *Entomologica Americana* 44:1–193.

Syrajämäki, J., and I. Ulmanen. 1970. Further experiments on male sexual behaviour in *Stictochironomus crassiforceps* (Kieff.) (Diptera, Chironomidae). *Annales Zoologici Fennici* 7:216–220.

Taber, S. 1954. The frequency of multiple mating of queen honeybees. *Journal of Economic Entomology* 47:995–998.

Tarshis, I. B. 1958. New data on the biology of *Stilbometopa impressa* (Bigot) and *Lynchia hirsuta* Ferris (Diptera: Hippoboscidae). *Annals of the Entomological Society of America* 51:95–105.

Tauber, M. J. 1968. Biology, behavior, and emergence rhythm of two species of *Fannia* (Diptera: Muscidae). *University of California Publications in Entomology* 50:1–86.

Tauber, M. J., and C. A. Toschi. 1965. Bionomics of *Euleia fratria* (Loew) (Diptera: Tephritidae). I. Life history and mating behavior. *Canadian Journal of Zoology* 43:369–379.

Tavolga, W. N. 1969. *Principles of animal behavior.* New York: Harper and Row.

Taylor, F. 1981. Ecology and evolution of physiological time in insects. *American Naturalist* 117:1–23.

Taylor, O. R., Jr. 1967. Relationship of multiple mating to fertility in *Atteva punctella* (Lepidoptera: Yponomeutidae). *Annals of the Entomolgical Society of America* 60:583–590.

Tengö, J. 1979. Odour-released behaviour in *Andrena* male bees (Apoidea, Hymenoptera). *Zoon* 7:15–48.

Tengö, J., and G. Bergström. 1977. Cleptoparasitism and odor mimetism in bees: do *Nomada* males imitate the odor of *Andrena* females? *Science* 196:1117–19.

Teskey, H. J. 1969. On the behavior and ecology of the face fly, *Musca autumnalis* (Diptera: Muscidae). *Canadian Entomologist* 101:561–576.

Thibout, E. 1975. Analyse des causes de l'inhibition de la receptivité sexuelle et de l'influence d'une éventuelle seconde copulation sur la réproduction chcz la teigne du poireau, *Acrolepia assectella* (Lepidoptera: Pluttellidae). *Entomologia Experimentalis et Applicata* 18:105–116.

Thomas, H. T. 1950. Field notes on the mating habits of *Sarcophaga* Meigen (Diptera). *Proceedings of the Royal Entomological Society of London (A)* 25:93–98.

Thornhill, R. 1975. Scorpionflies as kleptoparasites of web-building spiders. *Nature* 258:709–711.

——— 1976a. Sexual selection and paternal investment in insects. *American Naturalist* 110:153–163.

——— 1976b. Reproductive behavior of the lovebug, *Plecia nearctica* (Diptera: Bibionidae). *Annals of the Entomological Society of America* 69:843–847.

——— 1976c. Sexual selection and nuptial feeding behavior in *Bittacus apicalis* (Insecta: Mecoptera). *American Naturalist* 110:529–548.

——— 1977. The comparative predatory and sexual behavior of hangingflies (Mecoptera: Bittacidae). *Occasional Papers of the Museum of Zoology, University of Michigan* 677:1–43.

——— 1978a. Some arthropod predators and parasites of adult Mecoptera. *Environmental Entomology* 7:714–716.

——— 1978b. Sexually selected predatory and mating behavior of the hangingfly, *Bittacus stigmaterus* (Mecoptera: Bittacidae). *Annals of the Entomological Society of America* 71:597–601.

——— 1979a. Male and female sexual selection and the evolution of mating systems in insects. In *Sexual selection and reproductive competition in insects,* ed. M. S. Blum and N. A. Blum. New York: Academic Press.

——— 1979b. Male pair-formation pheromones in *Panorpa* scorpionflies (Mecoptera: Panorpidae). *Environmental Entomology* 8:886–889.

———— 1979c. Adaptive female-mimicking behavior in a scorpionfly. *Science* 295:412–414.

———— 1980a. Competitive, charming males and choosy females: was Darwin correct? *Florida Entomologist* 63:5–30.

———— 1980b. Sexual selection within mating swarms of the lovebug *Plecia nearctica* (Diptera: Bibionidae). *Animal Behaviour* 28:405–412.

———— 1980c. Mate choice in *Hylobittacus apicalis* (Insecta: Mecoptera) and its relation to some models of female choice. *Evolution* 34:519–538.

———— 1980d. Sexual selection in the black-tipped hangingfly. *Scientific American* 242:162–172.

———— 1980e. Rape in Panorpa scorpionflies and a general rape hypothesis. *Animal Behaviour* 28:52–59.

———— 1980f. Competition and coexistence among *Panorpa* scorpionflies (Mecoptera: Panorpidae). *Ecological Monographs* 50:179–197.

———— 1981. *Panorpa* (Mecoptera: Panorpidae) scorpionflies: systems for understanding resource-defense polygyny and alternative male reproductive efforts. *Annual Review of Ecology and Systematics* 12:355–386.

———— Forthcoming. Alternative hypotheses for traits presumed to have evolved in the context of sperm competition. In *Sperm competition and the evolution of animal mating systems*, ed. R. L. Smith. New York: Academic Press.

———— Manuscript a. Fighting and assessment in *Harpobittacus* scorpionflies.

———— Manuscript b. Cryptic female choice in the scorpionfly *Harpobittacus nigriceps* and its implications.

Tiemann, D. L. 1967. Observations on the natural history of the western banded glowworm, *Zarhipis integripennis* (Le Conte) (Coleoptera: Phengodidae). *Proceedings of the California Academy of Sciences* 35:235–264.

Tinbergen, N. 1951. *The study of instinct.* Oxford: Clarendon Press.

Tinbergen, N., B. J. D. Meeuse, L. K. Boerema, and W. W. Varossieau. 1942. Die Balz des Samtfallters, *Eumenis* (*Satyrus*) *semele* (L.). *Zeitschrift für Tierpsychologie* 5:182–226.

Tobin, E. N., and J. G. Stoffolano. 1973. The courtship of *Musca* species found in North America. 1. The house fly, *Musca domestica. Annals of the Entomological Society of America* 66:1249–57.

Tokunaga, M. 1935. Chironomidae from Japan (Diptera). V. Supplementary report on the Clunioninae. *Mushi* 8:1–20.

Tompkins, L., and J. C. Hall. 1981. The different effects on courtship of volatile compounds from mated and virgin *Drosophila* females. *Journal of Insect Physiology* 27:17–21.

Tompkins, L., J. C. Hall, and L. M. Hall. 1980. Courtship-stimulating volatile compounds from normal and mutant *Drosophila. Journal of Insect Physiology* 26:689–697.

Toschi, C. A. 1965. The taxonomy, life histories and mating behavior of the green lacewings of Strawberry Canyon (Neuroptera: Chrysopidae). *Hilgardia* 36:391–431.

Tostowaryk, W. 1971. Life history and behavior of *Podisus modestus* (Hemiptera: Pentatomidae) in boreal forest in Quebec. *Canadian Entomologist* 103:662–674.

Tozer, W. E., U. H. Resh, and J. O. Solem. 1981. Bionomics and adult behavior of a lentic caddisfly, *Nectopsyche albida* (Walker). *American Midland Naturalist* 106:133–144.

Triesman, M. 1976. The evolution of sexual reproduction: a model which assumes individual selection. *Journal of Theoretical Biology* 60:421–431.

Trivers, R. L. 1972. Parental investment and sexual selection. In *Sexual selection and the descent of man, 1871–1971,* ed. B. Campbell. Chicago: Aldine.

———— 1976. Sexual selection and resource accruing abilities in *Anolis garmani. Evolution* 30:253–269.

Trivers, R. L., and H. Hare. 1976. Haplodiploidy and the evolution of the social insects. *Science* 191:249–263.

Trivers, R. L., and D. E. Willard. 1973. Natural selection of parental ability to vary the sex ratio of offspring. *Science* 179:90–92.

Trottier, R. 1966. The emergence and sex ratio of *Anax junius* Drury (Odonata: Aeshnidae) in Canada. *Canadian Entomologist* 98:794–798.

Tschinkel, W. R. 1970. Chemical studies on the sex pheromone of *Tenebrio molitor* (Coleoptera: Tenebrionidae). *Annals of the Entomological Societey of America* 63:626–627.

Tschinkel, W. R., C. D. Wilson, and H. A. Bern. 1967. Sex pheromone of the mealworm beetle, *Tenebrio molitor. Journal of Experimental Zoology* 164:81–85.

Tullock, G. 1979. On the adaptive significance of territoriality: comment. *American Naturalist* 113:772–775.

Turell, M. J. 1976. Observation of the mating behavior of *Anthidiellum notatum* and *Anthidiellum perplexum. Florida Entomologist* 59:55–62.

Turner, N. 1960. The effect of inbreeding and crossbreeding on numbers of insects. *Annals of the Entomological Society of America* 53:686–688.

Tuskes, P. M., and L. P. Brower. 1978. Overwintering ecology of the monarch butterfly, *Danaus plexippus* L., in California. *Ecological Entomology* 3:141–153.

Tychsen, P. H. 1977. Mating behavior of the Queensland fruit fly, *Dacus tryoni,* in field cages. *Journal of the Australian Entomological Society* 16:459–465.

Ubukata, H. 1975. Life history and behavior of a corduliid dragonfly, *Cordulia aenea amurensis* Selys. II. Reproductive period with special reference to territoriality. *Journal of the Faculty of Science, Hokkaido University, Series VI, Zoology* 19:812–833.

Uéda, T. 1979. Plasticity of the reproductive behaviour in a dragonfly, *Sympetrus parvulum* Barteneff, with reference to the social relationships of males and the density of territories. *Researches on Population Ecology* 21:135–152.

Ulagaraj, S. M., and T. J. Walker. 1973. Phonotaxis of crickets in flight: attraction of male and female crickets to male calling songs. *Science* 182:1278–79.

Velthuis, H. H. W., and J. M. F. de Camargo. 1975a. Observations on male territories in a carpenter bee, *Xylocopa* (*Neoxylocopa*) *hirsutissima* Maidl (Hymenoptera, Anthophoridae). *Zeitschrift für Tierpsychologie* 38:409–418.

———— 1975b. Further observations on the function of male territories in the carpenter bee *Xylocopa* (*Neoxylocopa*) *hirsutissima* Maidl (Anthophoridae, Hymenoptera). *Netherlands Journal of Zoology* 25:516–528.

Velthuis, H. H. W., and D. Gerling. 1980. Observations on territoriality and mating behaviour of the carpenter bee *Xylocopa sulcatipes*. *Entomologia Experimentalis et Applicata* 28:82–91.

Villavaso, E. J. 1975. Functions of the spermathecal muscle of the boll weevil, *Anthonomus grandis*. *Journal of Insect Physiology* 21:1275–78.

Vité, J. P., and W. Francke. 1976. The aggregation pheromones of bark beetles: progress and problems. *Naturwissenschaften* 63:550–555.

Vité, J. P., and D. L. Williamson. 1970. *Thanasimus dubius*: prey perception. *Journal of Insect Physiology* 16:233–239.

Voelker, J. 1968. Untersuchungen zur Ernährung, Fortpflanzungsbiologie und Entwicklung von *Limnogeton fieberi* Mayr (Belostomatidae, Hemiptera) als Beitrag zur Kenntnis von natürlichen fienden tropischer Süsswasserschnecken. *Entomologische Mitteilungen* 3:1–24.

Waage, J. K. 1973. Reproductive behavior and its relation to territoriality in *Calopteryx maculata* (Beauvois) (Odonata: Calopterygidae). *Behaviour* 47:240–256.

———— 1975. Reproductive isolation and the potential for character displacement in the damselflies, *Calopteryx maculata* and *C. aequabilis* (Odonata: Calopterygidae). *Systematic Zoology* 24:24–36.

———— 1979a. Dual function of the damselfly penis: sperm removal and transfer. *Science* 203:916–918.

———— 1979b. Adaptive significance of postcopulatory guarding of mates and nonmates by male *Calopteryx maculata* (Odonata). *Behavioral Ecology and Sociobiology* 6:147–154.

———— 1979c. Reproductive character displacement in *Calopteryx* (Odonata: Calopterygidae). *Evolution* 33:104–116.

———— 1980. Adult sex ratios and female reproductive potential in *Calopteryx* (Zygoptera: Calopterygidae). *Odonatologica* 9:217–230.

———— Forthcoming. Sperm competition and the evolution of odonate mating systems. In *Sperm competition and the evolution of animal mating systems,* ed. R. L. Smith. New York: Academic Press.

Wade, M. J. 1976. Group selection among laboratory populations of *Tribolium*. *Proceedings of the National Academy of Science U.S.A.* 73:4604–7.

———— 1977. An experimental study of group selection. *Evolution* 31:134–153.

———— 1978. A critical review of the models of group selection. *Quarterly Review of Biology* 53:101–114.

———— 1979. Sexual selection and variance in reproductive success. *American Naturalist* 114:742–746.

———— 1980. Group selection, population growth rate and competitive ability in the flour beetles *Tribolium* spp. *Evolution* 61:1056–64.

Wade, M. J., and S. J. Arnold. 1980. The intensity of sexual selection in relation to male sexual behaviour, female choice, and sperm precedence. *Animal Behaviour* 28:446–461.

Waldbauer, G. P., and J. G. Sternburg. 1979. Inbreeding depression and a behavioral mechanism for its avoidance in *Hyalophora cecropia*. *American Midland Naturalist* 102:204–208.

Walker, T. J. 1964. Experimental demonstration of a cat locating orthopteran prey by the prey's calling song. *Florida Entomologist* 47:163–165.

———— 1969. Acoustic synchrony: two mechanisms in the snowy tree cricket. *Science* 166:891–894.

———— 1974a. *Gryllus ovisopis* n. sp.: a taciturn cricket with a life cycle suggesting allochronic speciation. *Florida Entomologist* 57:13–22.

———— 1974b. Character displacement and acoustic insects. *American Zoologist* 14:1137–50.

———— 1978. Post-copulatory behavior of the two-spotted tree cricket, *Neoxabea bipunctata*. *Florida Entomologist* 61:39–40.

———— 1980. Reproductive behavior and mating success of male short-tailed crickets: differences within and between demes. *Evolutionary Biology* 13:219–260.

———— Forthcoming a. Mating modes and female choice in short-tailed crickets (*Anurogryllus arboreus*). In *Orthopteran mating systems: sexual competition in a diverse group of insects*, ed. D. T. Gwynne and G. K. Morris. Boulder, Colorado: Westview Press.

———— Forthcoming b. Diel patterns of calling in nocturnal Orthoptera. In *Orthopteran mating systems: sexual competition in a diverse group of insects*, ed. D. T. Gwynne and G. K. Morris. Boulder, Colorado: Westview Press.

Walker, T. J., and D. Dew. 1972. Wing movements in calling katydids: fiddling finesse. *Science* 178:174–176.

Walker, W. F. 1980. Sperm utilization strategies in nonsocial insects. *American Naturalist* 115:780–799.

Wallace, A. R. 1889. *Darwinism*. 3rd ed., 1923. London: MacMillan.

Wallace, B. 1954. Genetic divergence of isolated populations of *Drosophila melanogaster*. *Caryologia* 6 (suppl.):761–764.

Weatherhead, P. J., and Robertson, R. J. 1979. Offspring quality and the polygyny threshold: "the sexy son hypothesis." *American Naturalist* 113:201–208.

Weber, H. 1930. *Biologie der Hemipteren*. Berlin: Springer.

———— 1954. *Grundriss der Insectkunde*. Stuttgart: Gustav Fischer.

Weissman, D. B. 1979. Assessment of mating status of female grasshoppers. *Entomological News* 90:121–124.

Weissman, D. B., and E. French. 1980. Autecology and population structure of *Trimerotropis occidentalis* (Bruner) (Orthoptera: Acrididae: Oedipodinae), a grasshopper with a reproductive dormancy. *Acrida* 9:145–157.

Werren, J. H. 1980. Sex ratio adaptations to local mate competition in a parasitic wasp. *Science* 208:1157–59.

Werren, J. H., M. R. Gross, and R. Shine. 1980. Paternity and the evolution of male parental care. *Journal of Theoretical Biology* 82:619–631.

West, M. J., and Alexander, R. D. 1963. Sub-social behavior in a burrowing cricket *Anurogryllus muticus* (DeGeer): Orthoptera: Gryllidae. *Ohio Journal of Science* 63:19–24.

West-Eberhard, M. J. 1979. Sexual selection, social competition and evolution. *Proceedings of the American Philosophical Society* 123:222–234.

———— Manuscript. Sexual selection, social competition, and speciation.

White, M. J. D. 1973. *Animal cytology and evolution*. 3rd ed. London: Cambridge University Press.

Whitham, T. G. 1978. Habitat selection by *Pemphigus* aphids in response to resource limitation and competition. *Ecology* 59:1164–76.

———— 1979. Territorial behavior of *Pemphigus* gall aphids. *Nature* 279: 324–325.

Wickler, W. 1968. *Mimicry*. London: Weidenfeld and Nicolson.

Wiklund, C. 1977. Courtship behaviour in relation to female monogamy in *Leptidea sinapis* (Lepidoptera). *Oikos* 29:275–283.

Wiklund, C., and T. Fagerström. 1977. Why do males emerge before females? A hypothesis to explain the incidence of protandry in butterflies. *Oecologia* 31:153–158.

Wilcox, R. S. 1979. Sex discrimination in *Gerris remigis:* role of a surface wave signal. *Science* 206:1325–27.

Wiley, R. H. 1973. Territoriality and nonrandom mating in sage grouse, *Centrocercus urophasianus. Animal Behaviour Monographs* 6:87–169.

Wiley, R. H., and D. G. Richards. 1978. Physical constraints in acoustic communication in atmosphere—implications for evolution of animal vocalizations. *Behavioral Ecology and Sociobiology* 3:69–94.

Willey, R. B., and R. L. Willey. 1970. The behavioral ecology of desert grasshoppers. I. Presumed sex role reversal in flight displays of *Trimerotropis agrestis. Animal Behaviour* 18:473–477.

Williams, G. C. 1966. *Adaptation and natural selection.* Princeton: Princeton University Press.

———— 1975. *Sex and evolution.* Princeton: Princeton University Press.

———— 1980. Kin selection and the paradox of sexuality. In *Sociobiology: beyond nature/nuture?*, ed. G. W. Barlow and J. Silverberg. Boulder, Colorado: Westview Press.

Williams, G. C., and J. B. Mitton. 1973. Why reproduce sexually? *Journal of Theoretical Biology* 39:545–554.

Willson, M. F. 1979. Sexual selection in plants. *American Naturalist* 113:777–790.

Wilson, D. S. 1975. A theory of group selection. *Proceedings of the National Academy of Sciences U.S.A.* 72:143–146.

———— 1977. Structured demes and the evolution of group-advantageous traits. *American Naturalist* 111:157–185.

———— 1980. *The natural selection of populations and communities.* Menlo Park: Benjamin/Cummings.

Wilson, D. S., and R. K. Colwell. 1981. Evolution of sex-ratio in structured demes. *Evolution* 35:882–897.

Wilson, E. O. 1957. The organization of the nuptial flight of the ant *Pheidole sitarches* (Wheeler). *Psyche* 64:46–50.

———— 1971. *The insect societies.* Cambridge, Massachusetts: Harvard University Press.

———— 1975. *Sociobiology: the new synthesis.* Cambridge, Massachusetts: Harvard University Press.

Wilson, E. O., and Bossert, W. H. 1963. Chemical communication among animals. *Recent Progress in Hormone Research* 19:673–716.

Withycombe, C. L. 1922. Notes on the biology of some British Neuroptera (Planipennia). *Transactions of the Entomological Society of London* 70:501–594.

Wittenberger, J. F. 1981a. *Animal social behavior.* Boston: Duxbury Press.

———— 1981b. Male quality and polygyny: the "sexy son" hypothesis revisited. *American Naturalist* 117:329–342.

Wood, D. L. 1973. Selection and colonization of ponderosa pine by bark beetles. *Symposium of the Royal Entomological Society of London* 6:101–117.

Wood, D. L., and W. D. Bedard. 1976. The role of pheromones in the population dynamics of the western pine beetle. *Proceedings of the XV International Congress of Entomology.*

Woodhead, A. P. 1981. Female dry weight and female choice in *Chauliognathus pennsylvanicus. Evolution* 35:192–194.

Woyke, J. 1955. Multiple mating of the honeybee queen (*Apis mellifera* L.) in one nuptial flight. *Bulletin de l'Academie Polonaise des Séances* 23:175–180.

Wright, C. G. 1960. Biology of the southern *Lyctus* beetle, *Lyctus planicollis. Annals of the Entomological Society of America* 53:285–292.

Wynne-Edwards, V. C. 1962. *Animal dispersion in relation to social behaviour.* New York: Hafner.

Young, A. M. 1980. Observations on the aggregation of adult cicadas (Homoptera: Cicadidae) in tropical forests. *Canadian Journal of Zoology* 58:711–722.

———— 1981. Temporal selection for communicatory optimization: the dawn-dusk chorus as an adaptation in tropical cicadas. *American Naturalist* 117:826–829.

Zahavi, A. 1975. Mate selection—a selection for a handicap. *Journal of Theoretical Biology* 53:205–214.

———— 1977a. The cost of honesty (further remarks on the handicap principle). *Journal of Theoretical Biology* 67:603–605.

———— 1977b. Reliability in communication systems and the evolution of altruism. In *Evolutionary ecology,* ed. B. Stonehouse and C. Perrins. Baltimore: University Park Press.

Zelazny, B. 1975. Behavior of young rhinoceros beetles, *Oryctes rhinoceros. Entomologia Experimentalis et Applicata* 18:135–140.

Author Index

Subject Index

(Last name,) First . int -

ital

Jornal 67: 65 - 67.